Analytical and
Approximate Methods
in Transport Phenomena

MECHANICAL ENGINEERING
A Series of Textbooks and Reference Books

Founding Editor

L. L. Faulkner

*Columbus Division, Battelle Memorial Institute
and Department of Mechanical Engineering
The Ohio State University
Columbus, Ohio*

1. *Spring Designer's Handbook*, Harold Carlson
2. *Computer-Aided Graphics and Design*, Daniel L. Ryan
3. *Lubrication Fundamentals*, J. George Wills
4. *Solar Engineering for Domestic Buildings*, William A. Himmelman
5. *Applied Engineering Mechanics: Statics and Dynamics*, G. Boothroyd and C. Poli
6. *Centrifugal Pump Clinic*, Igor J. Karassik
7. *Computer-Aided Kinetics for Machine Design*, Daniel L. Ryan
8. *Plastics Products Design Handbook, Part A: Materials and Components; Part B: Processes and Design for Processes*, edited by Edward Miller
9. *Turbomachinery: Basic Theory and Applications*, Earl Logan, Jr.
10. *Vibrations of Shells and Plates*, Werner Soedel
11. *Flat and Corrugated Diaphragm Design Handbook*, Mario Di Giovanni
12. *Practical Stress Analysis in Engineering Design*, Alexander Blake
13. *An Introduction to the Design and Behavior of Bolted Joints*, John H. Bickford
14. *Optimal Engineering Design: Principles and Applications*, James N. Siddall
15. *Spring Manufacturing Handbook*, Harold Carlson
16. *Industrial Noise Control: Fundamentals and Applications*, edited by Lewis H. Bell
17. *Gears and Their Vibration: A Basic Approach to Understanding Gear Noise*, J. Derek Smith
18. *Chains for Power Transmission and Material Handling: Design and Applications Handbook*, American Chain Association
19. *Corrosion and Corrosion Protection Handbook*, edited by Philip A. Schweitzer
20. *Gear Drive Systems: Design and Application*, Peter Lynwander
21. *Controlling In-Plant Airborne Contaminants: Systems Design and Calculations*, John D. Constance
22. *CAD/CAM Systems Planning and Implementation*, Charles S. Knox
23. *Probabilistic Engineering Design: Principles and Applications*, James N. Siddall

Analytical and Approximate Methods in Transport Phenomena

Marcio L. de Souza-Santos

CRC Press
Taylor & Francis Group
Boca Raton   London   New York

CRC Press is an imprint of the
Taylor & Francis Group, an **informa** business

CRC Press
Taylor & Francis Group
6000 Broken Sound Parkway NW, Suite 300
Boca Raton, FL 33487-2742

© 2008 by Taylor & Francis Group, LLC
CRC Press is an imprint of Taylor & Francis Group, an Informa business

First issued in paperback 2019

No claim to original U.S. Government works

ISBN 13: 978-0-367-45288-9 (pbk)
ISBN 13: 978-0-8493-3408-5 (hbk)

Library of Congress Cataloging-in-Publication Data

Souza-Santos, Marcio L. de.
  Analytical and approximate methods in transport phenomena / Marcio L. de Souza-Santos.
      p. cm. -- (Mechanical engineering)
    Includes bibliographical references and index.
    ISBN-13: 978-0-8493-3408-5
    ISBN-10: 0-8493-3408-X
    1. Transport theory. I. Title. II. Series.

  TP156.T7S78 2007
  621.01′530138--dc22   2007015322

Visit the Taylor & Francis Web site at
http://www.taylorandfrancis.com

and the CRC Press Web site at
http://www.crcpress.com

Dedication

To
my parents
Margarida and Americo

Table of Contents

Preface

Some people consider science arrogant—especially when it purports to contradict beliefs of long standing or when it introduces bizarre concepts that seem contradictory to common sense. Like an earthquake that rattles our faith in the very ground we're standing on, challenging our accustomed beliefs, shaking the doctrines we have grown to rely upon can be profoundly disturbing. Nevertheless, I maintain that science is part and parcel humility. Scientists do not seek to impose their needs and wants on Nature, but instead humbly interrogate Nature and take seriously what they find.

Carl Sagan

Over the years, I noticed a lacuna in engineering textbooks dedicated to transport phenomena. Most of them do not describe the details to arrive at solutions of specific examples, at least from an analytical point of view. Of course, the main reasons for this are the limitations in scope or space availability of the manuscript. However, even when the solutions are described, the texts do not generalize applied mathematical methods, therefore failing to provide a more solid foundation regarding the application of important mathematical tools. On the other hand, many texts preoccupied with mathematical methods in engineering and physical sciences fall into one of the following categories:

- Not covering examples of applications regarding the three phenomena, i.e., momentum, energy, and mass transfer
- Applying numerical methods when analytical solutions are possible without great difficulty
- Not applying approximate methods, which are very useful alternatives to numerical procedures, when analytical techniques are not able to provide a solution

There is also a false opinion on the importance of analytical methods. Some argue that numerical methods are more generally applicable than analytical or approximated methods. Of course, this is correct; however, if an analytical solution is possible, the advantage in understanding the phenomena or process cannot be overstressed. Once the equations describing the temperature, concentration, and velocity profiles are achieved, it is possible

- To observe the equation and learn the role of variables and several physical and chemical parameters regarding the behavior of a given process. In other words, it allows verifying the behavior, shapes, and limits provided by temperature, concentration, and velocity profiles. This very task improves the mathematical modeling skills of those involved with it.

- To use the solution to deduce important process parameters such as rates of heat, mass, and momentum transfers, among others.
- To plot graphs using the above results, which greatly improves the comprehension of the process at hand.

On the other hand, numerical methods depend on the capability of an available machine and accessibility of proper commercial routines, or require laborious development of such routines. Even newly developed numerical methods apply analytical solutions for certification of convergence robustness and precision. In addition, those intending to work in the area of numerical solutions need basic training on analytical solutions. This is not only a matter of obtaining a solid grounding in the field but also the ability to recognize when a numerical solution is the only alternative.

This book intends to fill some of the more common shortcomings mentioned above by seeking the following objectives:

1. To provide a source of analytical and approximate methods applicable to day-to-day problems on heat, mass, and momentum transfers.
2. To produce a text that can be used as basic or auxiliary material in courses on transport phenomena and applied mathematics.
3. To exemplify the applications and demonstrate how the solutions can be used to interpret a good range of real situations.
4. To allow using the text either for regular graduate or undergraduate courses. Hence, the sequence of problems should depart from simpler to more complex situations. Exercises are also included.
5. To allow sporadic consults by students, researchers, and professionals without too much time for reading complete texts and even less for consulting several publications in search of a method that might be valuable to solve a problem. In this respect, the book should be useful as a manual as well, where those interested in a specific type of problem may easily find the method or methods applicable to the problem. Therefore, a simple but precise classification of problems ought to be set to help those with a specific interest.

The basic governing equations of transport phenomena can be described in differential form. The tasks of anyone seeking the solution of a problem are as follows:

- Depart from the complete differential equations and after establishing reasonable assumptions, obtain the simplest equation or set of equations that govern the problem
- Properly set the boundary conditions

- Solve the differential equation to arrive at the general result
- Apply the boundary conditions to achieve the proper solution
- Interpret the result and apply to the physical situation
- Recognize possible ways to improve the process having in mind its applications

It is important to stress that the first step is the most critical. One should not refrain from making bold assumptions to arrive at a first model. Once the solution is achieved, the validity of assumptions might be verified by comparisons with real situations. If deviations against measurable values, for instance temperatures, velocities, and concentrations, are above acceptable levels, the model can be improved and the process starts again.

Having all the above in mind, the book is divided into chapters covering transport phenomena and methods to solve ordinary and partial differential equations. Each chapter deals with situations classified according to a code with three numbers, as follows:

1. The first indicates the number of independent variables involved in the differential equation. Three space coordinates and one for time are possible. Therefore, this group ranges from 1 to 4. Despite this, it has been considered that a maximum of three independent variables would suffice to illustrate the use of a wide range of mathematical methods and techniques applied to a relatively large variety of situations.
2. The second indicates the order of the differential equation (ordinary or partial) involved in the class of transport phenomena problems presented in the chapter. I have decided to limit the range to second-order equations because almost all engineering problems fall within this field.
3. The third indicates the highest class of boundary conditions involved in the problem. A possible classification for boundary conditions includes three main possibilities, as follows:
   - First-kind boundary conditions are those that set values for the dependent variable (velocity, temperature, or concentration) at certain values of the independent variable or variables (time or space coordinates). An example is when the temperature at position $x = a$ is known, or $T(x = a) = T_a$. Another is the so-called initial condition, which sets the value of temperature at the origin and represented by $T(t = 0) = T_0$.
   - Second-kind boundary conditions are those that set values for the derivatives of the dependent variable at given positions or time. This sort of condition is usually derived from the value of transport flux or flow at certain points. An example is the condition

$$\frac{dT}{dx}\bigg|_{x=0} = C$$

where C is a constant.

- Third-kind boundary conditions are those that set a relationship between values of flux and the transport variable at given positions or instants, for instance, when the heat fluxes by conduction and convection are equal at the interface between a solid and a fluid

$$-k\frac{dT}{dx}\bigg|_{x=0} = \alpha[T(x=0)-T_\infty]$$

Consider the following examples:

- Problems class 111: The chapter presents problems with time or one space-coordinate as independent variable. It deals with first-order differential equation. In addition, a first-kind boundary condition is set. If, for instance, ϕ is the transport variable (velocity, temperature, or concentration) and ω the variable time or space, this class of problems could be represented by the following equation:

$$f\left(\phi,\omega,\frac{d\phi}{d\omega}\right) = 0$$

- Problems class 223: Chapter 12 presents examples where two independent variables are involved with a second-order partial differential equation. Additionally, at least one third-kind boundary condition is found. Following the above notation, this class of problems could be represented by

$$f\left(\phi,\omega_i,\omega_j,\frac{\partial^2\phi}{\partial\omega_i^2},\frac{\partial\phi}{\partial\omega_i\partial\omega_j},\frac{\partial\phi}{\partial\omega_i}\right)$$

where ω_i and ω_j symbolize the space coordinates (rectangular, cylindrical, or spherical) or time.

Each chapter includes several examples. After the introduction to the problem, the simplifying assumptions are stated. This is very important because they provide training on modeling, i.e., they show how to arrive at a simple mathematical representation of a physical phenomenon. Then the differential equation or equations representing mass, energy, and momentum general balances are stated as well as the respective boundary conditions. The details to arrive at the solution by at least one mathematical method are shown. Of course, whenever possible or convenient, the problem is also solved by several methods, including using approximate methods. General considerations concerning the methods are described in appendixes. Once the method or methods are applied, the solution is commented upon. This allows exploring the mathematical result and shows how to appreciate its physical consequences and possible applications. Among the applications, one may also find the opportunity for process optimization.

Obviously, the possibility of an analytical solution for any physical problem is not guaranteed. As the number of variables increases and the boundary conditions become more complex, the relative number of cases for which an analytical solution is achievable decreases. The solutions become more elaborate and possible only at very particular circumstances. Not only that, but the method to obtain an analytical or even an approximate solution could lead to expressions that are too complex, and hence are difficult to apply. In these cases, numerical methods might be preferable.

It should be stressed that a preoccupation in presenting feasible situations with correlation to industrial production has been maintained throughout the book. However, in some instances, somewhat artificial situations are set to illustrate important features of mathematical tools.

The text does not present a thorough discussion on mathematical aspects. For instance, no demonstrations of theorems are shown, but the interested reader can find these in the listed references. The objective here is just to illustrate how few mathematical techniques can be applied to achieve solutions for a reasonably wide range of transport phenomena problems. Moreover, the text does not aim to be a comprehensive work on analytical or approximate methods. It advances to a point where the compromise between complexity and applicability of these techniques are considered reasonable.

Marcio L. de Souza-Santos

Acknowledgments

I am extremely grateful to my beloved wife, Marinalva, for her support and encouragement during the development of this book.

My sincere appreciation to the CRC Press staff, especially to Jill Jurgensen, Michael Slaughter, and Jonathan Plant, and also to Suryakala Arulprakasam of SPi for their kindness and helpful assistance.

It is important to express my appreciation to John Corrigan, for his encouragement during the contract phase.

I am also very grateful for the encouragement from various colleagues at the Faculty of Mechanical Engineering. The support of UNICAMP should also be acknowledged.

List of Symbols

NOMENCLATURE

a	parameter (dimension depends on the utilization)
A	area (m^2)
A_n	parameter (dimension depends on the utilization)
b	parameter (dimension depends on the utilization)
B	parameter (dimension depends on the utilization)
c	parameter (dimension depends on the utilization)
C	constant or parameter (dimension depends on the usage)
C_p	specific heat at constant pressure ($J\ kg^{-1}\ K^{-1}$)
d	diameter (m)
D_{AB}	diffusivity of component A in component B (or vice versa) ($m^2\ s^{-1}$). In a few situations, the second index may indicate the phase in which A diffuses.
D_T	thermal diffusivity ($m^2\ s^{-1}$)
$\tilde{E}$	reaction activation energy ($J\ kmol^{-1}$)
F	mass flow ($kg\ s^{-1}$) or independent variable
g	gravitational acceleration ($m\ s^{-2}$)
G	mass flux ($kg\ m^{-2}\ s^{-1}$) or independent variable
h	specific enthalpy ($J\ kg^{-1}$)
k_i	coefficient for the rate of reaction "i" (units depend on the order of reaction)
K	constant or parameter (unit depends on the usage)
L	length or distance (m) or Laplace transform
m	mass (kg)
n_j	rate of production (if negative: consumption) of component "j" per unit of volume of the reacting media ($kmol\ m^{-3}\ s^{-1}$)
N_j	mass flux of the component "j" indicated at the subscript ($kg\ m^{-2}\ s^{-1}$)
N_{Bi}	number of Biot (dimensionless)
N_{Fo}	number of Fourier (dimensionless)
N_{Pr}	number of Prandtl (dimensionless)
N_{Sh}	number of Sherwood (dimensionless)
p	pressure (Pa)
q	heat flux ($W\ m^{-2}$)
Q	heat (J) or independent variable
r	radial coordinate (m)
$\tilde{r}_i$	rate of reaction i indicated in the subscript ($kmol\ m^{-3}\ s^{-1}$)
$\tilde{R}$	universal gas constant ($8314.2\ J\ kmol^{-1}\ K^{-1}$)
$R_{M,j}$	rate of production (if >0) or consumption (if <0) of component "j," indicated at the subscript, per unit of volume ($kg\ m^{-3}\ s^{-1}$)

R_Q	rate of production (if >0) or consumption (if <0) of energy per unit of volume (W m^{-3})
s	parameter (unit depends on the usage)
t	time (s)
T	temperature (K)
u	parameter or velocity (m/s)
U	mass transfer resistance (s m^{-2})
v	velocity (m s^{-1})
V	volume (m^3)
w	mass fraction
W	weight function
x	coordinate (m)
y	coordinate (m)
z	coordinate (m)

GREEK LETTERS

α	convective heat transfer coefficient (W m^{-2} K^{-1})
β	coefficient of mass transfer (dimension depends on the utilization)
γ	parameter (units depend on the specific utilization)
ε	radiative heat transfer emissivity coefficient
ρ	density (kg m^{-3}). If subscript is present, it indicates the mass concentration of that component.
θ	angular coordinate (rad) or dimensionless temperature
ν	kinematic viscosity (m^2 s^{-1})
ν_{ij}	stoichiometric coefficient for component "j" in the reaction "i" (dimensionless)
ϕ	angular coordinate (rad) or sphericity (dimensionless)
Φ	Thiele modulus (dimensionless)
λ	thermal conductivity (W m^{-1} K^{-1})
τ	If with two coordinates as indexes it is the shear-stress tensor (Pa), otherwise it is dimensionless time
σ	Stefan–Boltzmann constant (5.67×10^{-8} W m^{-2} K^{-4})
Λ	parameter or residue function
μ	dynamic viscosity (kg m^{-1} s^{-1})
Ξ	parameter or variable
Ψ	parameter or variable
ω	general independent variable (time or space) (dimension depends on the use)
κ	general parameter (dimension depends on the use)
Ω	parameter (dimension depends on the use)

SUPERSCRIPTS

P	particle or particulate phase
U	unreacted core

X	exposed core
•	per unit of time or indication of time rate
~	molar basis

SUBSCRIPTS

A	ash layer
eq	at equilibrium condition
G	gas or gas boundary layer
N	nucleus or core
x	in the x direction
y	in the y direction
z	in the z direction
θ	in the θ direction
ϕ	in the ϕ direction
∞	far from the surface or at the middle of the continuous phase

GREEK ALPHABETS

A α	Alpha
B β	Beta
Γ γ	Gamma
Δ δ	Delta
E ϵ	Epsilon
Z ζ	Zeta
H η	Eta
Θ θ ϑ	Theta
I ι	Iota
K κ	Kappa
Λ λ	Lambda
M μ	Mu
N ν	Nu
Ξ ξ	Xi
O o	Omicron
Π π	Pi
P ρ	Rho
Σ σ	Sigma
T τ	Tau
Υ υ	Upsilon
Φ ϕ φ	Phi
X χ	Chi
Ψ ψ	Psi
Ω ω	Omega

1 Problems 111; One Variable, 1st Order, 1st Kind Boundary Condition

1.1 INTRODUCTION

This chapter presents methods to solve problems with one independent variable involving first-order differential equation and first-kind boundary condition. Mathematically, this class of problems can be summarized as $f\left(\phi,\omega,\frac{\mathrm{d}\phi}{\mathrm{d}\omega}\right)$, first-kind boundary condition.

1.1.1 MODEL AND REALITY

As shown throughout the text, even for apparently simple problems presented in this chapter, the solutions are possible only after a series of assumptions. Therefore, every mathematical model just approximately reproduces the relationships among the involved variables during real processes or phenomena. On the other hand, in science and engineering, there is always the need of relying on models to design any equipment or system or to predict the behavior of processes.

Thus, modeling starts with assumptions made about the behavior of real processes. There are several levels of assumptions, which in turn reflect the level of the model complexity. Of course, a model may or may not be a reasonable representation of reality. What is considered reasonable is a matter of conventions or criteria, but all those are based on comparisons between model and reality. Such comparisons are possible by measuring the variables involved in the real phenomena against the respective predicted values computed by the theoretical solution. Usually in engineering applications, deviations below 5% between the measured and predicted values are acceptable. This may not be the case for several other more critical applications.

Let us take, for instance, the case presented in the following section and illustrated by Figure 1.1.

A body exchanges heat with surrounding environment and the temperature inside that body has to be determined against time. Therefore, the temperature in

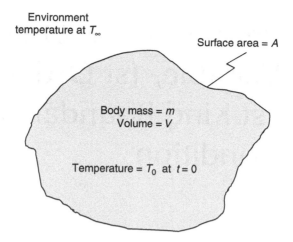

FIGURE 1.1 A solid body exposed to environment at a different temperature.

the body would constitute the main variable of the problem. In any case, the simple question of what and how comparisons would take place requires examination. For instance, during the process of heat exchange between the body and environment, temperature will vary along positions inside the body. Consequently, the physical and chemical properties of the material of the body and the immediate environment would change as well. For instance, the thermal conductivity will change with time from one point to another in the body. Concerning the environment, the influences of temperature variation may be noticeable as well. Let us imagine that air is surrounding the solid body. As the temperature of the body surface changes with time, so will the temperature of the air layer near that body. Hence, properties such as thermal conductivity of the air, density, and convective heat transfer coefficient will change, and these might significantly influence the rate of heat exchange between the body and the environment.

The problem would become even more complex if chemical transformations start occurring as a function of temperature variations. Even though such transformations are negligible, almost all solid materials go through such chemical changes. However, if they occur, modifications in chemical and physical properties would become even more important. This would severely influence the result and the temperature would deviate even more from the value predicted by a simple model. Normally, chemical reactions are either exothermic or endothermic. Therefore, they might become important sources or sinks of energy, and terms related to them should be added in the basic equations. Such relations are usually complex and introduce nonlinear terms to the differential equation or equations describing the problem.

Now, one wonders how a proposed model and its solution can be compared with the real process. As described above, if one sets a series of thermocouples along the body interior, the registered values will change among the thermo-couples with time. A possible simplification may assume that temperature is a weak function of space or position inside the body. According to such a simpli-fication, the temperatures of all thermocouples would be close to each other or to an average value among them. Therefore, this average will vary only with time. Let us assume that this is the case. Thus, if the model provides the average temperature of the solid against time, the comparisons between computed and measured average temperature would be simple. Two situations may occur:

1. The differences between real and model-computed values always stay below an acceptable deviation, for instance 2%.
2. The above does not occur, at least to one or few instants.

If the second situation is verified, one should revisit the simplification hypothesis and analyze which one could be the most critical.

A broader picture of the model quality can be obtained by plotting the measured and computed values, respectively, in the ordinate and abscissa axes of a graph and connecting the points related to each respective pair. The model would be perfect if a straight line with 45° inclination is obtained. Deviations from this would provide an indication of the model quality.

As seen, depending on the adopted level of rigor, even apparently simple situations might fall out of the range where analytical methods are applicable. Beyond that, only numerical models might solve the problem.

As a methodology applied to every problem is presented throughout the text, the assumptions are listed and commented. Such a procedure is strongly recom-mended to everyone trying to model a given situation for several reasons. Some of the reasons include

- It allows easy consultation to the assumptions made to model a given process.
- It facilitates the understanding of range and restrictions implicit to the achieved solution.
- It presents comments that might demonstrate how critical each assump-tion is.
- It allows future development not only by the person or team studying the problem but also by others on the same road.
- It helps to withdraw one or more assumptions, if the comparisons between the model solution and reality were not within the acceptable level of deviation.

In the examples throughout this book, the problem is presented and, after listing the assumptions, the fundamental differential equations of transport phe-nomena are applied. These are listed in Appendix A and can be found in almost all texts on the subject [1–3].

1.1.2 Verification of Solutions

Normally, arriving at an analytical solution representing any physical phenomena demands work and creativity. In addition, during the process of solution, one is involved with several equations to which various mathematical techniques may be applied. Most of the time, algebraic manipulations are necessary, and there is always a high probability of mistakes. Some of them are not easy to spot. Therefore, it is important to check every solution to verify its correctness. To decrease the probability of mistakes, after solving a problem the following steps are recommended:

1. Verify the dimensional consistency of the solution. For instance, the terms on both sides of equation should have the same dimension.
2. Verify if the solution satisfies the boundary condition or conditions.
3. Verify the physical consistency of the solution. This is possible by looking how each independent variable influences the dependent one. Ask yourself if these influences make sense.
4. Test if the original differential equation can be reproduced from the solution by properly derivating the dependent variables with regard to the independent ones. Even though this is the ultimate proof, it is not always easy to accomplish.
5. Whenever possible, plot graphs representing the solution. These would help to verify the consistency of the achieved results as well as help in understanding of the phenomena at hand.

1.1.3 Use of Dimensionless Variables

Obviously, the solution of problems does not require the transformation of variables into dimensionless variables. However, this is advisable whenever possible because of the following:

- It usually simplifies the handling of the boundary value problem to be solved.
- Result does not depend on the particular system of units.
- Solution of a problem can be generalized and applied to others—even not related to the one already solved—if those could set at the same dimensionless form.
- Graphs are also general and condense several effects into a single plotting. This also facilitates the understanding of interplay among the variables involved in a given situation.

Whenever possible, the transformation should lead to normalized variables, i.e., with variables ranging between zero and one. This allows a more comprehensive observation of the phenomena within the complete ranges of possible values that can be taken by the involved variables.

1.2 HEATING OF A SOLID

As shown in Figure 1.1, let a solid body be initially at temperature T_0 and suddenly exposed to the environment (air) at temperature T_∞.

Of course, heat transfer will occur from the body or to the body, depending on whether the temperature T_0 is greater or less than T_∞.

The heat transfer may occur by several mechanisms:

- Conduction
- Radiation
- Convection

Actually, the basic mechanisms for heat transfer are just conduction and radiation. Convection is a conduction process between the surface of the body and a moving fluid. Therefore, convection is not a basic phenomenon but a combination of conduction and momentum transfers.

Radiation also occurs between the body surface and the air and between the body and surfaces of the surrounding ambiance. Even when the body is exposed to open space, radiative heat transfer from the body will take place.

As recommended before, let us list the simplifications and assumptions made for the solution intended here:

1. Rate of heat transfer by convection is assumed much higher than the rate by radiation transfer to other surfaces or to open space. This is correct for moderate differences of temperature between the body and the surroundings. Radiative heat transfer depends on the fourth power of the temperatures, whereas convective heat transfer depends on the first power. On the other hand, in the case of radiative heat transfer, the temperature differences should be multiplied by the Stefan–Boltzmann constant $(5.67 \times 10^{-8} \text{ W m}^{-2} \text{ K}^{-4})$, which is a very small number. Therefore, the radiative heat transfer becomes competitive with convective heat transfer only for relatively large differences of temperature, usually above 500 K. It is also interesting to notice that, in most cases, the surrounding atmosphere can be considered transparent to the radiation at frequencies involved in thermal exchanges. Of course, this is an approximation because several gases, such as water and carbon dioxide, absorb a good fraction of thermal radiation.
2. Surrounding air has a mass such that its temperature—at least for positions far from the body surface—remains constant and equal to T_∞.
3. No chemical reaction or any other form of energy source is present in the body.
4. Within the range of temperatures involved in the process, no phase change is verified. Therefore, the body remains as a solid, thus no velocity field is present.
5. Volume of the body is small and its thermal conductivity is relatively high and does not change too much within the range of temperature of

the process. The high conductivity ensures fast heat transfer by conduction and small dimensions allow fast equalization of temperature throughout the body. Of course, this is an approximation, which might be more or less rude, depending on the mass, volume, and physical properties of the body as well as on external conditions. However, even if this approximation is applicable for most of the internal volume of the solid, it cannot be true for the layers near its surface. Of course, if no temperature gradient exists, at least at some part of solid, no heat transfer is possible, and therefore, the temperature would not vary at all, or the derivative of temperature against time is zero.

Let us take a small layer near the surface. The application of the hypothesis described earlier and Equation A.36 (Appendix A) leads to

$$\rho C_p \frac{\partial T}{\partial t} = R_Q \qquad (1.1)$$

The right-hand side of that equation represents a uniformly distributed rate of generation (+) or consumption (−) of energy throughout the considered volume of the body (W m^{-3}). This is due to the internal or external sources or sinks, for instance, uniformly distributed electrical heating and exothermic or endothermic chemical reactions that occur inside the control volume. Usually, the energy input by heat transfer through the surface is not included in this term because it is a localized process. However, in the particular situation where the temperature of the body is assumed uniform, the energy gained or lost through heat transfer by convection with air should be assumed to be immediately transferred to the entire body. In this way, the energy source or sink will be represented by the power input or output because of that heat transfer divided by the volume of the control volume, or

$$R_Q = \alpha A(T_\infty - T)/V \qquad (1.2)$$

One should be careful not to use this approach when space dimensions in the direction of the heat transfer are involved.

Once again, Equation 1.1 could be written here only because spatial variation of temperature is neglected, or the temperature of body is assumed uniform at any instant.

Substituting Equation 1.2 into Equation 1.1 leads to

$$\frac{dT}{dt} = -\frac{\alpha A}{\rho V C_p}(T - T_\infty) \qquad (1.3)$$

As a precaution against possible mistakes, it is advised to check the coherence of units at each side of the equation. In Equation 1.3, both sides have units of K s^{-1}. It is also important to analyze the physical significance of the differential equation to be solved. For example, the temperature should decrease with time (negative value for the derivative on the left) if the temperature T of the body is greater than the temperature of environment T_∞. This is correct in the present case due to negative sign in the right-hand side of the equation. Such procedure should become a habit in all problems. Of course, this is no guarantee against mistakes made during the solution; nonetheless, it decreases the probability of their occurrence. In addition, this habit is educational and serves as training for understanding the meaning of the differential equation and the behavior of the physical phenomena.

1.2.1 Solution by Separation of Variables

The above differential equation is a typical example of the class covered by this chapter. Furthermore, as seen in Appendix B, Equation 1.3 is separable. This allows placing the dependent and independent variables at distinct sides of the equation and integrating them leads to

$$\ln(T - T_\infty) = -\frac{\alpha A}{\rho V C_p} t + \ln C \qquad (1.4)$$

Here C is a constant. Therefore

$$T = T_\infty + C \exp\left(-\frac{\alpha A}{\rho V C_p} t\right) \qquad (1.5)$$

A condition needs to be set to determine the constant C.

If, for instance, the temperature of the body at time t_0 is T_0, the equation finally becomes

$$\frac{T - T_\infty}{T_0 - T_\infty} = \exp\left[-\frac{\alpha A}{\rho V C_p}(t - t_0)\right] \qquad (1.6)$$

The above describes the progress of average temperature in the body as a function of time. It is easily observed that temperature tends to approach the environmental one asymptotically.

The rate of heat transfer between the body and air is given by

$$\dot{Q} = \alpha A(T - T_\infty) = \alpha A(T_0 - T_\infty) \exp\left[-\frac{\alpha A}{\rho V C_p}(t - t_0)\right] \qquad (1.7)$$

The usual notation for the sign of heat transfer is employed, i.e., negative if the body energy decreases due to heat transfer to the ambiance, and positive if the body energy increases due to heat transfer from the environment. In addition, the total exchanged heat is given by integrating Equation 1.7 between time t_0 and t, or

$$Q = mC_p(T - T_0) = \rho V C_p(T_0 - T_\infty)\left\{1 - \exp\left[-\frac{\alpha A}{\rho V C_p}(t - t_0)\right]\right\} \qquad (1.8)$$

It is interesting to notice that the above equations could provide some indication of the validity and not of the approximations made. This could help to guide the decision to use these approximations. For this, let the following be the dimensionless variables:

$$N_{Bi} = \frac{\alpha L}{\lambda} \qquad (1.9)$$

and

$$N_{Fo} = \frac{D_T(t - t_0)}{L^2} \qquad (1.10)$$

where

$$D_T = \frac{\lambda}{\rho C_p} \qquad (1.11)$$

D_T is usually called thermal diffusivity. The parameter L in Equation 1.10 is the characteristic length, and in the present case is given by the ratio between the body volume and its surface area, or

$$L = \frac{V}{A} \qquad (1.12)$$

The Biot number (N_{Bi}) is a relation between the internal resistance to heat transfer by conduction inside the body and the resistance for this transfer by convection at its surface. The Fourier number is a measure of heat transfer inertia.

Let the following equation be the form of dimensionless temperature:

$$\theta = \frac{T - T_\infty}{T_0 - T_\infty} \qquad (1.13)$$

Using Equation 1.13, Equation 1.6 can now be written as

$$\theta = \exp(-N_{Bi}N_{Fo}) \qquad (1.14)$$

One should notice that assumption 5 (in Section 1.2) could only be assumed if the internal resistance for heat transfer is negligible when compared with the external one. The relative importance between the resistances is represented by the Biot number and can be put in quantitative terms. It is usually assumed that if the Biot number is smaller than 0.1, the above treatment can be used and it is called lumped analysis. Therefore, solution given by Equation 1.14 gives a reasonable description of the average temperature in the body against time. Dropping hypothesis 5 and developing the solution for the temperature profile inside the body can verify this. For $N_{Bi} < 0.1$, the error between the two calculations is below the usual precision for temperature measurements. Therefore, the lumped analysis may be a useful approach for various situations.

Figure 1.2 illustrates the role of each dimensionless parameter on the temperature progress.

To better observe the effect of each parameter, the graph includes values of Biot number above the limit of 0.1. However, it should be understood that the solution given by Equation 1.14 does not represent well for values above that limit.

The graph shows how the temperature approaches the environment temperature (T_∞) (or θ approaches zero) for larger values of time or higher values of N_{Fo}. The same happens for larger values of heat transfer coefficient (α) or larger N_{Bi}. One should also notice that although N_{Bi} and N_{Fo} depend upon the thermal conductivity of the body, this last property plays no role in the development of

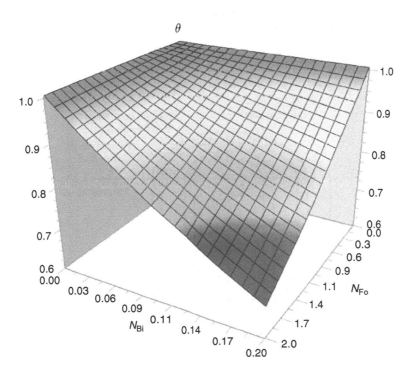

FIGURE 1.2 Dimensionless temperature (θ) as function of N_{Bi} and N_{Fo}.

average temperature of the body, as shown by Equation 1.5. Thermal conductivity is introduced here just to confirm the dimensionless variables N_{Bi} and N_{Fo}. The independence of average temperature of the body regarding its conductivity is a consequence of assumption 5. The supposition of uniform temperature throughout the body at any instant is equivalent as to assume instantaneous heat transfer by conduction. Therefore, it implies on infinite or very high thermal conductivity.

Parameter L is a part of both N_{Bi} and N_{Fo}. However, in N_{Fo} it is squared, which leads to a final inverse influence of L in the dimensionless temperature. In other words, larger values of L provide smaller values of product between N_{Bi} and N_{Fo}. Therefore, larger values of L leads to larger thermal inertia or to greater values of θ, which in turn is equivalent to larger periods to approach temperatures near that of the ambiance.

1.3   FLOW BETWEEN TWO DRUMS

The vertical cross sections of two concentric and static drums are shown in Figure 1.3.

Fluid continually passes from the internal to the external drum by crossing the permeable wall between them. From there, it passes through another permeable wall and flows to the external environment. These walls may be made of porous ceramic or any other permeable material. One is interested in determining the velocity and pressure profiles in the region between the permeable walls.

The following are assumed:

1. Steady-state regime.
2. Uniform temperature.

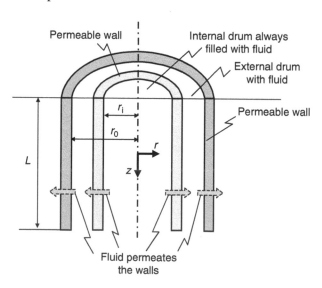

FIGURE 1.3 Flow between concentric drums.

3. Fluid is Newtonian with constant density and viscosity.
4. Permeable or porous walls distribute the fluid evenly.
5. Flow through the ring between the external and internal drums is laminar.
6. Studied region of external drum is far from its top and bottom walls. In other words, the height L is large enough and the flow around the center ($z = 0$) is not affected by those walls. Therefore, in such a region it is possible to assume radial velocity (v_r) as a function only of the radial component (r).

The departure from the total mass continuity is represented by Equation A.2, and using assumptions 1 and 6 one reaches at

$$\frac{d}{dr}(rv_r) = 0 \qquad (1.15)$$

This is a first-order separable equation and integration is immediate to give

$$rv_r = C \qquad (1.16)$$

Here C is a constant.

Equation 1.16 provides the velocity profile as

$$v_r = C\frac{1}{r} = v_r(r = r_i)\frac{r_i}{r} = \frac{F}{2\pi\rho L}\frac{1}{r} \qquad (1.17)$$

Here F is the mass flow (kg s^{-1}) of fluid injected into the internal drum, which is the same as that passing to the external drum.

The r-component of the momentum conservation equation for Newtonian fluid with constant density and viscosity is given by Equation A.13. After the assumptions listed above, it becomes

$$\rho\left(v_r\frac{dv_r}{dr}\right) = -\frac{\partial p}{\partial r} + \mu\left[\frac{d}{dr}\left(\frac{1}{r}\frac{d(rv_r)}{dr}\right)\right] \qquad (1.18)$$

The angular component of momentum conservation (Equation A.14) leads to

$$\frac{\partial p}{\partial \theta} = 0 \qquad (1.19)$$

The momentum conservation at the z-component (Equation A.15) leads to

$$\frac{\partial p}{\partial z} = -\rho g \qquad (1.20)$$

Applying Equations 1.15 and 1.16 to Equation 1.17, one gets

$$\frac{\partial p}{\partial r} = \rho C^2 \frac{1}{r^3} \tag{1.21}$$

The total derivative of pressure is

$$dp = \frac{\partial p}{\partial r} dr + \frac{\partial p}{\partial z} dz = \rho C^2 r^{-3} \, dr - \rho g \, dz \tag{1.22}$$

Integrating Equation 1.22

$$p = -\frac{\rho C^2}{2} \frac{1}{r^2} + f_1(z) - \rho g z + f_2(r) \tag{1.23}$$

Notice that f_1 is a function of coordinate z, therefore is a "constant" regarding the integration on variable r. The reverse occurs for function $f_2(r)$ regarding the integration on variable z. Functions f_1 and f_2 can be obtained after setting the boundary conditions of the problem. Assuming the pressure at a given height ($z = 0$, for instance) and at the surface of internal porous wall ($r = r_i$) as p_i, the solution for the pressure profile is

$$p(z = 0) = p_i - \frac{1}{8\rho} \left(\frac{F}{\pi L} \right)^2 \left(\frac{1}{r^2} - \frac{1}{r_i^2} \right) \tag{1.24}$$

The following comments can be made at this point:

1. As expected, the radial velocity of a fluid is inversely proportional to the radius.
2. Inside the external drum, the pressure decreases with the available flow area or with square of the radius.

1.4 HEATING OF A FLUID IN A STIRRING TANK

The following also exemplifies the application of lumped analysis. Figure 1.4 shows a tank with a stirring device. The tank has no insulation and the stirring device ensures a good homogeneity of temperature of the fluid. Initially, the fluid is at temperature T_0 and the ambiance at T_∞. An electrical resistance exchanges heat with the fluid and the rate of electrical energy input to the resistance can be set as a function of time.

We have to deduce the average temperature of the fluid in the tank as a function of time.

Let us list the adopted assumptions:

1. Differences of temperature between the surface of the tank and the surrounding air are relatively small. In addition, the air around the reactor is transparent to the radiation at frequencies involved in the problem.

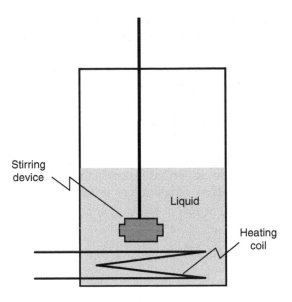

FIGURE 1.4 Stirring tank with heating device.

Therefore, radiative heat transfer between the tank and the surroundings is negligible and all transfers are described by convection.

2. Surrounding air has such a mass that its temperature—at least for positions far from the body surface—remains constant and equal to T_∞.

3. Because of the mixing, the temperature of the tank content can be considered uniform at any time. Of course, this is an approximation, which might be more or less rude, depending on the intensity of the mixing, mass, volume, and physical properties of the fluid as well as external conditions. Obviously, this assumption cannot be rigorously applied, at least for regions near the internal tank wall; otherwise, no heat transfer between the tank and the surrounding air would occur.

4. Fluid in the tank is Newtonian with constant density.

5. Work input due to the stirring device is negligible when compared with the energy input due to heating coil and loss to ambiance.

6. There is no increase in internal energy of fluid due to viscous dissipation.

Assumption 3 allows us to apply lumped analysis.

Owing to assumption 4, Equation A.34 (or Equation A.35) can be used. In addition, due to simplification 6, the terms preceded by the viscosity can be neglected. Therefore, it can be written as

$$\rho C_{\mathrm{p}} \frac{\mathrm{d}T}{\mathrm{d}t} = R_{\mathrm{Q}} \tag{1.25}$$

The same convention used at the previous section regarding heat losses (or gains) due to the convective exchange between the body (tank) and the environment (air) is applied. Moreover, the present case includes the source for energy produced by electrical resistance. As an example, let a simple relationship for the rate of energy delivery by the resistance and time be given by

$$R_Q = \frac{1}{V}\alpha A(T_\infty - T) + \frac{at}{V} \qquad (1.26)$$

Here parameter a is a constant.

To facilitate discussions as well as understanding of achieved solutions, the following set of dimensionless variables is proposed:

$$\theta = \frac{T - T_\infty}{T_0} \qquad (1.27)$$

and

$$\tau = t\frac{\alpha A}{\rho V C_p} \qquad (1.28)$$

Using the chain rule it is possible to see that

$$\frac{dT}{dt} = T_0\frac{d\theta}{d\tau}\frac{d\tau}{dt} = T_0\frac{\alpha A}{\rho V C_p}\frac{d\theta}{d\tau} \qquad (1.29)$$

Substituting Equation 1.29 into Equation 1.25, we get the following equation:

$$\frac{d\theta}{d\tau} = -\theta + b\tau \qquad (1.30)$$

The boundary condition is

$$\theta(0) = \theta_0 = \frac{T_0 - T_\infty}{T_0} \qquad (1.31)$$

Here

$$b = \frac{a\rho V C_p}{T_0(\alpha A)^2} \qquad (1.32)$$

Again, it is advisable to verify the coherence of units.

Since all variables of Equation 1.30 are dimensionless, the only care is to verify if the involved variables (θ, τ, and b) are indeed dimensionless. In addition, as heating is imposed, the last term on the right-hand side of Equation 1.30 is always positive.

1.4.1 VARIATION OF PARAMETERS

As seen, the ordinary differential equation (Equation 1.30) is not separable but can be solved by several methods. Among such methods, there is the method of variation of parameters described in Appendix B.

It starts by trying a solution with the following form:

$$\theta(\tau) = u_1(\tau)u_2(\tau) \tag{1.33}$$

Here, u_1 and u_2 are functions of τ. This maneuver does not impose any loss of generality.

Using Equations 1.33 and 1.30, it is possible to write

$$u_1(u_2' + u_2) + u_1'u_2 = b\tau \tag{1.34}$$

It is important to notice that Equation 1.33 introduces an extra degree of freedom. A convenient form of taking advantage of this is to impose a condition regarding the terms of Equation 1.34. On the other hand, such a condition should avoid restrictions to such a trivial solution θ, or lead to unreal results. Among the possibilities, it is always possible to set the first member of Equation 1.34 as zero. However, to avoid restrictions to solution θ, function u_1 cannot be imposed equal to zero, and therefore, the remaining possibility is to set

$$u_2' + u_2 = 0 \tag{1.35}$$

This leads to

$$u_2 = \exp(-\tau) \tag{1.36}$$

As the present problem requires just one boundary condition, this can be imposed through just one integration constant, either from integration of Equation 1.35 or other equations given below. For this reason, integration constant equal to zero has been conveniently chosen when writing Equation 1.36.

Substituting Equation 1.36 into Equation 1.34 one gets

$$u_1' = b\tau e^{\tau} \tag{1.37}$$

and

$$u_1 = be^{\tau}(\tau - 1) + C \tag{1.38}$$

Finally

$$\theta = b(\tau - 1) + Ce^{-\tau} \tag{1.39}$$

For the case where the initial temperature is equal to the ambiance temperature or $T_0 = T_\infty$, the boundary condition (Equation 1.31) becomes

$$\theta(0) = 0 \tag{1.40}$$

Therefore, $C = b$ and

$$\theta = b(e^{-\tau} + \tau - 1) \tag{1.41}$$

As seen, the average temperature increases indefinitely. Nonetheless, the rate of energy delivered by the heating device would decrease with time. This is illustrated by Figure 1.5.

Equation 1.41 shows that bath temperature increases due to an exponential and a linear term. However, for long periods, the exponential part tends to zero and the temperature increase would be linearly proportional to time. Of course, there are limitations for the temperature, such as the point at which phase change of the fluid begins. In addition, any given heating system may sustain increasing energy injection to the control volume just for a limited period or within a certain range of surrounding fluid temperature.

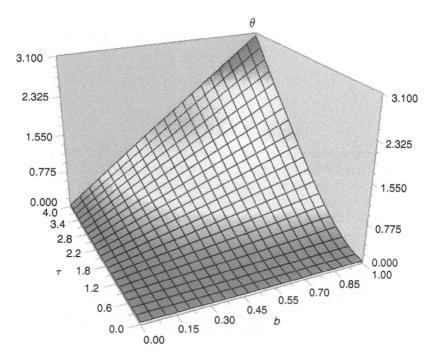

FIGURE 1.5 Dimensionless temperature (θ) as a function of dimensionless time (τ) and parameter b.

While observing Figure 1.5, it is also important to remember that the heating rate is proportional to parameter b. Additionally, from Equation 1.41, it is possible to verify that b increases for tanks with smaller A/V ratio or ratio between external area A and internal volume V. As expected, spherical tanks would provide the smallest rate of heat loss and fastest increase in the temperature of the stirred fluid.

1.5 HEATED BATCH REACTOR

Consider the same well-stirred tank studied in the previous section. In addition, a reaction is taking place in the vessel. This is similar to a temperature-controlled batch reactor, which is commonly used in the chemical and pharmaceutical industry.

Chemical reactions transform chemical potential difference between reactants and products into sensible internal energy of the mixture. If this difference is positive, the reaction is called exothermic, whereas, if negative, the reaction is endothermic. Therefore, exothermic reactions can be seen as a source of energy, which, if not ablated by a cooling system, would increase the enthalpy and consequently, the temperature of the reacting mixture.

This is a batch process; therefore, reactants are fed into the tank at once and reaction progresses until either the reactor content is withdrawn or the reaction reaches equilibrium. Of course, the rate of reaction is high at the beginning and decreases with time. In addition, the rate also depends upon the temperature.

For the sake of simplicity, it is assumed that the rate of energy production by the reaction is

$$R_{Q,\text{reaction}} = a_1 e^{-a_2 t} \tag{1.42}$$

Here a_i are constants. If a_1 were positive, an exothermic reaction would occur and the contrary for an endothermic reaction. Similarly to the last section, if the power delivered by the electrical heating is proportional to time (Equation 1.26), the total rate of energy generation by internal source should be added to the amount received by heat transfers. Hence, the overall energy balance for the reactor can be written as

$$\rho V C_p \frac{dT}{dt} = -\alpha A(T - T_\infty) + at + a_1 e^{-a_2 t} \tag{1.43}$$

Using the same change on variables as proposed by Equations 1.27 and 1.28, Equation 1.43 becomes

$$\frac{d\theta}{d\tau} = -\theta + b\tau + c e^{-\gamma \tau} \tag{1.44}$$

Here b is also given by Equation 1.32 and parameters c and γ by

$$c = \frac{a_1}{\alpha A T_0} \tag{1.45}$$

$$\gamma = a_2 \frac{\rho V C_p}{\alpha A} \tag{1.46}$$

1.5.1 SOLUTION BY LAPLACE TRANSFORM

Laplace transform is a powerful tool for solving ordinary as well as partial differential equations. The basic properties of this operator are described in Appendix D and the following exemplifies its application.

The Laplace transform of both sides of Equation 1.44 provides

$$s\varphi(s) - \theta(0) = -\varphi(s) + b\frac{1}{s^2} + c\frac{1}{s+\gamma} \tag{1.47}$$

Here, the boundary condition is set by assuming that the initial temperature in the tank equals the ambiance temperature. Thus, by Equation 1.31

$$\theta(0) = 0 \tag{1.48}$$

Therefore, Equation 1.47 becomes

$$\varphi(s) = \frac{b}{(s+1)s^2} + \frac{c}{(s+1)(s+\gamma)} \tag{1.49}$$

To facilitate the inversion, the following identity can be used to rewrite the first term of the right-hand side as

$$\frac{1}{(s+1)s^2} = \frac{B_1}{s+1} + \frac{B_2 s + B_3}{s^2} \tag{1.50}$$

As a rule, the numerator should be a polynomial one order below the respective denominator. From this, B_i ($i = 1, 2, 3$) can be found through polynomial identity to obtain the following:

$$\frac{b}{(s+1)s^2} = b\left(\frac{1}{s+1}\right) - b\left(\frac{s-1}{s^2}\right)$$

According to the exposed material at Appendix D, the inverse of that is

$$L^{-1}\left\{\frac{b}{(s+1)s^2}\right\} = be^{-\tau} - b(1-\tau) \tag{1.51}$$

Applying the inverse to the second term in the right-hand side of Equation 1.49 leads to

$$L^{-1}\left\{\frac{c}{(s+1)(s+\gamma)}\right\} = \frac{c}{1-\gamma}(e^{-\gamma\tau} - e^{-\tau}) \tag{1.52}$$

Finally, the solution is

$$\theta(\tau) = b(e^{-\tau} + \tau - 1) + \frac{c}{1-\gamma}(e^{-\gamma\tau} - e^{-\tau}) \tag{1.53}$$

It is interesting to compare the solution for this case with the previous solution given by Equation 1.41.

The last term on the right-hand side of Equation 1.53 shows the effect of the introduction of a reaction. Obviously, if endothermic ($c < 0$), the temperature would decrease and vice versa if exothermic ($c > 0$).

Figure 1.6 illustrates the play of variables in the process in case $c = 0.1$ whereas Figure 1.7 presents values for $c = 1$. In both cases, it has been assumed that $\gamma = 0.5$. In addition, Figure 1.8 shows the effect of increasing parameter γ, which is set as 5 while c is kept at 1.

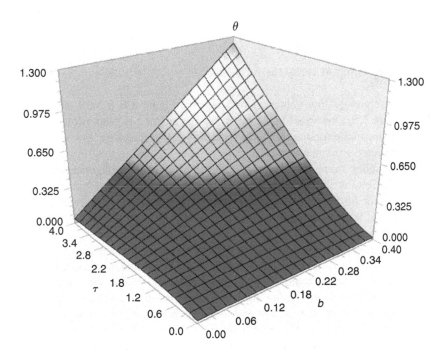

FIGURE 1.6 Dimensionless temperature (θ) as a function of dimensionless time (τ) and parameter b assuming $c = 0.1$ and $\gamma = 0.5$.

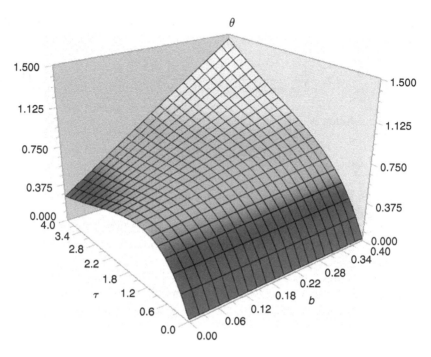

FIGURE 1.7 Dimensionless temperature (θ) as a function of dimensionless time (τ) and parameter b assuming $c = 1$ and $\gamma = 0.5$.

It is interesting to verify the following:

1. For relatively low values of parameter c, Figure 1.6 is very similar to Figure 1.5. One should remember that parameter c is proportional to the reaction pre-exponential factor (Equation 1.45); therefore, low values of c indicate slow reaction rates.
2. For higher values of parameter c, the effect of exothermic reaction starts to be felt, as shown by Figure 1.7 when compared with Figure 1.6. As seen, for faster reactions, temperature rises faster and, for a given instant (t or τ), it reaches higher values.
3. At the above figure it is possible to see the temperature profile for $b = 0$, i.e., when the heating device is not working and the temperature varies only because of exothermal reaction. At the beginning, as represented by Equation 1.42, the rate of energy release due to the reaction is relatively high. This is so due to the high concentrations of reactants. At the same time, heat is exchanged and lost to the environment. This is represented by the first term on the right-hand side of Equation 1.43. As the rate of energy release from reaction decreases due to decrease of reactant concentrations, the heat loss to environment begins to be felt and peak in the temperature is reached. After this, the rate of heat loss to

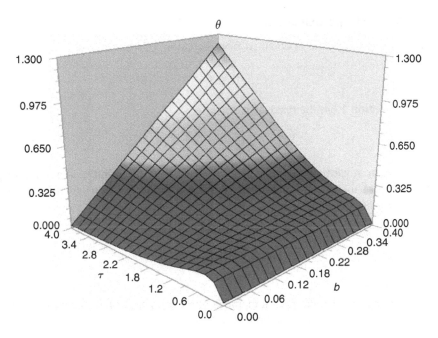

FIGURE 1.8 Dimensionless temperature (θ) as a function of dimensionless time (τ) and parameter b assuming $c = 1$ and $\gamma = 5$.

environment surpasses the rate of energy generation owing to the reaction and the temperature continues to decline.

4. Increases in parameter γ lead to slower reactions. Comparing Figures 1.7 and 1.8 (plotted for the same c), it is possible to verify that at any given time, higher values of γ lead to slower heating rates and lower temperatures.

1.5.2 SOLUTION BY WEIGHTED RESIDUALS

For some situations, it might be difficult or even impossible to find exact analytical solutions. In these cases, approximate methods may be employed. Among those, there is the method of weighted residuals (MWR), which is presented at Appendix E.

To show few details regarding the application of MWR as well as comparisons between various solutions, consider the same problem of the heated batch reactor.

Respecting boundary condition given by Equation 1.48, a possible trial function would be

$$\bar{\theta}_n = \sum_{j=1}^{n} C_j \tau^j \tag{1.54}$$

Here C_j are constants to be determined according to the following procedures.

1.5.2.1   First Approximation

The first approximation is obtained from Equation 1.54 as

$$\bar{\theta}_1 = C_1 \tau \tag{1.55}$$

Using Equation 1.44, the residual is

$$\Lambda_1 = C_1(1 + \tau) - b\tau - ce^{-\gamma\tau} \tag{1.56}$$

The domain regarding variable time is from zero to infinity; therefore, the integration of the residue multiplied by the weighting functions should be performed within this interval or

$$\int_0^\infty \Lambda_1 W_1 \, d\tau = 0 \tag{1.57}$$

As shown below, the various submethods differ from each other according to the weighting function applied in the integration.

1.5.2.1.1   Method of the Moments

From the exposed material at Appendix E, the weighting function for this method may be an orthogonal polynomial. However, to simplify, let us apply the following:

$$W_n = \tau^{n-1} \tag{1.58}$$

Therefore, for the first approximation, Equations 1.56 and 1.57 lead to

$$\int_0^\infty [C_1(1+\tau) - b\tau - ce^{-\gamma\tau}] \, d\tau = \left[C_1\tau + (C_1 - b)\frac{\tau^2}{2} + \frac{1}{\gamma}ce^{-\gamma\tau} \right]_{\tau=0}^\infty = 0 \tag{1.59}$$

Since τ might assume infinite value, it becomes impossible to determine C_1. Therefore, an alternative route should be tried.

1.5.2.2   Another Alternative

There are two possibilities that avoid such undefined values:

1. Solve the problem for an important time interval regarding the physical problem. Of course, it is unlikely that anyone would be interested in the reactor behavior throughout eternity. Even the steady-state regime would be approached after a finite period. Now, let us say that one would like to follow the process up to a time τ_f. This period may be, for instance, the one set for the batch period. Therefore, Equation 1.59 would be written as

$$\int_0^{\tau_f} (C_1 + C_1\tau - b\tau - ce^{-\gamma\tau})\, d\tau = \left[C_1\tau + (C_1 - b)\frac{\tau^2}{2} + \frac{1}{\gamma}ce^{-\gamma\tau} \right]_{\tau=0}^{\tau_f} = 0 \quad (1.60)$$

This would allow us to determine C_1 as

$$C_1 = \frac{\gamma b\tau_f^2 + 2c(1 - e^{-\gamma\tau_f})}{\gamma\tau_f(2 + \tau_f)} \quad (1.61)$$

2. The second alternative is to work with another trial function that could avoid infinite limits. Among them there is

$$\bar{\theta}_n = \sum_{j=1}^{n} C_j\left(1 - e^{-j\tau}\right) \quad (1.62)$$

Note that these trial functions obey the boundary condition given by Equation 1.48.

The first approximation would be

$$\bar{\theta}_1 = C_1(1 - e^{-\tau}) \quad (1.63)$$

Using Equation 1.44, the residue would become

$$\Lambda_1 = C_1 - b\tau - ce^{-\gamma\tau} \quad (1.64)$$

Of course, weighting functions should be carefully selected. If simple polynomials were chosen, terms with infinity limits would appear again. A simple choice to avoid is

$$W_n = e^{-n\tau} \quad (1.65)$$

For now, let us try a combination, i.e., apply trial functions given by Equation 1.63 and set a time τ_f as the upper limit of our domain.

1.5.2.2.1 Method of Collocation

According to the explanation at Appendix E, this method asks the residue to vanish at the collocation points. As a first attempt, let the middle of the interval be a collocation point, i.e., $\tau_f/2$. Therefore, from Equation 1.64 one gets

$$C_1 = \frac{b}{2}\tau_f + ce^{-\gamma\frac{\tau_f}{2}} \quad (1.66)$$

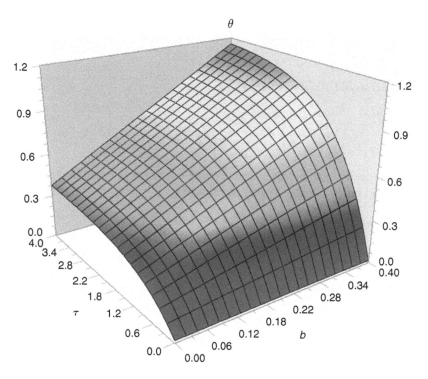

FIGURE 1.9 Dimensionless temperature (θ) as a function of dimensionless time (τ) and parameter b assuming $c = 1$ and $\gamma = 0.5$ obtained as first approximation using the collocation method.

Using Equation 1.66 in Equation 1.63, it is possible to write

$$\bar{\theta}_1 = \left(\frac{b}{2}\tau_f + ce^{-\gamma\frac{\tau_f}{2}} \right)\left(1 - e^{-\tau} \right) \tag{1.67}$$

Despite far from reproducing the exact solution (Equation 1.53), this first approximation already shows some promising similarity to it. To illustrate this, Equation 1.67 was computed choosing $c = 1$, $\gamma = 0.5$, and $\tau_f = 4$, which is the maximum period applied in previous graphs for the exact solution. The values are plotted leading to Figure 1.9, which can be compared with Figure 1.7 built for the same values of c and γ.

1.5.2.2.2 Method of Subdomain
Here, the integral of the residue should be equated to zero. Because of the choice made before for the weighting function at the method of moments (Equation 1.58), the integration for the first approximation by the method of subdomain is very simple and left as an exercise.

1.5.2.2.3　Method of Least Squares
As seen in Appendix E, this method sets the weighting function as

$$W_1 = \frac{\partial \Lambda_1}{\partial C_1} \tag{1.68}$$

Applying Equation 1.64, $W_1 = 1$ and the coefficient C_1 would be determined by

$$\int_0^{\tau_f} (C_1 - b\tau - ce^{-\gamma\tau}) \, d\tau = 0 \tag{1.69}$$

This leads to

$$C_1 = \frac{b}{2}\tau_f + \frac{c}{\gamma\tau_f}(1 - e^{-\gamma\tau_f}) \tag{1.70}$$

The reader is invited to use this in Equation 1.63 and compare the result with the exact solution (Equation 1.53), in the same way as performed after the application of collocation method.

1.5.2.2.4　Method of Galerkin
The Galerkin method requires the weighting function to be written as

$$W_1 = \frac{\partial \bar{\theta}_1}{\partial C_1} = 1 - e^{-\tau} \tag{1.71}$$

Therefore

$$\int_0^{\tau_f} (1 - e^{-\tau})(C_1 - b\tau - ce^{-\gamma\tau}) \, d\tau = 0 \tag{1.72}$$

This allows determining C_1. The details are left as exercise.

1.5.2.3　Second Approximation

It is possible to verify that almost no progress would be obtained if Equation 1.62 were applied to obtain a second approximation. To avoid that, one may think on changing Equation 1.62 and adopt the following:

$$\bar{\theta}_n = (1 - e^{-\tau}) \sum_{j=1}^{n} C_j \tau^j \tag{1.73}$$

The reader is asked to discuss this possibility and use the above to obtain a second approximation.

1.5.2.4 When to Stop

Of course, one would apply MWR only to problems in which an exact solution could not be found. Therefore, one would ask how many approximations would be reasonable to achieve an approximated solution within a given level of deviation. This important aspect is discussed at Section E.2.

1.6 REACTOR WITH A TIME-CONTROLLED RATE

The present problem exemplifies the solution of a differential equation with variable coefficients.

Consider the case of a well-stirred batch reactor as shown by Figure 1.4. Now, the rate of energy production by an exothermic reaction is a function of not only time but also temperature as follows:

$$R_{Q,\text{reaction}} = a_1(T - T_0)t \qquad (1.74)$$

If assumptions as stated at Section 1.4 continue to be valid and, again, the initial temperature in the tank is equal to the ambiance, the use of variables given by Equations 1.27 and 1.28 into Equation 1.25 gives

$$\frac{d\theta}{d\tau} = -\theta + b\tau + c\theta\tau \qquad (1.75)$$

Here b is given by Equation 1.32 and c is provided by

$$c = \frac{a_1 \rho V C_p}{(\alpha A)^2} \qquad (1.76)$$

As seen, the last term in the right-hand side of Equation 1.75 introduces a variable coefficient. This sort of differential equation is called linear with nonconstant coefficients. Despite the possibility of solving the present problem using the exact method (as shown below), one usually faces difficulties in finding analytical solutions for this class of differential equations. The MWR methods may be applied; however, let us now introduce the use of Picard's method, as explained in Appendix B.

As before, the boundary condition for this problem is

$$\theta(0) = 0 \qquad (1.77)$$

1.6.1 Solution by Picard's Method

According to Appendix B, the method can be applied to linear as well as non-linear differential equations as long as they can be put in the form given by Equation B.13 under the boundary condition that can be expressed by Equation B.14, or a first-kind boundary condition. In the present case, Equations 1.75 and 1.77 satisfy those requirements.

The first approximation can be written as

$$\theta_1 = \theta_0 + \int_0^\tau (-\theta_0 + b\tau + c\theta_0\tau)\, d\tau \qquad (1.78)$$

Owing to the boundary condition, which implies $\theta_0 = 0$, the following is obtained:

$$\theta_1 = \frac{b}{2}\tau^2$$

The second approximation becomes

$$\theta_2 = \int_0^\tau \left(-\frac{b}{2}\tau^2 + b\tau + \frac{bc}{2}\tau^3\right) d\tau = -b\left(\frac{\tau^3}{6} - \frac{\tau^2}{2} - \frac{c\tau^4}{8}\right) \qquad (1.79)$$

The third approximation is

$$\theta_3 = \int_0^\tau \left[b\left(\frac{\tau^3}{6} - \frac{\tau^2}{2} - \frac{c\tau^4}{8}\right) + b\tau - cb\left(\frac{\tau^4}{6} - \frac{\tau^3}{2} - \frac{c\tau^5}{8}\right)\right] d\tau$$

$$= b\left(\frac{\tau^4}{24} - \frac{\tau^3}{6} - \frac{c\tau^5}{40}\right) + \frac{b}{2}\tau^2 - cb\left(\frac{\tau^5}{30} - \frac{\tau^4}{8} - \frac{c\tau^6}{48}\right)$$

The fourth approximation becomes

$$\theta_4 = -b\left(\frac{\tau^5}{120} - \frac{\tau^4}{24} - \frac{c\tau^6}{240}\right) - \frac{b\tau^3}{6} + cb\left(\frac{\tau^6}{180} - \frac{\tau^5}{40} - \frac{c\tau^7}{336} + \frac{\tau^6}{144} - \frac{\tau^5}{30} - \frac{c\tau^7}{280}\right)$$

$$+ \frac{b}{2}\tau^2 + \frac{cb\tau^4}{8} - c^2b\left(\frac{\tau^7}{210} - \frac{\tau^6}{48} - \frac{c\tau^8}{384}\right) \qquad (1.80)$$

The process continues until the difference between the nth and $(n+1)$th approximation is equal or smaller than the desired deviation between model and reality, or below the expected error in measurements of temperature or other variables. For instance, let us assume 0.1 K as the maximum allowed deviation for temperature. Using Equation 1.27, it is easy to see that

$$d\theta = \frac{1}{T_0} dT \qquad (1.81)$$

Therefore, the desired deviation would represent variation of $0.1/T_0$ for θ. If, as an example, T_0 were 300 K, the maximum deviation for the dimensionless temperature would be 3.33×10^{-3}. In addition, let us take the case of a tank with 1 m diameter filled up to 1 m high with a fluid with properties similar to water. The heat transfer coefficient to still surrounding air is assumed as 2 W m^{-2} K^{-1} and constant a around 1 W s^{-1}. In this case, parameter b would be equal to 66.314.

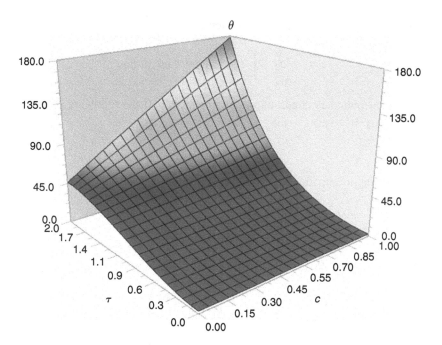

FIGURE 1.10 Second approximation of dimensionless temperature (θ) against dimensionless time (τ) and parameter (c).

Using these values, the approximations θ_2 and θ_4 at various values of dimensionless time τ and parameter c, which is related to the rate of energy delivery to the system by Equation 1.76, are presented by Figures 1.10 and 1.11, respectively.

From these, the following are noticeable:

- For c equal to zero, or cases with no extra heating, the reactor temperature tends to a limit.
- Temperature would increase faster against time for larger values of parameter c.

1.6.2 EXACT SOLUTION

Exceptionally, the present problem of nonlinear differential equation allows analytical exact solution.

Applying the method of parameter variation (see Appendix B), the following is written:

$$\theta = u_1(\tau)\, u_2(\tau) \tag{1.82}$$

Applying this on Equation 1.75 leads to

$$u_2(u_1' + u_1 - c\tau u_1) + u_2' u_1 = b\tau \tag{1.83}$$

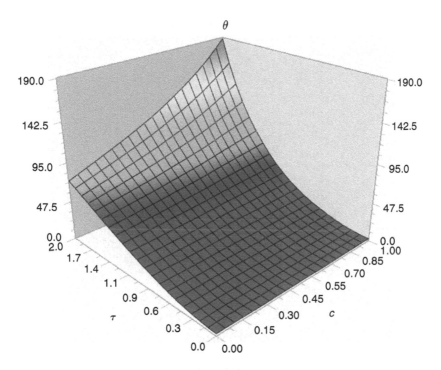

FIGURE 1.11 Fourth approximation of dimensionless temperature (θ) against dimensionless time (τ) and parameter (c).

Setting the term between parentheses as zero, one obtains

$$u_1 = \exp\left(\frac{c\tau^2}{2} - \tau\right) \tag{1.84}$$

Using Equation 1.84 in Equation 1.83, the following results:

$$u_2 = b\int_0^\tau \tau\exp\left(\tau - \frac{c\tau^2}{2}\right)d\tau \tag{1.85}$$

It is important to notice the definite integration, which complies with boundary condition given in Equation 1.77.

Finally

$$\theta = b\exp\left(\frac{c\tau^2}{2} - \tau\right)\int_0^\tau \tau\exp\left(\tau - \frac{c\tau^2}{2}\right)d\tau \tag{1.86}$$

Using the same value for parameter b (66.314) as before, the solution is illustrated by Figure 1.12.

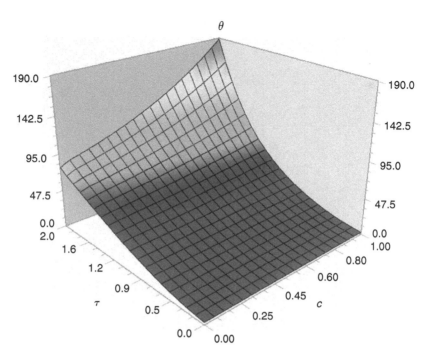

FIGURE 1.12 Exact solution of dimensionless temperature (θ) against dimensionless time (τ) and parameter (c).

Comparing Figure 1.12 with the Figures 1.10 and 1.11, it is easy to verify that the progress of approximations obtained by the Picard's method would tend to the exact solution.

The earlier observations based on the approximate solutions regarding the reactor behavior are valid.

1.7 PRESSURE IN A RESTING FLUID

Let us consider the simple example of the dependence of pressure within a fluid just due to the weight of its own column above the considered point. Figure 1.13 illustrates the situation.

The following are assumed:

1. Fluid is Newtonian with constant density and at rest inside a vertical container.
2. No movement is observed in the fluid.
3. Temperature of the fluid is uniform and is the same as that of the environment, which does not change as well.
4. Gravitational acceleration is constant throughout the entire region occupied by the fluid.

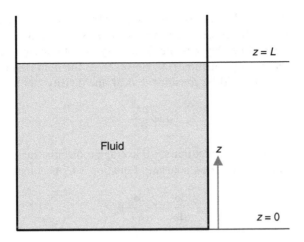

FIGURE 1.13 Stationary column of fluid.

The application of mass continuity to Equation A.1 leads to a zero identity, therefore to no information.

Applying the momentum Equation A.9 in the vertical direction, it is possible to write

$$\frac{dp}{dz} = \rho g_z \qquad (1.87)$$

Owing to assumption 4, the integration of Equation 1.87 from $z=0$ to $z=z$ is straightforward, or

$$p - p_0 = -\rho g z \qquad (1.88)$$

The negative sign on the right-hand side, or $g = -g_z$ is due to the upward orientation of coordinate z. Here p_0 is the pressure at the bottom of the vessel ($z=0$). This pressure can be put as function of the height if Equation 1.87 is integrated from $z=0$ to $z=L$ to give

$$p_L - p_0 = -\rho g L \qquad (1.89)$$

Therefore, it becomes

$$p = p_L + \rho g(L - z) \qquad (1.90)$$

The letter p represents the absolute pressure at a point z. If the pressure p_L were only due to atmosphere above the column, the relative pressure would be given just by the weight of the fluid above the considered point. This is a well-known result.

Let us assume now that the column is very high and the fluid is a gas (air, for instance). Therefore, the constant density hypothesis is no longer valid and the density would depend on the pressure. If density is not constant, Equation A.6 should be used instead of Equation A.9. However, as the fluid is at rest, Equation 1.87 holds. If the gas could be considered ideal, the density can be written as

$$\rho = \frac{pM}{\tilde{R}T} \tag{1.91}$$

Here M is the average molecular mass.* If average molecular mass as well as the temperature could be considered constant, Equation 1.87 becomes

$$\frac{dp}{dz} = -\frac{Mg}{\tilde{R}T}p \tag{1.92}$$

1.7.1 SOLUTION BY SEPARATION OF VARIABLES

If the other assumption could be kept as valid, Equation 1.92 is separable and the integration from $z = 0$ to $z = z$ leads to

$$p_z = p_0 e^{-\frac{Mg}{\tilde{R}T}z} \tag{1.93}$$

Here p_0 would be the pressure at $z = 0$ (or sea level, for instance).

As seen, the pressure decays exponentially with height in the atmosphere. Of course, this is still a great simplification because the temperature in the atmosphere varies with altitude as well. Let us now assume that the form of this dependence is given by

$$T = T_0 e^{-az} \tag{1.94}$$

T_0 is the temperature at $z = 0$ (sea level) and a is a positive constant. In this case, Equation 1.92 becomes

$$\frac{dp}{dz} = -\frac{Mg}{\tilde{R}T_0}e^{az}p \tag{1.95}$$

This is still a separable equation. The integration would lead to

$$p_z = p_0 \exp\left[\frac{Mg}{\tilde{R}T_0 a}(1 - e^{az})\right] \tag{1.96}$$

The next improvement would be to consider the variation of the gravitational field. This is left as an exercise.

---

* Actually, because of variation in the gravitational acceleration, a separation process takes place and the composition or molecular mass of air varies with altitude.

1.8 PRESSURE IN FLUID UNDER ROTATIONAL MOVEMENT

Let any fluid be placed between two very long vertical cylindrical drums. The outer one, with radius equal to r_o, rotates at constant angular velocity Ω and the inner one, with radius equal to r_i, remains stationary as shown by Figure 1.14.

It is desired to determine the velocity of the fluid at any radial position.

As usual, the list of assumptions follows:

1. Fluid between the drums is Newtonian with constant density.
2. Fluid viscosity is constant.
3. All other variables are kept constant, i.e., steady-state regime takes place.

Because of the obvious geometry, the cylindrical coordinates would be adopted.

As the only velocity component is the angular one, the continuity equation, Equation A.2, gives

$$\frac{\partial}{\partial \theta}(v_\theta) = 0 \tag{1.97}$$

In other words, the velocity does not vary with the angle. Of course, this is no surprise.

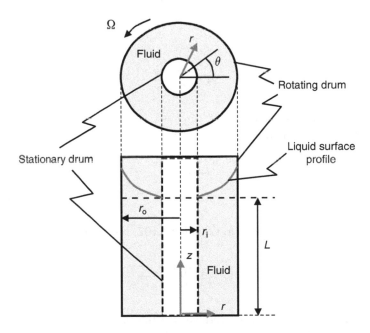

FIGURE 1.14 Scheme of concentric drums to impose a rotating field in a fluid between them.

The momentum continuity in the radial direction, or Equation A.13, provides

$$\frac{\partial p}{\partial r} = \rho \frac{v_\theta^2}{r} \tag{1.98}$$

It should be noticed that g_r is zero because the drums are in the vertical position. The momentum in the angular direction or Equation A.14 leads to

$$\frac{d}{dr}\left[\frac{1}{r}\frac{d(rv_\theta)}{dr}\right] = 0 \tag{1.99}$$

Finally, momentum in vertical direction or Equation A.15 allows writing

$$\frac{\partial p}{\partial z} = \rho g_z \tag{1.100}$$

1.8.1 SOLUTION BY SEPARATION OF VARIABLES

Equation 1.99 is separable, and

$$\frac{d(rv_\theta)}{dr} = ar \tag{1.101}$$

Here a is constant. Therefore

$$v_\theta = \frac{ar}{2} + \frac{b}{r} \tag{1.102}$$

where b is another constant.
The boundary conditions are

$$v_\theta(r_i) = 0 \tag{1.103}$$
$$v_\theta(r_o) = \Omega r_o \tag{1.104}$$

After applying these equations to Equation 1.102, it is possible to write

$$v_\theta = \Omega \frac{r_o^2}{r}\frac{r^2 - r_i^2}{r_o^2 - r_i^2} \tag{1.105}$$

The pressure profile against the vertical direction is the same as for the resting fluid. Actually, as the pressure varies with the height as well, the complete pressure profile can be obtained using the definition of total derivative, or

$$dp = \frac{\partial p}{\partial r}dr + \frac{\partial p}{\partial z}dz \qquad (1.106)$$

Applying Equations 1.98 and 1.100, one gets

$$dp = \rho\Omega^2 \frac{r_0^4}{r^3}\left(\frac{r^2 - r_i^2}{r_0^2 - r_i^2}\right)^2 dr - \rho g \, dz \qquad (1.107)$$

Since each right-side term depends only on the respective variable, the integrations can be performed without any problem. However, two integration constants would appear, or

$$p = \rho\Omega^2 r_0^4 \int \frac{1}{r^3}\left(\frac{r^2 - r_i^2}{r_0^2 - r_i^2}\right)^2 dr + p_r(z) - \rho g z + p_z(r) \qquad (1.108)$$

Functions $p_r(z)$ and $p_z(r)$ should be set. For instance, at the bottom of the space between the drums, or $z = 0$, the pressure might be known. Even better, if the pressure at the surface of the fluid wetting the inner drum ($z = L$) is the atmospheric pressure (p_0), then $p_r(L) = p_0$ and $p_z(r_i) = p_0$, and

$$p = p_0 + \rho\Omega^2 r_0^4 \int_{r_1}^{r} \frac{1}{r^3}\left(\frac{r^2 - r_i^2}{r_0^2 - r_i^2}\right)^2 dr - \rho g(L - z) \qquad (1.109)$$

Of course, the height (L) of the surface at the internal drum surface is a function of the geometry and rotational velocity, or $L = L(r_i, r_o, \Omega)$. Therefore, the present solution is not general and depends on that information.

The form of the liquid surface can be obtained by the reverse problem, i.e., finding the function describing the form when the pressure is atmospheric. To simplify, let us take the case when $r_i = 0$. Hence, there is no internal drum and the liquid rotates as in a revolving cup. Therefore, Equation 1.109 becomes

$$p = p_0 + \frac{1}{2}\rho\Omega^2 r^2 - \rho g(L - z) + p_z(r) \qquad (1.110)$$

At $z = L$ and $p = p_0$, the pressure should not depend on the radius; therefore, $p_z = L(r) = 0$. Therefore, the form of the surface would be given by

$$z = L + \frac{\Omega^2}{2g}r^2 \qquad (1.111)$$

A parabolic surface would form at the liquid surface, which is illustrated by Figure 1.15.

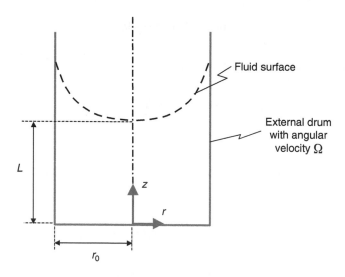

FIGURE 1.15 Form of a liquid surface in a rotating drum.

1.9 PLUG-FLOW REACTOR

Consider a tubular reactor as in Figure 1.16.

Once the temperature, pressure, velocity, and composition of the injected mixture are known, it is desired to obtain these properties for the leaving stream from the reactor.

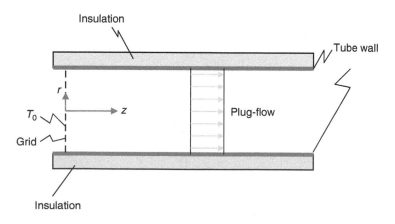

FIGURE 1.16 Plug-flow reactor.

Of course, depending on the degree of precision required and the complexity of the reaction or reactions taking place inside the reactor, the problem can become difficult to solve, at least to achieve an analytical solution. One can imagine that the reactions might be endothermic or exothermic, which will lead to decreasing or increasing temperatures for positions ahead of the entrance. This, combined with the changes in the composition from point to point, leads to changes in the physical and chemical properties of the traveling mixture. The complete solution for such a problem would require the use of momentum, energy, and mass conservation equations along with a considerable amount of auxiliary routines for computations of physical–chemical properties of mixtures as functions of temperature, pressure, and composition. In addition, routines to compute the rates of involved reactions should be provided. When solving the problem using a dimensional approach, even if laminar flow could be assured, one still has to consider variations of temperature, composition, and pressure in two or three directions. This last one would be mandatory if rotational flow occurs. In cases of unsteady-state regime, time would be added as another independent variable. If the problem involves turbulent flow, the problem becomes much more complex. Of course, such problems can only be solved by numerical methods, mainly by commercial computational fluid dynamics (CFD) programs. More details and considerations about modeling reactive systems can be found elsewhere [4].

Our objective here is much more modest, i.e., to obtain the temperature profile inside the reactor, when the following are assumed:

1. Steady-state regime.
2. At position $z = 0$, the fluid is at uniform temperature T_0.
3. Tube is perfectly insulated, i.e., reactor has adiabatic walls.
4. Velocity profile is flat, or does not depend on the radial direction. This model is also called plug-flow regime. Despite an approximation, it is very useful and often applied for a first attack to reactor design.
5. Fluid properties are constants. This is another approximation since the chemical species are reacting and the composition changes from point-to-point in the reactor.
6. Chemical reaction delivers a constant and uniform rate (R_Q) of energy per unit of volume of the fluid. This is very difficult to achieve because the reaction rate depends on the concentration of reactants, which in turn are decreasing for increasing values of coordinate z. Therefore, the objective of the present example is just to illustrate a solution of differential equation.

The above assumptions allow one to simplify Equation A.35 and to write

$$\rho C_p v_z \frac{\partial T}{\partial z} = \lambda \frac{\partial^2 T}{\partial z^2} + R_Q \qquad (1.112)$$

It is important to notice that no variation of temperature occurs in the radial coordinate. This is due to the combination of following assumptions:

- Assumption 3, which first imposes no heat flux in that direction at the wall, or

$$q_r|_R = -\lambda \frac{\partial T}{\partial r}\bigg|_R = 0 \qquad (1.113)$$

- Assumptions 4 and 6 guarantee evenly distributed energy release and flat temperature profile. Therefore, no heat flux will appear in the radial direction.

Even though Equation 1.112 is a second-order differential equation, it can be easily converted into a first-order equation by

$$\theta(z) = \frac{dT}{dz} \qquad (1.114)$$

Thus, Equation 1.112 becomes

$$\frac{d\theta}{dz} - a\theta + b = 0 \qquad (1.115)$$

Here

$$a = \frac{\rho C_p v_z}{\lambda} \qquad (1.116)$$

and

$$b = \frac{R_Q}{\lambda} \qquad (1.117)$$

One boundary condition is

$$T(0) = T_0 \qquad (1.118)$$

1.9.1 SOLUTION BY SEPARATION OF VARIABLES

As long as parameters a and b in Equation 1.115 remain constant, the solution for the separable equation is straightforward leading to

$$\theta = \frac{dT}{dz} = \frac{C_1}{a} e^{az} - \frac{b}{a} \qquad (1.119)$$

This can be integrated again to give

$$T = \frac{C_1}{a^2} e^{az} - \frac{b}{a} z + C_2 \tag{1.120}$$

As seen, apart from the condition given by Equation 1.118, another condition is necessary. A possibility is to set the derivative of temperature at $z=0$ as zero as well, or

$$\theta(0) = 0 \tag{1.121}$$

Of course, this is one possibility among various others. However, one would be sure after a careful examination of the real problem at hand. As the fluid mixture enters the reactor, it is reasonable to expect a zero derivative at entrance position either for exothermic or endothermic reactions. Therefore, the condition given by Equation 1.120 is a distinct possibility and its application leads to

$$T = \frac{b}{a^2} e^{az} - \frac{b}{a} z + C_2 \tag{1.122}$$

Finally, applying the condition given by Equation 1.118 provides

$$T = T_0 + \frac{b}{a^2} (e^{az} - 1) - \frac{b}{a} z \tag{1.123}$$

To facilitate and generalize the representations by graphs, Equation 1.123 may be written in a dimensionless form as

$$\Phi = \frac{T - T_0}{T_0} = c_1 \left(e^{c_2 \zeta} - 1 \right) - c_1 c_2 \zeta \tag{1.124}$$

Here

$$c_1 = \frac{b}{a^2 T_0} \tag{1.125}$$

$$c_2 = aR \tag{1.126}$$

$$\zeta = \frac{z}{R} \tag{1.127}$$

To discuss an example, let parameter c_2 be equal to 0.1. With this, Figure 1.17 shows the dimensionless temperature profile in the reactor for parameter c_1 varying between 0 and 1.

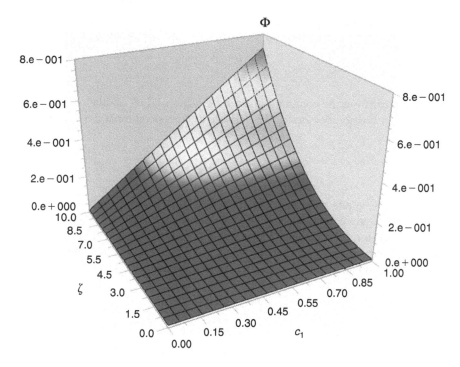

FIGURE 1.17 Graph of dimensionless temperature (Φ) against dimensionless length (ζ) and parameter (c_1) inside the reactor for parameter $c_2 = 0.1$.

1.9.2 COMMENTS

It is interesting to observe the following:

- As seen from Figure 1.17, the temperature increases for positions ahead of the entrance. The rate of increase is higher for larger values of parameter c_1, or larger values of b, which according to Equation 1.117 is proportional to the reaction rate.
- As the reaction is irreversible, the temperature does not know boundaries. Of course, this does not represent a possible real situation; nonetheless, it can be applied to very certain cases and within a given range of conditions. For instance, several combustion reactions can be considered irreversible within a relatively large range of conditions.
- Assumption 5 is among the most critical, mainly in cases of gaseous reacting media where severe changes of density and velocity usually occur. However, despite this, Equation 1.112 can be put into a very convenient form that allows its application even in such a situation. This is so because

$$F = \rho v_z A \tag{1.128}$$

where F is the mass flow and A is the reactor cross-sectional area. Both do not change throughout the reactor. Therefore, despite changes in the density and velocity, Equation 1.112 can be written as

$$FC_p \frac{\partial T}{\partial z} = A\lambda \frac{\partial^2 T}{\partial z^2} + AR_Q \tag{1.129}$$

This would lead to Equation 1.115, with just changing parameter a to

$$a = \frac{FC_p}{A\lambda} \tag{1.130}$$

The solution is the same as shown above.

- Owing to the decrease of reactants, assumption 6 is also critical. Actually, it is almost impossible to maintain a constant rate of energy delivery or source because of a reaction or reactions. One more acceptable relation for R_Q would be

$$R_Q = a_1 \exp\left(-\frac{a_2}{T}\right) \tag{1.131}$$

which reminds us of the classical Arrhenius equation for reaction rate. As the rate of energy delivery follows the reaction rate, the above equation might be reasonable as a first approximation. However, such an equation would not allow the technique described above, which transforms a second-order differential equation into two first-order equations.

1.10 HEAT CONDUCTION IN AN INDEFINITE WALL

Let us consider the classical problem of heat conduction through a wall of thickness L but with an infinite area and length, which is illustrated by Figure 1.18.

The temperatures at both faces are set and the temperature profile in the wall against the coordinate x at steady-state condition has to be determined.

Despite the apparent simplicity of this problem, several assumptions should be made to allow an analytical solution. Of course, depending on the assumptions, any problem can become cumbersome. The solution usually found in textbooks assumes thermal conductivity and other properties as constants. However, the real situation should include at least the thermal conductivity as a function of temperature. For most of the solid materials, the thermal conductivity decreases with increase in temperature. Let us solve the problem for such a situation. For clarity's sake, the following are assumed:

1. The wall has an indefinite dimension in a direction orthogonal to x. This allows assuming temperature variations only through the thickness of the wall.

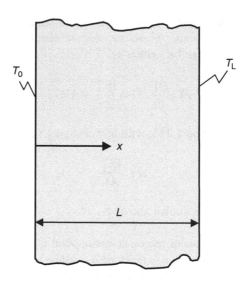

FIGURE 1.18 Conduction through infinite wall with constant temperature at each face.

2. Steady-state condition, i.e., the temperatures or any other property or variable of the problem modifies with time. In other words, the temperatures T_0 and T_L also remain constant.
3. Apart from thermal conductivity, all other properties of the wall material do not depend on the temperature. This might become a rough approximation, if the difference between the temperatures at each face increases.
4. Thermal conductivity is a function of temperature. This is discussed below.
5. Solid material of the wall does not go through any phase change or chemical transformation.

As the thermal conductivity is not constant, Equation A.34 cannot be used and Equation A.31 should be applied to set the basic equation. After the above assumptions, it can be simplified to

$$\frac{dq_x}{dx} = 0 \tag{1.132}$$

Using Equation A.22, the equation becomes

$$\lambda \frac{dT}{dx} = C_1 \tag{1.133}$$

Notwithstanding a second-order nature of the problem, Equation 1.132 leads to a first-order one. However, two boundary conditions should be set. Those are

$$T(0) = T_0 \tag{1.134}$$

$$T(L) = T_L \tag{1.135}$$

Usually, the conductivity of metals is predicted as a function of temperature by Perry and coworkers [5]:

$$\lambda = aT - bT^2 + \frac{c}{T} \tag{1.136}$$

Despite the dependence of parameters a, b, and c on the density, specific heat, and electrical conductivity of the metal, they would be assumed as constants. This is a reasonable approximation because the thermal conductivity usually is more sensitive to temperature variations than these other properties.

Combining Equations 1.135 and 1.136, one gets

$$\left(aT - bT^2 + \frac{c}{T}\right)\frac{\mathrm{d}T}{\mathrm{d}x} = C_1 \tag{1.137}$$

1.10.1 SOLUTION BY SEPARATION OF VARIABLES

Since Equation 1.137 is separable, the differential equation is easily integrated to give

$$a\frac{T^2}{2} - b\frac{T^3}{3} + c\ln T = C_1 x + C_2 \tag{1.138}$$

After applying conditions given by Equations 1.134 and 1.135, it is possible to write

$$\frac{\frac{a}{2}(T^2 - T_0^2) - \frac{b}{3}(T^3 - T_0^3) + c\ln\frac{T}{T_0}}{\frac{a}{2}(T_L^2 - T_0^2) - \frac{b}{3}(T_L^3 - T_0^3) + c\ln\frac{T_L}{T_0}} = \frac{x}{L} \tag{1.139}$$

Regardless of its transcendent nature, the temperature profile in the wall as well as the heat transfer rate through it can be computed.

1.11   PLATE-AND-CONE VISCOMETER

Plate-and-cone devices, as illustrated in Figure 1.19, are used for determinations of liquid viscosity. A steady rotation is imposed in the upper cone while the lower plate is kept static and the tested fluid is held by surface tension between these two parts. Therefore, a velocity gradient is established in the fluid. The viscosity is related to the torque necessary to maintain the constant rotation (Ω) of the upper cone.

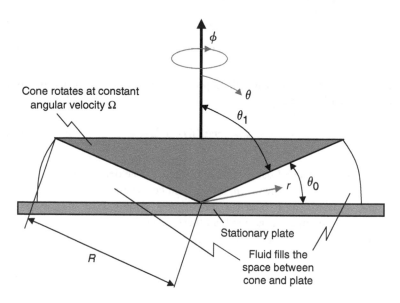

FIGURE 1.19 Vertical cut of the plate-and-cone viscometer.

The velocity profile in the fluid between the cone and the plate as well as an expression to provide the fluid viscosity as function of the angular velocity (Ω) and geometry of the viscometer has to be determined.

Let us clearly list the main assumptions:

1. Steady-state regime is established.
2. Temperature across the fluid is constant and uniform.
3. Fluid is Newtonian with constant physical properties.
4. Energy dissipation due to attrition is negligible.
5. There are no chemical reactions involved or any other source of energy present.
6. Flow is exclusively tangential; therefore, there are no velocity fields in the radial and angular θ directions, or $v_\theta = v_r = 0$. Because of the geometry of the device, this assumption is very reasonable as long as laminar flow is observed.

Departing from the continuity (Equation A.3) and using the assumptions listed before, it is possible to write

$$\frac{\partial v_\phi}{\partial \phi} = 0 \qquad (1.140)$$

Now, instead of working directly with the form involving velocities, let us use the stress forms. It will be shown that this is a more convenient form for the present situation. In this way, let us write the equations of the various components.

For the r-component, the stress form is given by Equation A.16. Having in mind, assumption 6 and Equation 1.140, one can apply Equations A.16a, A.17a, A.18a, A.19a, and A.20a to find that all stress involved in the above are zero. Therefore, it is possible to write

$$\rho \frac{v_\phi^2}{r} = \frac{\partial p}{\partial r} \qquad (1.141)$$

The component of velocity in θ direction can be obtained from the Equation A.17. Likewise, as before, one may verify that the above would lead to

$$\rho \cot \theta \frac{v_\phi^2}{r} = \frac{1}{r} \frac{\partial p}{\partial \theta} \qquad (1.142)$$

The ϕ-component is given by Equation A.18. Using the same procedure as before, Equation A.18 can be simplified to

$$\frac{1}{r^2} \frac{\partial r^2 \tau_{r\phi}}{\partial r} + \frac{1}{r} \frac{\partial \tau_{\theta\phi}}{\partial \theta} + \frac{\tau_{r\phi}}{r} + \frac{2 \cot \theta}{r} \tau_{\theta\phi} = 0 \qquad (1.143)$$

Notice that pressure does not vary with angle ϕ.

Despite the apparently difficult problem involving the above partial differential equation, it will be shown that the present problem leads to a simple ordinary differential equation. This is so because the symmetry allows the following assumption:

$$v_\phi(r,\theta) = \frac{r}{R} v_\phi(R,\theta) = rf(\theta) \qquad (1.144)$$

Of course, this is an approximation and it leads to large deviations near the fluid surface ($r = R$). This point is discussed at the end of this section.

The following are the boundary conditions of the present attack:

$$v_\phi\left(r, \frac{\pi}{2}\right) = 0, \quad 0 < r < R \qquad (1.145)$$

$$v_\phi(r,\theta_1) = r\Omega \sin \theta_1, \quad 0 < r < R \qquad (1.146)$$

$$v_\phi(0,\theta) = 0, \quad \theta_1 < \theta < (\theta_1 + \theta_0) \qquad (1.147)$$

Notice that Equation 1.144 satisfies boundary conditions given by Equations 1.146 and 1.147. If Equation 1.144 is true, Equation A.21a leads to

$$\tau_{r\phi} = 0 \qquad (1.148)$$

Therefore, Equation 1.143 becomes

$$\frac{d\tau_{\theta\phi}}{d\theta} = -2\cot\theta \; \tau_{\theta\phi} \tag{1.149}$$

The solution of which is

$$\tau_{\theta\phi} = \frac{C}{\sin^2\theta} \tag{1.150}$$

From Equations A.20a and 1.144 the following can be expressed:

$$\tau_{\theta\phi} = -\mu\left[\frac{\sin\theta}{r}\frac{\partial}{\partial\theta}\left(\frac{v_\phi}{\sin\theta}\right)\right] = -\mu\sin\theta\frac{d}{d\theta}\left(\frac{f(\theta)}{\sin\theta}\right) \tag{1.151}$$

Combining Equations 1.151 and 1.150, one arrives at

$$\frac{d}{d\theta}\left(\frac{f(\theta)}{\sin\theta}\right) = -\frac{C}{\mu\sin^3\theta} \tag{1.152}$$

As seen, it is a separable ordinary first-order differential equation. Its integration yields

$$v_\phi = \frac{Cr}{2\mu}\left(\cot\theta - \sin\theta\ln\tan\frac{\theta}{2}\right) + C_1 r\sin\theta \tag{1.153}$$

The condition given by Equation 1.145 forces $C_1 = 0$. The condition given by Equation 1.146 leads to

$$v_\phi = \Omega r\sin\theta_1\frac{\cot\theta - \sin\theta\ln\tan\dfrac{\theta}{2}}{\cot\theta_1 - \sin\theta_1\ln\tan\dfrac{\theta_1}{2}} \tag{1.154}$$

From Equation 1.150, the shear stress $\tau_{\theta\phi}$ becomes

$$\tau_{\theta\phi} = 2\mu\Omega\frac{\sin\theta_1/\sin^2\theta}{\cot\theta_1 - \sin\theta_1\ln\tan\dfrac{\theta_1}{2}} \tag{1.155}$$

The torque between the fluid and the stationary base can be computed by

$$\text{Torque} = \int_0^{2\pi}\int_0^R \tau_{\theta\phi}\bigg|_{\theta=\frac{\pi}{2}} r^2 dr\, d\phi = \frac{4\pi\mu\Omega R^3}{3}\frac{\sin\theta_1}{\cot\theta_1 - \sin\theta_1\ln\tan\dfrac{\theta_1}{2}} \tag{1.156}$$

To simplify the representation of velocity profiles, let us rewrite Equation 1.154 using the following dimensionless variables:

$$Y = \frac{v_\phi}{R\Omega \sin\theta_1} \qquad (1.157)$$

$$\zeta = \frac{r}{R} \qquad (1.158)$$

Therefore

$$Y = \zeta \frac{\cot\theta - \sin\theta \ln\tan\dfrac{\theta}{2}}{\cot\theta_1 - \sin\theta_1 \ln\tan\dfrac{\theta_1}{2}} \qquad (1.159)$$

Figure 1.20 illustrates the dimensionless profile against dimensionless position (ζ) and angle (θ). In the present case, the value for $\theta_0 = \pi/10$, or $\theta_1 = \pi/2 - \pi/10$. The following can be noticed:

- Maximum velocity ($Y = 1$) is obtained at the end of the film ($\zeta = 1$) and at the rotating cone surface ($\theta = \theta_1$).

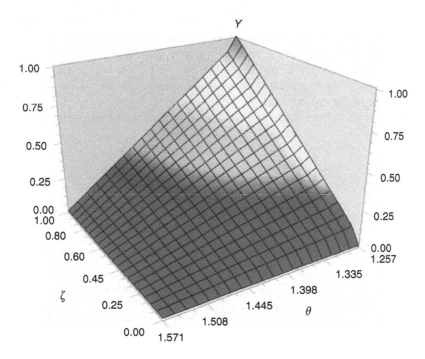

FIGURE 1.20 Fluid dimensionless velocity (Y) profile against angle (θ) and dimensionless radius (ζ).

- Velocity is zero at the base ($\theta = \pi/2$).
- Assumption of linear velocity profile against position r is reasonable because for positions near the surface meniscus ($r = R$, or $\zeta = 1$) the shear stress $\tau_{r\phi}$ should approach zero. Since v_r is zero, from Equation A.21a, the following can be written:

$$\left. \frac{\partial}{\partial r}\left(\frac{v_\phi}{r}\right)\right|_{r=R} = 0$$

Therefore, the dependence of velocity on radius should be linear.

A more rigorous approach of the present problem is presented in Chapter 11.

1.12 THERMOCOUPLE

Consider a thermocouple with a spherical extremity exposed to ambiance at temperature T_∞, as illustrated in Figure 1.21.

The thermocouple is composed of a wire and a solid sphere that surrounds it. The set, initially at an unknown relative low temperature, is inserted into an environment at higher temperature T_∞, such as the interior of a furnace. During the processes, heat transfer between the sphere and environment occurs by convection and radiation, whereas between that of the solid and the wire by conduction. Therefore, the average temperature (T) of the solid as well as the one at the wire (T_w) would be functions of time. T_w is also called "cold-end" of the thermocouple, while T remains as the "hot-end." The difference between T and T_w induces an electrical current at the wiring and allows measuring the temperature of the environment T_∞.

It is desired to have the relationship between the various temperatures against time.

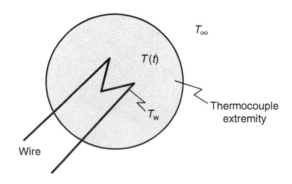

FIGURE 1.21 A thermocouple extremity exposed to ambiance at T_∞.

Let the following be the assumptions:

1. Spherical extremity is very small and metallic, which permits the approximation of uniform temperature throughout the whole solid. Actually, this can be securely assumed for such terminals because the Biot number defined by Equation 1.9 is always below 0.1.
2. As the temperature is uniform throughout the sphere, the heat transfers from the environment and to the wiring can be, respectively, seen as an uniformly distributed source and sink of energy.
3. No phase change takes place in the solid or in the wire, and therefore, no velocity field is present.
4. Air is transparent to the radiation at frequencies involved in the problem.
5. All involved bodies can be considered gray bodies for computations of heat transfer by radiation between the sphere and other bodies surrounding it.
6. Environment surrounding the sphere remains at constant temperature T_∞.
7. No chemical reaction or any other form of energy source is present in the sphere or in the air.
8. All physical properties of the sphere are constant. Obviously, this is another approximation and is valid for a relatively small range of temperature. As the thermocouples are usually applied to measure relatively high temperatures, such an assumption is among the most critical in the present treatment.

Once these conditions are accepted, the energy equation (Equation A.36) can be applied to the sphere surrounding the wire and can be written as

$$\rho C_P \frac{dT}{dt} = R_{Q,\text{envir.}} + R_{Q,\text{wire}} \tag{1.160}$$

On the other hand, a similar equation can be applied for the wire, or

$$\rho_w C_{Pw} \frac{dT_w}{dt} = -R_{Q,\text{wire}} \tag{1.161}$$

Here

$$R_{Q,\text{envir.}} = \frac{\alpha A(T_\infty - T)}{V} + \frac{\sigma(T_\infty^4 - T^4)A}{V\left(\frac{1-\varepsilon}{\varepsilon} + 1\right)} \tag{1.162}$$

$$R_{Q,\text{wire}} = -a(T - T_w) \tag{1.163}$$

The volume and area of the sphere are represented by V and A, respectively. The first term in the right-hand side of Equation 1.160 represents the energy per volume exchanged by convection between the sphere and surrounding air. The second term in the right-hand side represents the heat transfer by radiation between sphere and bodies in the environment. These basic aspects can be found in almost any reference dealing with radiative heat transfer [1,2,6–8].

Equations 1.161 and 1.162 would require a single boundary each. Of course, the most natural choice of boundary condition would be the one related to the initial temperature of the sphere. It is also reasonable that before being injected into the furnace, both sphere and wire were at the same temperature, say T_0. Therefore

$$T(0) = T_0 \tag{1.164}$$

and

$$T_w(0) = T_0 \tag{1.165}$$

The system formed by Equations 1.160 and 1.161 and boundary conditions given by Equations 1.164 and 1.165 can be solved by various methods. However, the solution of nonlinear systems is beyond the scope of the present text.

EXERCISES

1. Imagine a thin disk and a sphere, both with the same mass and made out of the same material. The lumped analysis would lead which one to smaller deviations? Justify your answer.
2. Repeat the treatment shown in Section 1.2 for a body where an exothermic chemical reaction takes place. Assume that the rate of energy generation due to chemical reactions R_Q remains constant.
3. From solution given at Section 1.4, deduce the relations to provide the instantaneous rate of heat transfer to the ambiance and the total heat transferred from the start to a given instant.
4. Repeat the treatment shown in Section 1.4 for the case where the coil provides a constant power input from $t=0$ until $t=t_1$ and stops operating from that time onward.
5. Observing Equation 1.44, an interesting situation arises when parameter γ approaches 1. Obtain the progress of the tank temperature for this case.
6. Demonstrate relations given by Equations 1.61 and 1.70.
7. Using the trial functions given by Equation 1.73, determine the first, second, and third approximations by any MWR method of the problem posed at Section 1.5. Compute the deviations (see Appendix E) among the successive approximations in order to decide, if the third approximation would be enough for a reasonable representation of the temperature inside the reactor as a function of time. Assume maximum acceptable deviation of 5%.

8. For the following values of parameters, $a = 1$ W s^{-1}, $A = 10$ m^2, $\alpha = 100$ W m^{-2} K^{-1}, $\rho = 1000$ kg/m^3, $V = 1$ m^3, $C_p = 4.2$ kJ kg^{-1} K^{-1}, $T_0 = 290$ K, $a_1 = 1000$ W, $a_2 = 0.1$ s^{-1}, and $\tau_f = 2\frac{\rho V C_p}{\alpha A}$ (or twice the time constant), compare the second approximation obtained in Problem 7 with the exact solution. This can be done by computing values for temperature at various instants between zero and τ_f.

9. Consider a cylindrical electrical wire with radius R and exposed to air at temperature T_a. The wire heats up when electrical current is imposed upon it. Assuming steady-state regime and constant rate (R_Q) of energy per unit of volume delivered to wire due to electrical energy dissipation, determine the
   (a) Temperature profile in the wire. The heat loss from the wire surface to ambiance due to convection transfer is given by $\alpha[T(r = R) - T_a]$, where α is constant and r is the radial coordinate ($r = 0$ at the center line of the wire). Answer: $T - T_a = \frac{R_Q}{4\lambda}(R^2 - r^2) + \frac{R_Q}{2\alpha}R$
   (b) Temperature at the surface of the wire. Answer: $T = T_a + \frac{R_Q}{2\alpha}R$
   (c) Rate (W m^{-2}) of heat transfer per unit of surface area to air. Answer: $R_Q\frac{R}{2}$

10. Try to solve Equation 1.75 using the method of variation of parameters, Laplace transform, and MWR (any type until second approximation). Compare that approximate solution with the ones obtained by other methods.

11. Using the method that better fits or leads to easier solution, solve the problem shown at Section 1.4 (Equations 1.25 and 1.26, and $T(0) = T_0 = T_\infty$) assuming now that the electrical resistance delivers a power to the fluid at a rate equal to
    (a) aT^2, where a is constant.
    (b) ae^{-ct}, where a and c are constants.
    (c) Repeat the above cases adding energy delivered by a chemical reaction following an Arrhenius form: $a\exp\left(-\frac{\tilde{E}}{\tilde{R}T}\right)$, where a and $\tilde{E}$ are constants ($\tilde{R}$ is the gas constant).
    (d) Repeat the previous problem using the method of Picard.

12. Develop a further improvement on the relation described in Section 1.7 by adding the influence of a varying gravitational field according to the famous law of decrease with the square of the distance.

13. Try to solve the problem presented at Section 1.7 assuming a gas with a state described by Van der Waals equation.

14. Imagine if in the problem of Section 1.8, the temperatures of inner and outer cylinders were different and kept constants at T_i and T_0, respectively. In addition, assume that an ideal gas is between the cylinders. Answer the following questions:
    (a) Would it be reasonable to assume a linear profile for the temperature in the gas between the cylinders? Justify your answer.
    (b) Determine the temperature profile between the cylinders.

15. Obtain the temperature profile for the fluid in Section 1.9 as a function of the derivative of temperature at $z = 0$.

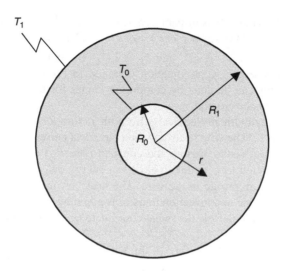

FIGURE 1.22 Conduction through a spherical shell with constant temperature at each face.

Answer: $T - T_0 = [a\theta(0) + b]\frac{1}{a^2}(e^{az} - 1) - \frac{b}{a}z$, where $\theta(0)$ is the derivative of temperature at $z = 0$.

16. In the former problem, what would the physical condition to obtain a linear temperature profile for the temperature? Discuss the significance of that situation.

17. Solve the temperature profile for the plug-flow reactor, described in Section 1.9 using Laplace transform.

18. Solve the problem presented in Section 1.9 when the rate of energy delivery by chemical reaction is given by $R_Q = a_1 \exp(-a_2 z)$ where a_1 and a_2 are constants.

19. Using the result presented in Section 1.10, determine the rate of heat transfer to the wall.

20. Repeat problem presented in Section 1.10 for the case of a sphere with cavity (or spherical shell), as shown by Figure 1.22. The external surface is kept at temperature T_1 while the internal one at T_0. Use the correlation shown by Equation 1.136 to describe the dependence of thermal conductivity on the temperature.

21. Using the results presented in Section 1.11, find the pressure profile as function of radius and angle (θ) within the viscometer fluid.

REFERENCES

1. Bird, R.B., Stewart, W.E., and Lightfoot, E.N., *Transport Phenomena*, John Wiley, New York, 1960.
2. Slattery, J.C., *Momentum, Energy, and Mass Transfer in Continua*, Robert E. Kriefer (Ed.), Huntington, New York, 1978.

3. Brodkey, R.S., *The Phenomena of Fluid Motions*, Dover, New York, 1967.
4. de Souza-Santos, M.L., *Solid Fuels Combustion and Gasification: Modeling, Simulation and Equipment Operation*, Marcel Dekker, New York, 2004.
5. Perry, J.H. et al., *Chemical Engineers' Handbook*, 4th ed., McGraw-Hill, New York, 1963, pp. 3–224.
6. Incropera, F.P. and DeWitt, D.P., *Fundamentals of Heat and Mass Transfer*, John Wiley, New York, 1996.
7. Schmidt, F.W., Henderson, R.E., and Wolgemuth, C.H., *Introduction to Thermal Sciences*, 2nd ed., John Wiley, New York, 1993.
8. Brewster, M.Q., *Thermal Radiative Transfer and Properties*, John Wiley, New York, 1992.

2 Problems 112; One Variable, 1st Order, 2nd Kind Boundary Condition

2.1 INTRODUCTION

This chapter presents methods to solve problems with one independent variable involving the first-order differential equation and the second-kind boundary condition. Mathematically, this class of cases can be summarized as $f\left(\phi, \omega, \frac{d\phi}{d\omega}\right)$, second-kind boundary condition.

2.2 HEATING OF A SOLID

Let us consider the same problem shown in Section 1.2. However, instead of the initial body temperature T_0, the derivative of its temperature is known at this instant or at any other instant (a). Again, one seeks to predict the temperature in the body against time.

The same assumptions made in Section 1.2 are valid here; therefore, the problem would lead to the same final differential equation as in Section 1.2, or

$$\frac{dT}{dt} = -\frac{\alpha A}{\rho V C_p}(T - T_\infty) \qquad (2.1)$$

Nevertheless, now the boundary condition is given by

$$\left.\frac{dT}{dt}\right|_{t=a} = b \qquad (2.2)$$

Here parameters a and b are known constants.

2.2.1 Solution by Separation of Variables

As in Section 1.2.1, the above differential equation is separable and can be easily integrated to give

$$\ln(T - T_\infty) = -\frac{\alpha A}{\rho V C_p} t + \ln C \qquad (2.3)$$

Here C is a constant. Thus

$$T = T_\infty + C \exp\left(-\frac{\alpha A}{\rho V C_p} t\right) \qquad (2.4)$$

Its derivative is

$$\frac{dT}{dt} = -\frac{\alpha A}{\rho V C_p} C \exp\left(-\frac{\alpha A}{\rho V C_p} t\right) \qquad (2.5)$$

Using condition given by Equation 2.2, the constant C is obtained, and Equation 2.4 becomes

$$T = T_\infty - \frac{b \rho V C_p}{\alpha A} \exp\left[\frac{\alpha A}{\rho V C_p}(a - t)\right] \qquad (2.6)$$

Notice that if $b > 0$, the body is heating up, and therefore the temperature T is smaller than that of the environment (T_∞). Of course, if $b < 0$ the contrary occurs.

The initial temperature of the body is found for $t = 0$, leading to

$$T_0 = T_\infty - \frac{b \rho V C_p}{\alpha A} \exp\left[\frac{\alpha A}{\rho V C_p} a\right] \qquad (2.7)$$

Finally, the relationship between T and T_0 can be obtained by combining the above two equations to give

$$T = T_0 + \frac{b \rho V C_p}{\alpha A} \exp\left(\frac{\alpha A}{\rho V C_p} a\right) \left[1 - \exp\left(-\frac{\alpha A}{\rho V C_p} t\right)\right] \qquad (2.8)$$

Combining Equations 2.6 and 2.7 it is possible to write

$$\frac{T - T_\infty}{T_0 - T_\infty} = \exp\left(-\frac{\alpha A}{\rho V C_p} t\right) \qquad (2.9)$$

This is exactly the same as Equation 1.6 if $t_0 = 0$.

2.3   HEAT CONDUCTION IN A SPHERICAL SHELL

Let us consider a problem similar to that presented in Section 1.10; however, for the case of a spherical shell and with a second-kind boundary condition, as illustrated in Figure 2.1.

The temperature at the central cavity surface $(r = R_0)$ is set as T_0 while a constant heat flux q_1 is maintained on the external surface $(r = R_1)$. It is desired to determine the temperature profile inside the spherical shell at steady-state conditions. The thermal conductivity of the shell material is a function of temperature.

The assumptions are as follows:

1. Steady-state condition, i.e., the temperatures or any other property or variable of the problem does not change with time.
2. Apart from thermal conductivity, all other properties of the sphere material do not depend on temperature. This might become a rough approximation if the gradient of temperature inside the sphere increases.
3. Thermal conductivity is a function of temperature and is given by Equation 1.136 [1].
4. Solid material of the shell does not go through any phase change or chemical transformation.

Again, because of variations of thermal conductivity, Equation A.36 cannot be applied and Equation A.33 should be used instead. Because of the assumptions made above and because of no transfer in any other direction but r, this equation simplifies to

$$\frac{d(r^2 q_r)}{dr} = 0 \qquad (2.10)$$

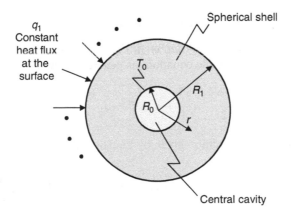

FIGURE 2.1 Conduction through a spherical shell with constant heat flux at external surface and constant temperature at internal face.

and

$$r^2 q_r = C_1 \tag{2.11}$$

The boundary conditions are

$$T(R_0) = T_0 \tag{2.12}$$

and

$$q_1 = \left(-\lambda \frac{dT}{dr}\right)_{r=R_1} = \text{constant} \tag{2.13}$$

This last condition can be immediately applied to Equation 2.11 to give constant C_1 or

$$C_1 = R_1^2 q_1 \tag{2.14}$$

Using Equation 2.14, Equation 2.11, and Equation A.28, one gets

$$r^2 \lambda \frac{dT}{dr} = -R_1^2 q_1 \tag{2.15}$$

Applying the correlation for thermal conductivity against temperature (see Equation 1.136), Equation 2.15 becomes

$$\left(aT - bT^2 + \frac{c}{T}\right) \frac{dT}{dr} = -q_1 \frac{R_1^2}{r^2} \tag{2.16}$$

2.3.1 SOLUTION BY SEPARATION OF VARIABLES

Despite being a second-order differential, Equation 2.11 led to Equation 2.14, which is a first-order separable equation. Its general solution is

$$a\frac{T^2}{2} - b\frac{T^3}{3} + c \ln T = q_1 \frac{R_1^2}{r} + C_2 \tag{2.17}$$

After applying the condition given by Equation 2.12, the solution becomes

$$\frac{a}{2}(T^2 - T_0^2) - \frac{b}{3}(T^3 - T_0^3) + c \ln \frac{T}{T_0} = q_1 R_1^2 \left(\frac{1}{r} - \frac{1}{R_0}\right) \tag{2.18}$$

This allows the determination of the temperature profile at the spherical shell according to heat flux at the surface [q_1 (W)] and is illustrated in Figure 2.2.

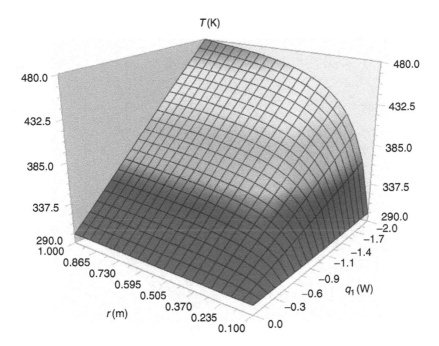

FIGURE 2.2 Temperature profile through a spherical shell with constant heat flux at external surface and constant temperature at internal face; case for varying thermal conductivity.

The heat flux assumes negative values because it is in an opposite direction to that of the radial coordinate. In this case, the following parameters were assumed: $R_0 = 0.1$ m, $R_1 = 1$ m, $T_0 = 298$ K, $a = 0.1$ W m^{-1} K^{-2}, $b = 2 \times 10^{-4}$ W m^{-1} K^{-3}, and $c = 0.1$ W m^{-1}. The numbers chosen for the parameters of conductivity relationship do not relate to any specific material and are used here just as an example.

It should be noticed that the above parameters a, b, and c lead to conductivities around 12 W m^{-1} K^{-1} at 298 K, which is similar to values found in metals. The above parameters have been chosen to produce an increased effect on the thermal conductivity. For instance, at 400 K it is around 7 W m^{-1} K^{-1}.

For the case of constant thermal conductivity, the solution to the temperature profile is given by

$$T = T_0 + \frac{q_1 R_1^2}{\lambda}\left(\frac{1}{r} - \frac{1}{R_0}\right) \tag{2.19}$$

Figure 2.2 can be compared with Figure 2.3, illustrated for the case of constant thermal conductivity (12 W m^{-1} K^{-1}) obtained at 298 K from Equation 1.136. In both cases, the same geometry and heat flux were imposed.

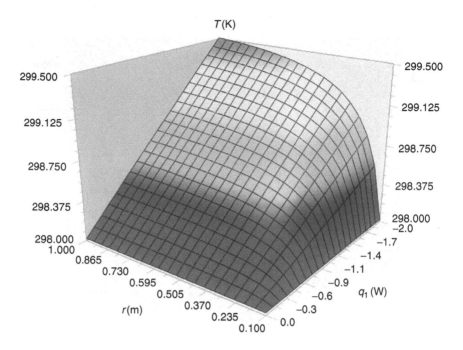

FIGURE 2.3 Temperature profile through a spherical shell with constant heat flux at external surface and constant temperature at internal face; case for constant thermal conductivity.

Of course, the higher conductivity used to chart Figure 2.3 led to faster heat transfer; therefore, preventing the temperature to reach as high as in the case represented in Figure 2.2.

2.4 BATCH REACTOR

Consider a well-stirred batch reactor, where a chemical species A is to be produced by the decomposition of another component B according to the simple following reaction:

$$B \rightarrow A \tag{2.20}$$

The reaction is first-order reversible and the rate is given by

$$\tilde{R}_{M,A} = k_d \tilde{\rho}_B - k_i \tilde{\rho}_A \tag{2.21}$$

Here, the reaction coefficients for the direct and for the inverse reactions are constant and represented by the symbols k_d and k_i, respectively.

Assume that the production rate (a) of component A at a given instant b is known and given by

$$\left.\frac{d\tilde{\rho}_A}{dt}\right|_{t=b} = a \tag{2.22}$$

The concentration of A at any instant can be determined. It represents a second-kind boundary condition.

As always, the assumptions should be clearly stated as follows:

1. Reactor is well stirred; therefore, the concentration of reactants and products do not vary from point to point inside the reacting vessel.
2. Reactor has good heating and refrigeration systems that ensure constant temperature throughout the entire reacting vessel.
3. Pressure inside the reactor is maintained constant as well.
4. Temperature and pressure inside the reactor are known.
5. Fluid in the reactor is Newtonian with constant density.

Because of assumptions 4 and 2, the equation of energy conservation would not be necessary.

The equations for conservation of species are to be used. Applying the above assumptions, Equation A.40 can be simplified to

$$\frac{d\rho_A}{dt} = R_{M,A} \tag{2.23}$$

Using Equation 2.21, this equation can be written at molar basis as

$$\frac{d\tilde{\rho}_A}{dt} = \tilde{R}_{M,A} = k_d\tilde{\rho}_B - k_i\tilde{\rho}_A \tag{2.24}$$

Since the temperature is constant, the kinetic coefficients do not vary.

The stoichiometry of Equation 2.20 allows writing

$$\frac{d\tilde{\rho}_A}{dt} = -\frac{d\tilde{\rho}_B}{dt} \tag{2.25}$$

which

$$\frac{d\tilde{\rho}_A}{dt} + \frac{d\tilde{\rho}_B}{dt} = \frac{d(\tilde{\rho}_A + \tilde{\rho}_B)}{dt} = \frac{d\tilde{\rho}}{dt} = 0 \tag{2.26}$$

Thus, the total molar concentration is constant. This is only possible in the particular case of Equation 2.20, where the number of moles produced is equal to the number of moles that react; therefore, the total remains constant. Equation 2.24 becomes

$$\frac{d\tilde{\rho}_A}{dt} = k_d\tilde{\rho} - (k_i + k_d)\tilde{\rho}_A \tag{2.27}$$

2.4.1 Solution by Separation of Variables

Equation 2.27 is separable and the solution can be written as

$$\tilde{\rho}_A = \frac{k_d}{k}\tilde{\rho} + C_1 e^{-kt} \tag{2.28}$$

where

$$k = k_d + k_i \tag{2.29}$$

Using Equations 2.28 and 2.26, the derivative of concentration of A is given by

$$\frac{d\tilde{\rho}_A}{dt} = -kC_1 e^{-kt} \tag{2.30}$$

Applying boundary condition given by Equation 2.22, it is possible to determine C_1 and to write Equation 2.28 as

$$\tilde{\rho}_A = \frac{k_d}{k}\tilde{\rho} - \frac{a}{k}e^{k(b-t)} \tag{2.31}$$

Notice that Equation 2.31 is dimensionally and physically coherent because the concentration of A increases with time.

2.4.2 Solution by Laplace Transform

The problem formed by Equation 2.27 with boundary condition given by Equation 2.22 can also be solved by Laplace transform.

Let the transform be

$$L\{\tilde{\rho}_A(t)\} = \varphi(s) \tag{2.32}$$

This when applied to Equation 2.26 yields

$$s\varphi(s) - \tilde{\rho}_A(0) = \frac{k_d\tilde{\rho}}{s} - k\varphi(s) \tag{2.33}$$

where k is given by Equation 2.29.

An apparent difficulty arises here because the initial value of concentration is not known. However, it is a constant and can be determined later. At present, let us write Equation 2.33 in the following form:

$$\varphi(s) = \frac{\tilde{\rho}_A(0)}{k+s} + \frac{k_d \tilde{\rho}}{s(k+s)} \tag{2.34}$$

The inverse is

$$\tilde{\rho}_A(t) = \tilde{\rho}_A(0)e^{-kt} + \tilde{\rho}\frac{k_d}{k}\left(1 - e^{-kt}\right) \tag{2.35}$$

Equation 2.35 leads to

$$\frac{d\tilde{\rho}_A}{dt} = -(k\tilde{\rho}_A(0) - k_d\tilde{\rho})e^{-kt} \tag{2.36}$$

Using boundary condition given by Equation 2.22 one gets

$$\tilde{\rho}_A(0) = \frac{k_d\tilde{\rho} - ae^{kb}}{k} \tag{2.37}$$

This when applied to Equation 2.35 reproduces Equation 2.31.

2.4.3 Discussion

Equation 2.31 can be easily put under a dimensionless form, or

$$x_A = b_1 - b_2 e^{b_3 - \tau} \tag{2.38}$$

where

$$b_1 = \frac{k_d}{k}, \quad b_2 = \frac{a}{k}, \quad b_3 = kb, \quad \tau = kt \tag{2.39}$$

The derivative of molar concentration (x_A) of product A in the reactor is

$$\frac{dx_A}{d\tau} = b_2 e^{b_3 - \tau} \tag{2.40}$$

From these equations, it is clear that the molar concentration of species A depends on the following:

1. Parameter b_1, which is the ratio between the direct reaction coefficient and the sum of direct and inverse reaction coefficients. Of course, this agrees with the fact that concentration of A should increase with faster direct reaction when compared with the inverse reaction.

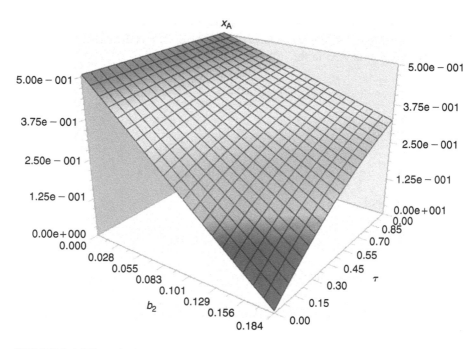

FIGURE 2.4 Dimensionless concentration in the reactor against time.

2. Parameter b_2, which is the ratio between the instantaneous rate (a) of A production and the sum of reaction coefficients. Higher b_2 leads to higher rates of A production. However, as the second derivative of x_A against time is negative, a maximum in the concentration should be expected. Nonetheless, the point of maximum cannot be reached because Equation 2.40 does not have a root. This is a consequence of simplified—and somewhat artificial—boundary condition given by Equation 2.22. In addition, the concentration of A will increase faster for $\tau < b_3$ (or $t < b$) and slower for $\tau > b_3$ (or $t > b$). In Chapter 3, the same problem is solved using a third-kind boundary condition, which allows a more realistic approach.

3. Attention should be paid to representations because of the limitation in molar concentration. An example is presented in Figure 2.4, where the following values have been assumed: $b_1 = 0.5$ and $b_3 = 1.0$. From this, it is easy to see the progress of species A concentration against time and the decrease of achieved molar fraction for larger values of b_2.

2.5 PLUG-FLOW REACTOR

Consider the problem of determining the temperature profile inside a tubular plug-flow reactor, similar to the one presented in Section 1.9 and illustrated in Figure 1.16. However, at this time, the temperature T_0 at the reactor entrance is not known. Instead, the derivative of temperature at this position is known. This situation might occur if, for instance, a constant flux of energy is applied by the grid to the passing fluid. The controlled power input to the electrical resistance grid would ensure an initial positive derivative of temperature at the reactor entrance.

To leave the conditions clear, let us list the assumptions:

1. Steady-state regime.
2. At the position $z = 0$, the fluid experiences a given and known constant increase of temperature in the ϕ direction. In other words, a positive constant derivative of temperature is maintained at the grid.
3. Tube is perfectly insulated, e.g., a reactor with adiabatic walls.
4. Plug-flow regime. Of course, this is a strong approximation but useful and often applied for a first attack to reactor design.
5. Fluid properties are constants. Since the chemical species are reacting and the composition as well as temperature change from point to point in the reactor; this is another severe approximation.
6. Reaction delivers a constant and uniform rate (R_Q) of energy per unit of volume. This is very difficult or even impossible to achieve because the reaction depends on the concentration of reactants, which in turn are decreasing for increasing values of coordinate z. Therefore, the objective of the present example is just to illustrate how to solve a differential equation. Situations with a more plausible picture are present in chapters ahead.

The initial development is the same as shown in Section 1.9. Therefore, calling

$$\theta(z) = \frac{dT}{dz} \tag{2.41}$$

Equation 1.112 becomes

$$\frac{d\theta}{dz} - a\theta + b = 0 \tag{2.42}$$

where constants a and b are given by Equations 1.116 and 1.117.

The boundary condition is

$$\left.\frac{dT}{dz}\right|_{z=0} = \theta(0) = \phi \tag{2.43}$$

where ϕ is a constant.

2.5.1   SOLUTION BY SEPARATION OF VARIABLES

As before, the solution is straightforward because Equation 2.42 is a separable differential equation, leading to

$$\theta = \frac{dT}{dz} = \frac{C_1}{a}e^{az} - \frac{b}{a} \tag{2.44}$$

On applying the condition in Equation 2.43, it provides

$$C_1 = a\phi + b \tag{2.45}$$

and

$$\theta = \frac{dT}{dz} = \left(\phi + \frac{b}{a}\right)e^{az} - \frac{b}{a} \tag{2.46}$$

The above can be integrated again to give

$$T = \left(\phi + \frac{b}{a}\right)\frac{1}{a}e^{az} - \frac{b}{a}z + C_2 \tag{2.47}$$

As seen earlier, apart from the condition given by Equation 2.43, another one is necessary. For instance, if the temperature of the fluid entering the reactor is known, the remaining condition is translated by

$$T(0) = T_0 \tag{2.48}$$

Of course, this is one possibility among several, and leads to

$$T = T_0 + \left(\phi + \frac{b}{a}\right)\frac{1}{a}(e^{az} - 1) - \frac{b}{a}z \tag{2.49}$$

In order to facilitate and generalize the representations by graphs, Equation 2.49 is represented by the following dimensionless form:

$$\Phi = \frac{T - T_0}{T_0} = (c_1 + c_3)\left(e^{c_2\zeta} - 1\right) - c_1 c_2 \zeta \tag{2.50}$$

where

$$c_1 = \frac{b}{a^2 T_0} \tag{2.51}$$

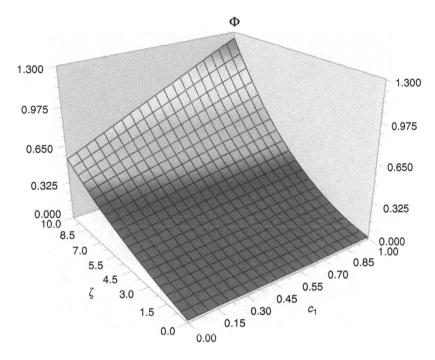

FIGURE 2.5 Dimensionless temperature against position in the tubular reactor for $c_2 = 0.1$ and $c_3 = 0.3$.

$$c_2 = aR \tag{2.52}$$

$$c_3 = \frac{\phi}{aT_0} \tag{2.53}$$

$$\zeta = \frac{z}{R} \tag{2.54}$$

Apart from the introduction of a new parameter c_3, the solution given by Equation 2.50 is similar to the one provided by Equation 1.124. The parameter c_3 only adds to c_1 and Figure 2.5 can be seen as a particular case of Figure 1.17, at which c_1 is always greater than zero. Figure 2.5 is plotted for the particular values of c_2 and c_3 (0.1 and 0.3, respectively). It is easy to see that

1. Because of the exothermic reaction, the temperature will increase for positions ahead the entrance. Notice that, according to Equation 1.117, the reaction rate affects parameter b, and therefore parameter c_1. Thus, higher values of c_1 lead to higher temperatures.
2. As imposed by boundary condition given by Equation 2.43, the derivative of temperature remains constant at $z = 0$ or $\zeta = 0$.

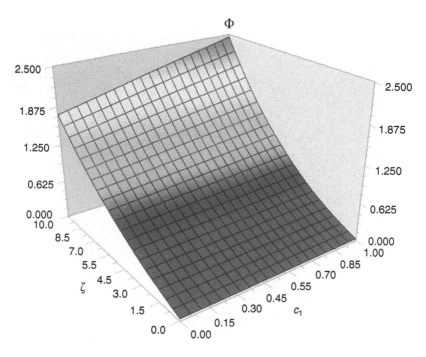

FIGURE 2.6 Dimensionless temperature against position in the tubular reactor for $c_2 = 0.1$ and $c_3 = 1.0$.

3. Temperature derivative is always positive, even when $c_1 = 0$.
4. Increases in the imposed temperature derivative (or increases in parameter c_3) lead to higher temperature across the reactor. This is illustrated in Figure 2.6 for the case where c_2 remains equal to 0.1, but c_3 is set as 1.0.

EXERCISES

1. Solve the problem presented in Section 2.3 by replacing the spherical shell with a cylindrical one.
2. Solve the case of a well-stirred reactor (Section 2.4) if the reaction were

$$B \rightarrow 2A$$

Again, assume a first-order reversible reaction occurring at constant temperature and pressure.
3. Solve the case of well-stirred tank (as shown in Section 2.4) for a second-order reaction, or $\tilde{R}_{M,A} = k_d \tilde{\rho}_B^2 - k_i \tilde{\rho}_A^2$. Continue to assume all other hypotheses listed in Section 2.4.

4. Solve the problem presented in Section 2.4, using collocation and Galerkin methods. Suggest a reference time for the interval in which the solution is to be sought.
5. Solve the problem presented in Section 2.5 using any branch of the method of weighted residuals. Find, at least, two levels of approximations.

REFERENCE

1. Perry, J.H. et al., *Chemical Engineers' Handbook*, 4th ed., McGraw-Hill, New York, 1963, pp. 3–22.

3 Problems 113; One Variable, 1st Order, 3rd Kind Boundary Condition

3.1 INTRODUCTION

This chapter presents methods to solve problems with one independent variable involving first-order differential equations and third-kind boundary conditions. Mathematically, this class of cases can be summarized as $f\left(\phi,\omega,\frac{\mathrm{d}\phi}{\mathrm{d}\omega}\right)$, third-kind boundary condition.

3.2 HEATING OF A SOLID WITH CONTROLLED HEAT TRANSFER RATE

Let us consider a similar problem as shown in Sections 1.2 and 2.2. Instead knowing the initial body temperature or assuming a constant energy flux at the surface, the rate of heat transfer is now controlled according to a given function of its temperature or time.

Basically, the same assumptions made in Section 1.2 are valid.

The problem would lead to the same final differential equation, i.e., Equation 1.3, or

$$\frac{\mathrm{d}T}{\mathrm{d}t} = -\frac{\alpha A}{\rho V C_{\mathrm{p}}}(T - T_\infty) \tag{3.1}$$

The boundary condition related to the heat transfer is represented by

$$\left.\frac{\mathrm{d}T}{\mathrm{d}t}\right|_{t=a} = f(t = a) \tag{3.2}$$

Here $f(t)$ is any continuous function. This is an example of a third-kind condition.

3.2.1 SOLUTION BY SEPARATION OF VARIABLES

As in Section 1.2.1, the above differential equation is separable and can be easily integrated to give

$$\ln(T - T_\infty) = -\frac{\alpha A}{\rho V C_p} t + \ln C \tag{3.3}$$

where C is a constant. Therefore,

$$T = T_\infty + C \exp\left(-\frac{\alpha A}{\rho V C_p} t\right) \tag{3.4}$$

Its derivative is

$$\frac{dT}{dt} = -\frac{\alpha A}{\rho V C_p} C \exp\left(-\frac{\alpha A}{\rho V C_p} t\right) \tag{3.5}$$

Using the boundary condition given by Equation 3.2, the constant C is obtained and Equation 3.4 becomes

$$T = T_\infty - f(a)\frac{\rho V C_p}{\alpha A} \exp\left[\frac{\alpha A}{\rho V C_p}(a - t)\right] \tag{3.6}$$

The initial temperature of the body is deduced for $t=0$, leading to

$$T_0 = T_\infty - f(a)\frac{\rho V C_p}{\alpha A} \exp\left[\frac{\alpha A}{\rho V C_p}a\right] \tag{3.7}$$

Finally, a relation between T and T_0 can be obtained by combining Equations 3.6 and 3.7 to give

$$T = T_0 + f(a)\frac{\rho V C_p}{\alpha A} \exp\left(\frac{\alpha A}{\rho V C_p}a\right)\left[1 - \exp\left(-\frac{\alpha A}{\rho V C_p}t\right)\right] \tag{3.8}$$

3.3 TEMPERATURE-CONTROLLED BATCH REACTOR

Almost all chemical reactors require temperature control, either to avoid accidents or to ensure maximum productivity or both. In cases of exothermic reactions, safety is a major concern and this demands precise control of the reactor temperature. In some cases, even a control of the temperature derivative is required to take action before any surge occurs.

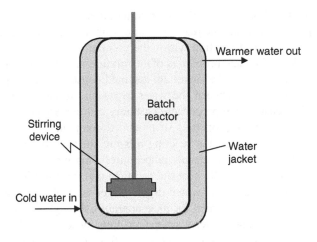

FIGURE 3.1 Scheme of a water-jacketed and well-stirred batch reactor.

Figure 3.1 presents a scheme to illustrate the present case. An exothermic reaction takes place in a well-stirred reactor, which is cooled by a water jacket. A sensor measures the temperature inside the reactor and its derivative is imposed according to a given relationship. This is accomplished by varying the flow of cool water through the jacket. One is interested in the profile of reactor temperature against time.

To follow the usual procedure, let us list the assumptions applied in the present problem:

1. Given amounts of two liquid reactants A and B are introduced in the reactor and an exothermic process takes place where a liquid product C is generated. Within the residence time of reacting species inside the vessel, conditions are far from equilibrium. Therefore, the reaction can be considered irreversible.
2. Pressure inside the reactor is high enough to maintain the mixture as a liquid, even at relatively high temperatures.
3. As explained in Section 1.5, exothermic reactions increase the internal energy or enthalpy of the reacting mixture. In the present case, it is assumed that the rate of energy generation per unit of reactor volume is a function of the temperature itself and given by the following simple relation:

$$R_{Q,\ reaction} = aT \tag{3.9}$$

where parameter a is constant. Of course, this is just an example. In more realistic situations, the reaction rate follows the Arrhenius relation. This is discussed ahead.

4. Reactor is well stirred and the average temperature (T)—as well as any other parameter inside the reactor—varies only against time; therefore, it is not a function of position within the volume.
5. All physical–chemical properties of the mixture inside the reactor as well as of the water inside the jacket are assumed constant, including the heat transfer coefficient between the water jacket and reactor interior. Obviously, the constancy of properties constitutes a strong assumption because it would require the properties of produced components to be equal or approximately equal to the reacting mixture. In addition, all properties need to be weak functions of temperature. This may be a reasonable approximation for specific heat but a critical one for viscosities.
6. Temperature variation of water in the jacket is small, and an average value (T_W) may be assumed for its temperature throughout the jacket. In other words, the reactor derivative-controlling system would not require significant variations of the cold-water flow to the point so that its average temperature might be assumed constant. Of course, that simplification might be criticized if large variation of the water temperature or phase change occurs.
7. Rate of temperature increase of the reacting mixture against time is controlled. This is accomplished by a system that measures its derivative and, by increasing or decreasing the mass flow of cooling water, forces the derivative to follow a prescribed relationship against time. Let us assume here that such a relationship, at a given instant t_b is given by

$$\left.\frac{dT}{dt}\right|_{t=t_b} = B[T(t_b)]^{-1} \tag{3.10}$$

where B is a constant. This is a typical third-kind boundary condition. Notice that parameter B reflects how well the cooling jacket helps to control the inevitable increase on the reacting mixture inside the internal vessel. Larger values of B means that the cooling is not too effective and the temperature would increase faster than for lower values of B.

Using the above assumptions and considering the heat transfer by convection to the water in the jacket, one may apply the energy balance Equation A.31 to arrive at

$$\frac{dT}{dt} = \frac{\alpha A(T_W - T)}{V\rho C_p} + \frac{a}{\rho C_p}T \tag{3.11}$$

Here A is the area of the surface between the reactor and the water jacket, and V is the internal volume of the reactor. The density and specific heat refer to the mixture inside the reactor, which are assumed constant.

In the present case, it should be noticed that the source and sink terms include the energy input or output because of heat transfer through the reactor surface.

This can only be done because of assumption 4, which ensures uniform temperature inside the reactor at any time.

Equation 3.11 can be put in the form

$$\frac{dT}{dt} = bT + c \tag{3.12}$$

where

$$b = \frac{aV - \alpha A}{\rho C_p V} \tag{3.13}$$

and

$$c = \frac{\alpha A T_W}{\rho C_p V} \tag{3.14}$$

3.3.1 Solution by Separable Equation

Equation 3.12 is separable with the following solution:

$$\ln\left(T + \frac{c}{b}\right) = bt + \ln C_1 \tag{3.15}$$

or

$$T = \frac{C_1}{b} e^{bt} - \frac{c}{b} \tag{3.16}$$

In order to facilitate the application of boundary condition given by Equation 3.10, let us take the case when it is an initial one, or $t_b = 0$. This would lead to

$$C_1 = \frac{c + \sqrt{c^2 + 4bB}}{2} \tag{3.17}$$

Finally, the temperature progress against time would be described by

$$T = \frac{c + \sqrt{c^2 + 4bB}}{2b} e^{bt} - \frac{c}{b} \tag{3.18}$$

As seen, the temperature will be a function of imposed parameter B. On the other hand, the temperature has limits. First, its absolute value should not only be positive but also fall within the range dictated by the validity (or at least approximate validity) of the assumptions listed before. For instance, if the reactants are in liquid phase, the temperature inside the reactor cannot become lower than the freezing value or higher than the boiling point of any phase of the mixture. Additionally, the temperature value is a real number; therefore

$$c^2 + 4bB \geq 0$$

It is interesting to notice that

$$\frac{dT}{dt} = \frac{c + \sqrt{c^2 + 4bB}}{2} e^{bt} \tag{3.19}$$

Therefore, the derivative decreases with b. According to Equation 3.13, this can be achieved for increasing values of heat transfer coefficient (α) or higher cooling rates by the jacket. If b becomes negative, there is a possibility of achieving a minimum derivative value given by

$$\frac{dT}{dt} = \frac{c}{2} e^{b_{min}t} \tag{3.20}$$

where

$$b_{min} = -\frac{c^2}{4B} \tag{3.21}$$

To illustrate one possible situation, let us assume the following values:

- Reacting part of the vessel is 2 m high and 1 m in diameter ($A = 6.2831 \text{ m}^2$, $V = 1.5708 \text{ m}^3$).
- Heat transfer coefficient (α) is around 100 W m^{-2} K^{-1}.
- Average temperature of water (T_W) in the jacket is equal to 298 K.
- Average density of the liquid mixture (ρ) in the reactor is equal to 1000 kg m^{-3}.
- Average specific heat of the liquid mixture (C_p) is equal to 4200 J kg^{-1} K^{-1}.

It should be noticed that the average temperature in the reactor would only become stable or controllable for negative values of parameter b, or for $\alpha A > aV$. In other words, the rate of heat transfer by convection between reactor and cooling jacket should surpass the rate of energy generation by the exothermic reaction. In the case of the example, the rate of reaction should be below 400 W m^{-3} K^{-1}. As an example, let $a = 100$ W m^{-3} K^{-1}.

The above values lead to parameter b equal to -7.143×10^{-5} s^{-1} and c equal to 2.838×10^{-2} K s^{-1}. According to the consideration related to possibilities of temperature value, B should be smaller than 2.819 K^2 s^{-1}. With these values, Equation 3.18 provides the values of temperature against time. In addition, if we assume that the freezing temperature of the mixture inside the vessel is $T_f = 270$ K, this value should be used as a boundary for possible ones computed using Equation 3.18.

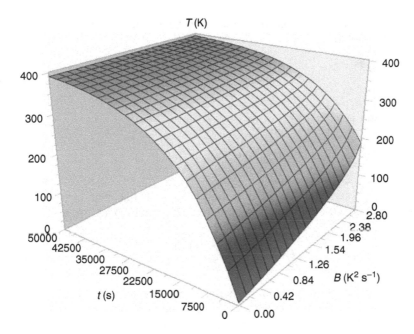

FIGURE 3.2 Temperature as function of time and process parameter inside a batch reactor.

The temperature against time and controlling parameter B is illustrated in Figure 3.2.

It becomes clear that

- Gradient temperature increases in the reactor against variations of parameter B.
- Because of the negative value of b, the temperature is controllable and tends to limit around 400 K.

3.3.2 SOLUTION BY LAPLACE TRANSFORM

The problem given by Equations 3.12 and 3.10 can be solved by Laplace transform as well.

Let us consider the following transform:

$$\varphi(s) = L\{T(t)\} \tag{3.22}$$

The transform of Equation 3.12 becomes

$$s\varphi(s) - T(0) = b\varphi(s) + \frac{c}{s} \tag{3.23}$$

The initial condition $T(0)$ is not known; however, it may be determined later. For now, let us write

$$\varphi(s) = \frac{T(0)}{s-b} + \frac{c}{s(s-b)} \tag{3.24}$$

Its inversion is given by

$$T(t) = T(0)e^{bt} + \frac{c}{b}(e^{bt} - 1) \tag{3.25}$$

If this solution is introduced into Equation 3.10, one gets a second-degree polynomial equation. The only feasible (positive) root would lead to

$$T(0) = \frac{c + \sqrt{c^2 + 4bB}}{2b}e^{-bt_b} - \frac{c}{b} \tag{3.26}$$

Substituting Equation 3.26 into Equation 3.25 reproduces Equation 3.18.

The reader is invited to verify the limitations for parameters in order to allow a meaningful real problem or $T(0)$ greater or equal to zero.

3.3.3 ARRHENIUS RELATION

A more realistic equation for the energy source because of the reaction is described by the Arrhenius formulation, or

$$R_{Q,\text{ reaction}} = a_1 \exp\left(-\frac{\tilde{E}}{\tilde{R}T}\right) \tag{3.27}$$

Thus, the differential energy balance becomes

$$\frac{dT}{dt} = \frac{\alpha A(T_W - T)}{V\rho C_p} + \frac{a_1}{\rho C_p}\exp\left(-\frac{\tilde{E}}{\tilde{R}T}\right) \tag{3.28}$$

The boundary condition given by Equation 3.10 should be observed.

The above differential equation is not separable and cannot be solved by Laplace transform. Among the possible methods to achieve a solution there is the weighted residual method.

3.3.4 SOLUTION BY WEIGHTED RESIDUALS

As seen in Appendix E, there are several available weighted residual methods, which may be applied to solve the boundary value problem formed by Equation 3.28 and the boundary condition given by Equation 3.10. Moreover, these methods are easier applied when variables are normalized and dimensionless. For such purposes, let us define

$$\theta = \frac{T}{T_{\rm L}} \tag{3.29}$$

and

$$\tau = \frac{t}{t_{\rm res}} \tag{3.30}$$

Here, $T_{\rm L}$ is a limiting temperature observed for the process and $t_{\rm res}$ is the reactor residence time, which is usually defined by the ratio between the reactor volume and the injection volume flow of reactants. Such parameters are assumed known for the given process. Let us seek approximated solutions for the problem within the ranges $0 \le \theta \le 1$ and $0 \le \tau \le 1$.

Using the dimensionless variables, Equation 3.28 becomes

$$\frac{d\theta}{d\tau} = b_1 - b_2\theta + b_3 \exp\left(-b_4\theta^{-1}\right) \tag{3.31}$$

where

$$b_1 = \frac{\alpha A t_{\rm res} T_{\rm W}}{V \rho C_{\rm p} T_{\rm L}}, \quad b_2 = \frac{\alpha A t_{\rm res}}{V \rho C_{\rm p}}, \quad b_3 = \frac{a_{\rm l} t_{\rm res}}{\rho C_{\rm p} T_{\rm L}}, \quad \text{and } b_4 = \frac{\tilde{E}}{R T_{\rm L}} \tag{3.32a,b,c,d}$$

The boundary condition given by Equation 3.10 becomes

$$\left.\frac{d\theta}{d\tau}\right|_{\tau=\tau_{\rm b}} = \beta[\theta(\tau_{\rm b})]^{-1} \tag{3.33}$$

where

$$\beta = B\frac{t_{\rm res}}{T_{\rm L}^2} \quad \text{and } \tau_{\rm b} = \frac{t_{\rm b}}{t_{\rm res}} \tag{3.34a,b}$$

3.3.4.1 Choosing the Form of Approximations

According to Section E.1, the best approach is to have approximations in the form

$$\bar{\theta}_n = \phi_0 + \sum_{j}^{n} C_j \phi_j \tag{3.35}$$

with functions ϕ_0 and ϕ_j $(1 \leq j \leq n)$ that satisfy the boundary condition or conditions. For instance, the function

$$\phi_0 = (2\beta\tau)^{1/2} \tag{3.36}$$

satisfies the boundary condition given by Equation 3.33.

Finding functions ϕ_j $(1 \leq j \leq n)$ that satisfy that same condition is not a trivial task. In such cases, it is best to set functions that equal zero at the boundary point, leaving just function ϕ_0, which does satisfy the condition. Therefore, let the following be the approximation function:

$$\bar{\theta}_n = (2\beta\tau)^{1/2} + \sum_{j}^{n} C_j(\tau - \tau_b)^j \tag{3.37}$$

where coefficients C_j can be found through the application of one or more weighted residual methods, as follows.

3.3.4.2 First Approximation

The first approximation would be

$$\bar{\theta}_1 = (2\beta\tau)^{1/2} + C_1(\tau - \tau_b) \tag{3.38}$$

Its respective residue is obtained by substituting the above equation in Equation 3.31, or

$$\Lambda_1 = \frac{d\bar{\theta}_1}{d\tau} - b_1 + b_2\bar{\theta}_1 - b_3 \exp\left(-b_4\bar{\theta}_1^{-1}\right) \tag{3.39}$$

3.3.4.3 Collocation Method

If the residual is to be made equal to zero at, for instance, the middle point of the range in which we are interested in $(\tau = \frac{1}{2})$, then the following results:

$$\beta^{1/2} + C_1 - b_1 - b_2 C^{1/2} - b_2 C_1\left(\frac{1}{2} - \tau_b\right) - b_3 \exp\left[-\frac{b_4}{\beta^{1/2} + C_1\left(\frac{1}{2} - \tau_b\right)}\right] = 0 \tag{3.40}$$

The presence of an exponential function does not allow obtaining an explicit form for parameter C_1. Nonetheless, this difficulty is easily overcome by any simple convergence procedure. In order to facilitate discussions, let us assume a case where t_b is half of the residence time t_{res}. In this case, $\tau_b = \frac{1}{2}$ and C_1 can be written as

$$C_1 = b_1 + \beta^{1/2}(b_2 - 1) + b_3 \exp\left(-b_4\beta^{-1/2}\right) \tag{3.41}$$

The simplest set of collocation points would be provided by evenly distributed points. Therefore, the first approximation can be computed using Equation 3.38.

3.3.4.4 Second Approximation

The second approximation is

$$\bar{\theta}_2 = (2\beta\tau)^{1/2} + C_1(\tau - \tau_b) + C_2(\tau - \tau_b)^2 \tag{3.42}$$

Following the same steps, two equations can be obtained by equating the residual to zero at two collocation points. Now, let us use the roots of an orthogonal polynomial, such as Legendre. As seen in Section E.4, the range of such a function rests between 0 and 1. This illustrates why the normalization of independent variable is a useful procedure.

The second-order Legendre polynomial is $\frac{1}{2}(3x^2 - 1)$, and therefore the roots are $\pm\sqrt{1/3}$. However, there are two roots in the 0–1 domain. For this, the fourth-order polynomial, or $\frac{1}{8}(35x^4 - 30x^2 + 3)$ would give (approximate) roots $\tau_1 = 0.33998104$ and $\tau_2 = 0.86113631$. The residue for each root τ_i is

$$\frac{1}{2}(2\beta)^{1/2}\tau_i^{-1/2} + C_1 + 2C_2(\tau_i - \tau_b) - b_1 + b_2(2\beta\tau_i)^{1/2} + b_2C_1(\tau_i - \tau_b)$$

$$+ b_2C_2(\tau_i - \tau_b)^2 - b_3\exp\left[-\frac{b_4}{(2\beta\tau_i)^{1/2} + C_1(\tau_i - \tau_b) + C_2(\tau_i - \tau_b)^2}\right] = 0$$

$$\tag{3.43}$$

Parameters can now be obtained through the application of any convergence method to solve the system given by Equation 3.43 at two roots τ_1 and τ_2.

Section 3.3.1 provides values to illustrate a plausible situation. In addition, the following are considered:

- Residence time in the reactor (t_{res}): 50,000 s
- Limiting temperature (T_L): 400 K
- Ratio between reaction activation energy and universal gas constant $(\tilde{E}/\tilde{R})$: 1×10^4 K

The above values have been chosen to approximately match the reaction rate given by Equation 3.9 at 400 K. With this, it is possible to calculate parameters b_1–b_4 in Equation 3.32. Then the values of C_1 and C_2 can be obtained, and therefore, the approximated solution for temperature as a function of time given by Equation 3.42 and illustrated in Figure 3.3.

Comparison between Figures 3.3 and 3.2 shows that even a second approximation is enough to approximately represent the exact solution.

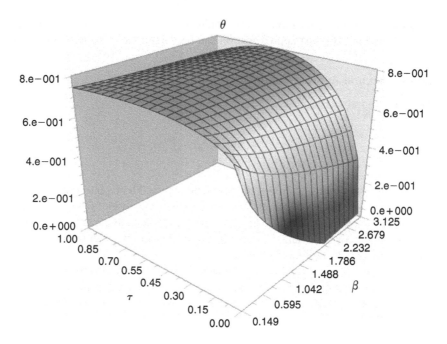

FIGURE 3.3 Second approximation of dimensionless temperature as function of time and process parameter inside a batch reactor.

3.4 BATCH REACTOR

This is similar to the problem seen in Section 2.4, i.e., a well-stirred batch reactor where chemical species A is to be produced by the decomposition of another component B according to the simple reaction:

$$B \rightarrow A \tag{3.44}$$

Again, the reaction is of first order and the rate is given by

$$\tilde{R}_{M, A} = k_d \tilde{\rho}_B - k_i \tilde{\rho}_A \tag{3.45}$$

As before, the reaction coefficients for the direct and the inverse reactions are assumed constant and represented by the symbols k_d and k_i, respectively.

On the other hand, at a given instant (b) the production rate of component A depends on the concentration of the reactant at the same instant, or

$$\frac{d\tilde{\rho}_A}{dt}\bigg|_{t=b} = a\tilde{\rho}_B(b) \tag{3.46}$$

This is a more realistic picture for batch reactors than the one described in Section 2.4. Notice that as the concentration of reactant B decreases with time so does the production rate of component A. As Equation 3.45 links the concentration of reactant with the concentration of product, Equation 3.46 represents a third-kind boundary condition. Therefore, the difference between the situation presented earlier and here resides on the type of imposed boundary condition.

The assumptions made in Section 2.4 are still valid; therefore, one would arrive at Equation 2.27, or

$$\frac{d\tilde{\rho}_A}{dt} = k_d \tilde{\rho} - (k_i + k_d)\tilde{\rho}_A \tag{3.47}$$

3.4.1 Solution by Separation of Variables

Equation 3.47 is separable and the solution can be written as

$$\tilde{\rho}_A = \frac{k_d}{k}\tilde{\rho} + C_1 e^{-kt} \tag{3.48}$$

where

$$k = k_d + k_i \tag{3.49}$$

Using Equations 3.48 and 2.26, the derivative of concentration of A is given by

$$\frac{d\tilde{\rho}_A}{dt} = -kC_1 e^{-kt} \tag{3.50}$$

In order to apply the boundary condition given by Equation 3.46, it is necessary to use Equation 2.26, which shows a constant total concentration, therefore

$$\tilde{\rho}_B = \tilde{\rho} - \tilde{\rho}_A \tag{3.51}$$

With this, Equation 3.46 becomes

$$\left.\frac{d\tilde{\rho}_A}{dt}\right|_{t=b} = a[\tilde{\rho} - \tilde{\rho}_A(b)] \tag{3.52}$$

This allows determining the integration constant C_1 and to write Equation 3.48 as

$$\frac{\tilde{\rho}_A}{\tilde{\rho}} = \frac{k_d}{k} - \frac{a\left(\frac{k_d}{k_i} - 1\right)}{k - a} e^{k(b-t)} \tag{3.53}$$

The left-hand side of Equation 3.53 represents the molar fraction of component A in the reactor and the whole equation can be easily put in the dimensionless form or

$$x_A = b_1 - b_2 e^{b_3 - \tau} \tag{3.54}$$

where

$$b_1 = \frac{k_d}{k}, \quad b_2 = \frac{a\left(\frac{k_d}{k_i} - 1\right)}{k - a}, \quad b_3 = kb, \quad \tau = kt \tag{3.55a,b,c,d}$$

The form of Equation 3.54 is similar to Equation 2.38, as well as the respective derivative, as given by Equation 2.40. The only difference remains in the definition of parameter b_2. In the present case, this parameter includes the ratio between direct and inverse reaction coefficients.

For the following discussions, it is worthwhile to repeat the derivative

$$\frac{dx_A}{d\tau} = b_2 e^{b_3 - \tau} \tag{3.56}$$

Its second derivative is

$$\frac{d^2 x_A}{d\tau^2} = -b_2 e^{b_3 - \tau} \tag{3.57}$$

Keeping in mind that k is always greater than parameter a, the following can occur:

1. If $k_d > k_i$, b_2 would be positive and therefore the time derivative of molar fraction of A (Equation 3.56) would also be positive, but its second derivative Equation 3.57 would remain negative. Differently from Section 2.4, the value of derivative may reach zero when k_d is equal to k_i. Therefore, a maximum concentration of component A might be reached, which is a plausible picture for a batch reactor. However, this is not really a maximum because b_2 would be equal to zero and the concentration of A would remain constant. Figure 2.4 is a possible representation for this case.

2. If $k_d < k_i$, the time derivative of the molar fraction of component A will be negative and the second derivative will be positive. Such a situation only will be feasible if some product A is present at the start of the reaction. Its concentration would decrease to reach a minimum. This case is illustrated in Figure 3.4 for $b_1 = 0.5$ and $b_3 = 1.0$. Of course, the concentration of A would decrease faster for larger absolute values of parameter b_2 or smaller ratios k_d/k_i.

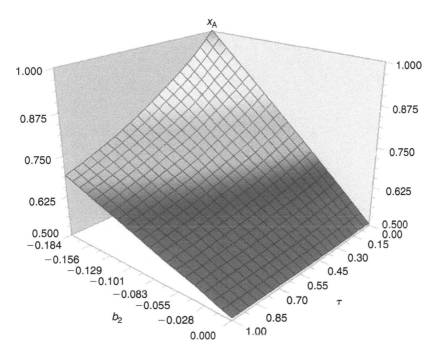

FIGURE 3.4 Dimensionless concentration in the reactor against time; case of reversing reaction.

EXERCISES

1. From the conditions set and the solution achieved in Section 3.2, deduce the rate of heat transfer between the body and the surrounding environment at instant $t = a$.

2. Redraw Figure 3.2 for the case when the minimum value for b (given by Equation 3.21) is used.

3. Develop a graphical representation of temperature as a function of time using the second approximation given by the collocation method. Apply the values suggested for an example in Section 3.3.4.4 and compare the graphical representation with Figure 3.2.

4. Develop the third approximation for the problem presented in Section 3.3.3 using the collocation method. Compare it with the previous first and second approximations.

5. Repeat the solution presented in Section 3.3.4 by the Galerkin method.

6. Solve the problem presented in Section 3.4 by Laplace transform.

4 Problems 121; One Variable, 2nd Order, 1st Kind Boundary Condition

4.1 INTRODUCTION

This chapter presents methods to solve problems with one independent variable involving second-order differential equations and first-kind boundary conditions. Mathematically, this class of cases can be summarized as $f\left(\phi, \omega, \frac{d\phi}{d\omega}, \frac{d^2\phi}{d\omega^2}\right)$, first-kind B.C.

Likewise, almost all classes of differential equations, second-order ordinary equations, can be classified by several methods. The most common classifications are

1. Linear
2. Nonlinear

Additionally, there are situations where second-order differential equations can be reduced to first-order equations. This chapter starts with such cases. Then, examples of linear and nonlinear second-order differential equations with first-kind boundary conditions are shown. If the reader is interested only on solutions of complete second-order differential equations, it is advisable to consult the problems presented after Section 4.8. More examples of such equations are shown in Chapters 5 and 6.

4.2 MASS TRANSFER THROUGH A CYLINDRICAL ROD

Consider two very large tanks connected by a cylindrical porous rod, as illustrated in Figure 4.1.

Both tanks contain liquid solutions of component A in one and component B in the other. The porous rod connecting both tanks allows diffusion of components A and B; however, the impermeable coat around it prevents any mass transfer to the external ambiance. The left tank is maintained at constant concentration ρ_{A0}, while

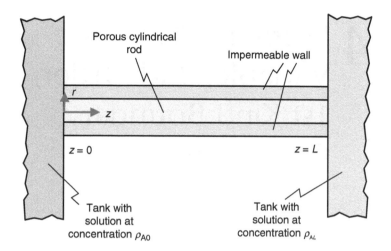

FIGURE 4.1 Scheme for mass transfer through a rod.

the tank on the right holds a solution at constant concentration ρ_{AL}. This is possible because both tanks are very large and small amounts of A or B transferred through the rod would not significantly affect the concentrations in the tanks.

One is interested in the profile of component A (or B) concentration in the porous rod as well as its rate of mass transfer from one tank to the other.

The following assumptions are made:

1. Steady-state regime.
2. Extremities are kept at constant and different concentrations.
3. Rod is made of homogenous porous material with an impermeable coating that avoids mass transfers to the external environment, or in the radial direction (r). In addition, no swirl flow is verified or imposed. Therefore, concentrations are a function of only coordinate z.
4. Components A and B do not react.
5. Temperature and pressure in the whole system remain constant.
6. Density of A and B mixtures can be considered independent of their concentrations.
7. Diffusivity of component A into component B is constant.
8. Rod is very porous and does not impose additional resistance to mass transfers between the two tanks.
9. No appreciable velocity field is observed in the rod, i.e., the densities of solutions in both tanks are almost equal.

Some comments may be necessary on the above assumptions:

• Rate of mass transfer of one component into another depends on various factors such as difference in concentration, diffusivity, temperature, pressure, etc. [1–4]. Of course, the dominant factor is difference in concentration. As seen, the total mass transfer of component A into component B can be expressed in mass or molar basis by Equations A.53 and A.54, respectively. The sum of mass fluxes N_A and N_B are connected to the overall velocity field (v) by Equation A.55. According to assumption 9 above, the overall mass flow is negligible and, therefore, the total velocity or velocity of mass center is zero. This can be understood by imagining that the molecules of A run, for instance, from left to right (or positive z direction) and B in the opposite direction. If their densities are similar, the total velocity is zero. This can also be seen by examining the overall continuity equation, which for cylindrical coordinates in the rod is given by Equation A.2. According to assumptions 1 and 3, this equation can be simplified to

$$\frac{\mathrm{d}}{\mathrm{d}z}(\rho v_z) = 0$$

Now, the overall density of a mixture is given by Equation A.45, and the assumption of similar densities for A and B leads to constant overall density. Therefore, the above equation shows that the overall velocity in the transfer direction must be zero.

• Mass transfer may be affected by several factors, among them temperature and pressure. These are special effects called Dufour and Soret, which depend on gradients of pressure and temperature, respectively. In view of assumption 5, these are neglected here.
• Diffusion coefficients either between two pure components or between any other combinations are functions of temperature, pressure, and concentration itself. For mass transfer problems on isothermal and isochoric ambiances with relatively small variations of concentration, diffusion coefficients can be taken as constants. This justifies assumption 7 above. In addition, the presence of a porous media through which A and B diffuse interferes in that phenomenon. Usually, this interference tends to decrease the diffusion coefficient found for pure components [5,6].

Having in mind the geometry of the rod, it is possible to start from Equation A.41, which in view of the assumptions listed before becomes

$$\frac{\mathrm{d}^2 \rho_A}{\mathrm{d}z^2} = 0 \qquad (4.1)$$

Owing to its simplicity, Equation 4.1 does not require any special method because it results in

$$\frac{\mathrm{d}\rho_A}{\mathrm{d}z} = C_1 \qquad (4.2)$$

Hence

$$\rho_A = C_1 z + C_2 \tag{4.3}$$

Using the boundary conditions

$$\rho_A(0) = \rho_{A0} \tag{4.4}$$

and

$$\rho_A(L) = \rho_{AL} \tag{4.5}$$

into Equation 4.3, the solution becomes

$$\rho_A = \rho_{A0} - (\rho_{A0} - \rho_{AL})\frac{z}{L} \tag{4.6}$$

The expected linear profile has been obtained.

It is worthwhile to deduce some important parameters related to the mass transfer phenomena, such as the following:

- Mass flux of component A. This can be obtained by applying Equation A.53. According to the discussion above, the sum N_A and N_B is zero; therefore, using Equation 4.6 this would give

$$N_A = \rho_A v_A = -D_{AB}\frac{d\rho_A}{dz} = D_{AB}\frac{\rho_{A0} - \rho_{AL}}{L} \tag{4.7}$$

Notice that if $\rho_{AL} < \rho_{A0}$, the mass flux would be positive, i.e., it would occur in the positive z direction. Of course, molecules of component B would travel in the opposite direction with flux given by

$$N_B = \rho_B v_B = -D_{AB}\frac{d\rho_B}{dz} = -D_{AB}\frac{\rho_{BL} - \rho_{B0}}{L} \tag{4.8}$$

Using Equation A.53 it is easy to verify that $N_A + N_B$ becomes zero.

4.3   MASS TRANSFER IN A ROD WITH VARIABLE DIFFUSIVITY

In Section 4.2, the diffusion coefficient of component A into B was assumed constant. This may be approximately true for cases where the temperature, pressure, and concentrations in the extremities of the slab are not too different. However, a more realistic situation involves diffusivity as a function of these variables. Of course, the solution for the concentration profile in the rod would depart from the linear behavior, and the problem may become somewhat difficult to solve.

To demonstrate this, let us assume diffusivity as a function of concentration itself. Therefore, the differential equation governing the problem is

$$\frac{d}{dz}\left[D_{AB}(\rho_A)\frac{d\rho_A}{dz}\right] = 0 \qquad (4.9)$$

with

$$\rho_A(0) = \rho_{A0} \qquad (4.10)$$

and

$$\rho_A(L) = \rho_{AL} \qquad (4.11)$$

As an example, let there be a linear dependence of diffusivity on concentration, or

$$D = D_0 + a(\rho_A - \rho_{A0}) \qquad (4.12)$$

where the symbol D represents D_{AB}. Parameters a and D_0 are constants. Of course, D_0 would be the diffusion coefficient between components A and B at the position $z=0$ or at concentration ρ_{A0}.

Consider the following change of variables:

$$\Omega = \frac{\rho_A - \rho_{A0}}{\rho_{AL} - \rho_{A0}} \qquad (4.13)$$

and

$$x = \frac{z}{L} \qquad (4.14)$$

In Equation 4.9, these lead to

$$\left[(-b + a\Omega)\Omega'\right]' = 0 \qquad (4.15)$$

or

$$(-b + a\Omega)\Omega'' + a(\Omega')^2 = 0 \qquad (4.16)$$

where

$$b = \frac{D_0}{\rho_{A0} - \rho_{AL}} \qquad (4.17)$$

As commented before, if the concentration of A at $z=0$ is greater than concentration at $z=L$, the mass flow of component A occurs in z direction, leading to positive parameter b.

The boundary conditions become

$$\Omega(0) = 0 \tag{4.18}$$

$$\Omega(1) = 1 \tag{4.19}$$

Despite being nonlinear, Equation 4.15 is another particular case where it is possible to reduce to a first-order equation. Therefore

$$(-b + a\Omega)\Omega' = C_1 \tag{4.20}$$

This is also a separable equation, and the following is derived after integrations:

$$-b\Omega + a\frac{\Omega^2}{2} = C_1 x + C_2 \tag{4.21}$$

Using Equations 4.18 and 4.19 one gets to

$$\gamma\Omega^2 - 2\Omega = (\gamma - 2)x \tag{4.22}$$

where

$$\gamma = \frac{a}{b} \tag{4.23}$$

which is a dimensionless parameter.

Keeping in mind Equation 4.12, it is easy to see that, at least for most cases, the value of parameter a should be relatively small if compared with the ratio between diffusivity and concentration difference. Actually, this parameter should be in the vicinity of parameter b, as defined by Equation 4.17. A simple estimation is possible.

Assume, for instance, the case of liquids, where diffusivities are around 10^{-9} m^2 s^{-1}. On the other hand, the concentrations of salts in liquids usually fall in the range of 10^2 kg m^{-3}. Therefore, in such situations, parameter b should have magnitudes around 10^{-11} m^5 kg^{-1} s^{-1}. In any situation, it is expected that parameter γ should vary from 0 to 1 or near that.

Keeping in mind that γ is a positive real number and that x and Ω vary between 0 and 1, the only solution that might lead to valid values is

$$\Omega = \frac{1 - [1 + \gamma(\gamma - 2)x]^{1/2}}{\gamma} \tag{4.24}$$

It should be observed that the boundary conditions given by Equations 4.18 and 4.19 are satisfied using the negative determination. Moreover, as $0 \leq \Omega \leq 1$, the possible range for parameter γ is $0 < \gamma \leq 1$.

4.3.1 Application of Method of Weighted Residuals

Of course, the above explicit solution was only possible for a simplified dependence of diffusivity on concentration. However, for more complex cases, other methods could be necessary or useful, such as the weighted residuals. Its application is illustrated below.

4.3.1.1 Choice of Trial Functions

As explained in Appendix E, it is very convenient if function ϕ_0 could satisfy the boundary conditions given by Equations 4.18 and 4.19. For the sake of simplicity (desired whenever it is viable), a possible choice is

$$\phi_0(x) = x \tag{4.25}$$

The other trial functions are set as

$$\phi_j(x) = x^{j+1} - x, \quad j = 1, 2, \ldots, n \tag{4.26}$$

Therefore, the approximation function becomes

$$\Omega_N = x + \sum_{j=1}^{N} C_j(x^{j+1} - x) \tag{4.27}$$

Using Equations 4.16 and 4.27, the residual is

$$\Lambda(x, \Omega_N) = (-1 + \gamma \Omega_N)\Omega_N'' + \gamma(\Omega_N')^2 \tag{4.28}$$

Let us now work on some approximations and apply the various weighted residual methods (WRM) introduced before.

4.3.1.2 First Approximation

Using Equation 4.26, the first approximation is

$$\Omega_1 = x + C_1(x^2 - x) \tag{4.29}$$

with

$$\Omega_1' = 1 + C_1(2x - 1) \tag{4.30}$$

and

$$\Omega_1'' = 2C_1 \tag{4.31}$$

Thus, the residual given by Equation 4.28 becomes

$$\Lambda_1(x,C_1) = -2C_1 + \gamma + C_1^2\gamma - 2C_1\gamma + 6C_1\gamma(1 - C_1)x + 6C_1^2\gamma x^2 \qquad (4.32)$$

4.3.1.2.1   Method of Moments

Let the Legendre polynomials (Appendix C) be the weighting functions. Therefore

$$W_1(x) = P_1(x) = x \qquad (4.33)$$

and

$$W_2(x) = P_2(x) = 1/2(3x^2 - 1) \qquad (4.34)$$

The method of moments requires that

$$\int_0^1 x\Lambda_1(x,C_1)\,dx = 0 \qquad (4.35)$$

The solution for the above equation provides

$$C_1 = \frac{\gamma}{2(1 - \gamma)} \qquad (4.36)$$

Therefore

$$\Omega_1 = x + \frac{\gamma}{2(1 - \gamma)}(x^2 - x) \qquad (4.37)$$

It becomes now clear that a linear profile would be reproduced if γ equals zero, which would be the same solution achieved in Section 4.2.

4.3.1.2.2   Collocation

As seen in Appendix E, the application of the collocation method requires choosing points inside the domain of variable x. Of course, the first approximation demands just one point at which the residue given by Equation 4.32 would be made equal to zero. If, within the frame of a simple choice, the point $x = 1/2$ is chosen, the following is obtained:

$$C_1 = \frac{\gamma - 2 + \sqrt{4 - 4\gamma - \gamma^2}}{\gamma} \qquad (4.38)$$

Clearly, the linear concentration profile should be reproduced for $\gamma = 0$. Although this value could just be approached as a limit, the only possible choice is the positive determination. In addition, to allow real values for C_1, parameter γ should range between 0 and $2\sqrt{2} - 2$. This has no special significance because other collocation points could be chosen. However, as higher levels of approximations are used, the solution should converge to the exact one, no matter what the choices of collocation points are. The difference is the efficiency, or the level of approximation needed to achieve an acceptably low deviation from the exact solution.

4.3.1.2.3 Sub-domain
Here it is only possible for one partition, or

$$\int_0^1 \Lambda(x,\theta_1)\,dx = 0 \tag{4.39}$$

The remaining is left to the reader as an exercise.

4.3.1.2.4 Least Squares
Here, the weighting function would be

$$W_1 = \frac{\partial \Lambda(x,\theta_1)}{\partial C_1} \tag{4.40}$$

The remaining details on how to obtain the approximate solution are also left as an exercise to the reader.

4.3.1.2.5 Galerkin
Observing Equation E.16, the weighting function is

$$W_1 = x^2 - x \tag{4.41}$$

Using Equation 4.32, the integration (Equation 4.35) leads to

$$C_1 = \frac{5(\gamma - 2) + \sqrt{145\gamma^2 + 100}}{2(6\gamma - 5)} \tag{4.42}$$

Again, the linear concentration profile should be reproduced for $\gamma = 0$. This is the reason for choosing the positive determination. Additionally, the above imposed no limiting values for parameter γ, except $5/6$, which equals the root of the function in the denominator.

4.3.1.2.6 Comparisons between Various Solutions
The various approximations can be compared with the exact solution given by Equation 4.24. Graphs representing each approximation and the exact solution are presented below.

Figure 4.2 shows the exact solution within the possible range for parameter γ and dimensionless coordinate x.

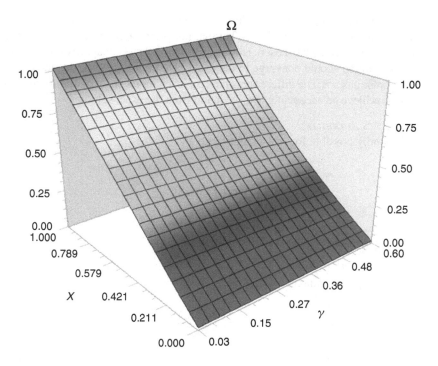

FIGURE 4.2 Dimensionless concentration profile against dimensionless coordinate and diffusion-correlated parameter: exact solution.

It is interesting to observe that the concentration profile deviates from the linear for larger values of γ. The range for parameter γ was kept between 0 and 0.6 only to allow possible real values for all studies based on the first approximations made here. The reader is invited to enlarge the plotting shown by Figure 4.2 for $0 < \gamma < 1$.

The representations of first approximations obtained using the method of moments are presented in Figure 4.3, and in Figures 4.4 and 4.5 in the cases of method of collocation and Galerkin, respectively.

It is easy to see that, at this particular approximation and case, the method of moments provides the best result. However, in general, the method of Galerkin usually works better.

Of course, if one is trying to use the method of weighted residuals (MWR), the exact solution is not available and the decision whether to go further or not is based on comparisons between subsequent approximations. More on this can be found in Section E.2.

4.3.1.3 Second Approximation

From Equation 4.27 it is possible to write the second approximation as

$$\Omega_2 = x + C_1 x^2 + C_2 x^3 - (C_1 + C_2) x^4 \qquad (4.43)$$

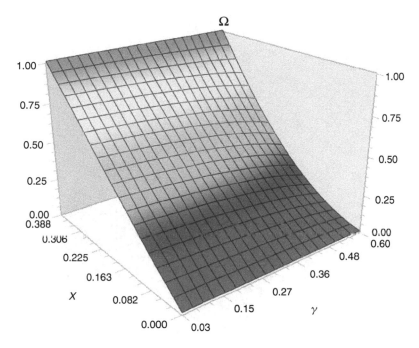

FIGURE 4.3 Dimensionless concentration profile against dimensionless coordinate and diffusion-correlated parameter: first approximation by method of momentum.

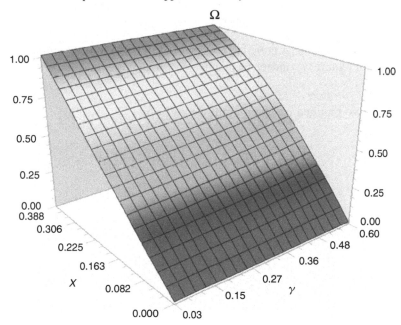

FIGURE 4.4 Dimensionless concentration profile against dimensionless coordinate and diffusion-correlated parameter: first approximation by method of collocation.

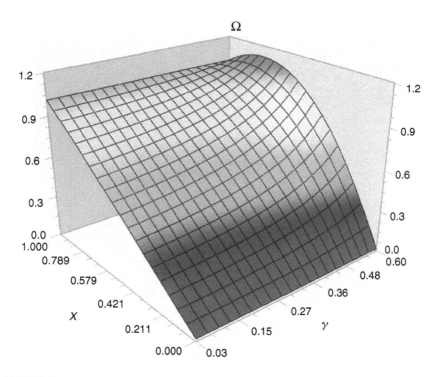

FIGURE 4.5 Dimensionless concentration profile against dimensionless coordinate and diffusion-correlated parameter: first approximation by method of Galerkin.

The respective residue $[\Lambda\,(x,\theta_2)]$ is obtained from Equation 4.28. After this, each method will require two independent equations for the determination of C_1 and C_2.

4.3.1.3.1 Method of Moments
Here, the two required equations could be given by

$$\int_0^1 \Lambda(x,\Omega_2)x\,\mathrm{d}x = 0 \qquad (4.44)$$

and

$$\int_0^1 \Lambda(x,\Omega_2)(3x^2 - 1)\,\mathrm{d}x = 0 \qquad (4.45)$$

4.3.1.3.2 Collocation
The two necessary equations for the determination of C_1 and C_2 require that two points (p_1 and p_2) within the domain D be chosen to give

$$\Lambda(p_1, \Omega_2) = 0 \tag{4.46}$$

and

$$\Lambda(p_2, \Omega_2) = 0 \tag{4.47}$$

It is possible to prove [7] that the best choice for these points is the roots of an orthogonal polynomial of Ω_2. This method is called orthogonal collocation.

For now, the equidistant collocation could be used, and the points would be given by

$$p_j = \frac{j}{N+1}(x_{max} - x_{min}) \tag{4.48}$$

Of course, x_{max} is equal to 1 and x_{min} is equal to 0. If $N=3$, the points will be $p_1 = 1/3$ and $p_2 = 2/3$.

4.3.1.3.3   Galerkin

Here the two weighting functions are

$$W_1 = \frac{\partial \Omega_2}{\partial C_1} = x^2 - x \tag{4.49}$$

and

$$W_2 = \frac{\partial \Omega_3}{\partial C_2} = x^3 - x \tag{4.50}$$

The remaining details are again left to the reader as an exercise.

4.4   CONDUCTION THROUGH A PIPE WALL

As illustrated by Figure 4.6, a fluid at average temperature T_i flows in a pipe with internal and external diameters equal to d_i and d_0, respectively. Its external surface is kept at constant temperature T_0, while the internal surface at temperature T_i. The temperature profile in the wall as well as the rate of heat transfer through it has to be determined.

The following is assumed:

1. Steady-state regime.
2. External surface is in contact with fast moving air at temperature T_0, and because of the high rate of heat transfer, this surface remains at the same temperature. Of course, this is only approximately possible because no matter how high the convective transfer at the surface, the temperature at the surface should differ from that of the environment.

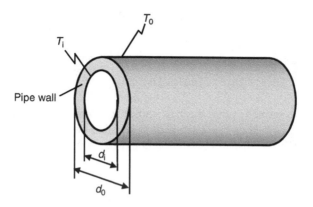

FIGURE 4.6 A pipe with insulation.

3. Flow rate of fluid inside the tube is so high that its temperature remains constant at T_i and the temperature of the tube wall is also at the same temperature. Similar to the situation before, this is an approximation; otherwise no heat transfer would take place.
4. Wall material has a constant thermal conductivity.
5. Pipe is very long and no border effects are present. In other words, developed flow regime is established at the heating region.

Departing from Equation A.35, it is possible to write

$$\frac{d}{dr}\left(r\frac{dT}{dr}\right) = 0 \tag{4.51}$$

which leads to

$$r\frac{dT}{dr} = C_1 \tag{4.52}$$

where C_1 is a constant.

As seen, Equation 4.51 is another example of a second-order equation that can be reduced one level, leading to a simple first-order equation with the following solution:

$$T = C_1 \ln r + C_2 \tag{4.53}$$

as the boundary conditions are

$$T(r_i) = T_i \tag{4.54}$$

and

$$T(r_0) = T_0 \tag{4.55}$$

After using the above conditions, the result is given by

$$\frac{T - T_0}{T_i - T_0} = \frac{\ln\left(\dfrac{r}{r_0}\right)}{\ln\left(\dfrac{r_i}{r_0}\right)} \tag{4.56}$$

Hence, if the internal and external conditions of a given geometry are known, the temperature profile would have the same form, no matter what the conductivity value. This is valid as long as the conductivity remains constant.

The rate of heat transfer is given by

$$\dot{Q} = A_i q_r\big|_{r=r_i} = -A\lambda \frac{dT}{dr}\bigg|_{r=r_i} = -A\lambda \frac{C_1}{r_i} = A\lambda \frac{T_i - T_0}{\ln\dfrac{r_0}{r_i}} \tag{4.57}$$

Here A is the internal area of the tube section in which the rate of heat transfer is computed. Of course, the rate would depend on the wall thermal conductivity.

4.5 ELECTRICALLY HEATED PIPE WALL

It is common to find a situation where the transport of hot fluids between two distant points requires systems to heat the pipe walls, not only to keep its high temperature due to process requirements but also, in some situations, to allow for transportation of the fluids. For instance, some fluids, such as asphalt, need to be heated to decrease their viscosity, otherwise more energy should be delivered by a pump or pumps. In several situations, this can be accomplished by an electric blanket wrapped around the pipe. The power delivered to the blanket should be enough to guarantee the fluid arriving at a desired temperature at the destination.

For several reasons, one being the safety of the heating system, it is interesting to determine the temperature profile inside the electric blanket around the pipe. Obviously, another important information, related to costs, is the heat transfer rate between the blanket and the pipe.

For the sake of simplification and to allow comparisons with the solution achieved in the last section, let us assume that an electric current is made to pass through the metal of the pipe shown in Figure 4.6. In other words, the pipe by itself a source of electric resistance and energy is dissipated by the Joule effect.

As above, the conditions and simplifying hypothesis should be clearly stated:

1. Steady-state regime.
2. External surface of the electrically heated pipe is kept at temperature T_0. This is approximately possible only if high rates of heat transfer were observed between the environment at T_0 and the surface.
3. Internal surface of the pipe remains at constant temperature T_i. This is also approximately possible for fast flowing fluid at constant temperature T_i.

Of course, this is convenient because it coincides with the desired effect, i.e., approximately constant fluid temperature throughout the pile length.

4. Thermal conductivity of the pipe material may be considered constant. This also constitutes a more or less strong approximation and it will be approximately valid for relatively small differences between the internal and external surfaces of the heated pipe.

5. Very long pipe and no border effects. In other words, accomplished flow regime is already established at the heating region.

The problem is the same as that presented in the previous section, but with the addition of an energy source. Therefore, after the above assumption, Equation A.35 is simplified to

$$\lambda \frac{1}{r} \frac{d}{dr} \left(r \frac{dT}{dr} \right) + R_Q = 0 \tag{4.58}$$

It should be remembered that the term R_Q represents the power input per unit of volume where it is delivered. Electric heating manufacturers inform the power or rate of energy delivered by unit of pipe length, or

$$a = \frac{\text{Power}}{L} \tag{4.59}$$

According to the treatment used here, the source should have the following form:

$$R_Q = \frac{\text{Power}}{V} \tag{4.60}$$

Consequently,

$$R_Q = \frac{aL}{V} = a \frac{1}{\pi(r_0^2 - r_i^2)} \tag{4.61}$$

In this case R_Q is a constant. Thus, Equation 4.58 can be rewritten as

$$r \frac{d^2T}{dr^2} + \frac{dT}{dr} + a_1 r = 0 \tag{4.62}$$

Here the constant a_1 is given by

$$a_1 = \frac{R_Q}{\lambda} = \frac{a}{\lambda \pi(r_0^2 - r_i^2)} \tag{4.63}$$

The boundary conditions given by Equations 4.54 and 4.55 are applied again.

Consider a new variable φ given by

$$\varphi(r) = \frac{dT}{dr} \tag{4.64}$$

With this, it is possible to write Equation 4.62 in the following form:

$$r\frac{d\varphi}{dr} + \varphi + a_1 r = 0 \tag{4.65}$$

As seen, Equation 4.62 is another example of a second-order equation that can be reduced to a first-order equation. This in turn can be solved by several methods, one among them being the variation of parameters (Appendix B), which assumes a solution in the form

$$\varphi(r) = u(r)v(r) \tag{4.66}$$

Applying this to Equation 4.65, one arrives at

$$v(r) = r^{-1} \tag{4.67}$$

and

$$u(r) = -a_1\frac{r^2}{2} + C_1 \tag{4.68}$$

where C_1 is an integration constant. Therefore

$$\varphi(r) = -a_1\frac{r}{2} + C_1 r^{-1} \tag{4.69}$$

Using Equation 4.64 and after another round of integration, it is possible to write

$$T = -a_1\frac{r^2}{4} + C_1 \ln r + C_2 \tag{4.70}$$

Boundary conditions given by Equations 4.54 and 4.55 can now be applied, leading to

$$\frac{T - T_i}{T_0 - T_i} = \frac{a_1}{4}\frac{r^2 - r_i^2}{T_0 - T_i}\left(1 - \frac{\ln\frac{r}{r_i}}{\ln\frac{r_0}{r_i}}\right) + \frac{\ln\frac{r}{r_i}}{\ln\frac{r_0}{r_i}} \tag{4.71}$$

The dimensionless form is

$$\theta = \frac{a}{4\pi\lambda(T_0 - T_i)}\frac{x^2 - x_i^2}{x_0^2 - x_i^2}\left(1 - \frac{\ln x}{\ln x_0}\right) + \frac{\ln x}{\ln x_0} \tag{4.72}$$

where

$$\theta = \frac{T - T_i}{T_0 - T_i}, \quad x = \frac{r}{r_i}, \quad x_i = 1, \quad x_0 = \frac{r_0}{r_i} \qquad \text{(4.73a,b,c,d)}$$

This equation provides the temperature profile in the pipe wall, and therefore can be used to obtain the rate of heat transfer to the pipe interior. This is illustrated by Figure 4.7 for a particular case where $T_i = 300$ K, $T_0 = 800$ K, $\lambda = 5$ W m^{-1} K^{-1}, $x_0 = 2$, $x_i = 1$, and parameter a varies from 0 to 2×10^5 W m^{-1}. It is interesting to observe that if $a = 0$, the above solution includes the solution for the case without the heating (Equation 4.56) effect. Equation 4.72 shows how increases of parameter a lead to higher temperatures inside the wall. Despite fixed end values of temperatures (T_i and T_0), larger values of parameter a lead to greater derivatives of temperature at the extremities, which represent higher rates of heat transfers delivered to the flowing fluid.

Usually, the electrical heating blanket is isolated from the ambiance to minimize energy losses. The reader is asked to apply Equation A.25 to find the relation for the rate of heat loss (W m^{-2}) from the external surface to ambiance in the case of no insulation.

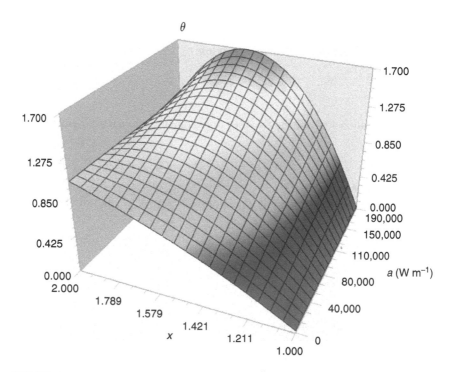

FIGURE 4.7 Dimensionless temperature profile in the electrically heated pipe wall.

4.6   HEAT TRANSFER IN A SPHERICAL SHELL

Consider a solid sphere with a central cavity, as shown by Figure 4.8. The internal and external surfaces of the shell can be kept at constant temperatures, T_1 and T_2, respectively. It is desired to determine the temperature profile in the shell as well as the rate of heat transfer between the external surface and surrounding environment.

The following are the main characteristics or assumptions:

1. Steady-state regime.
2. External surface of the sphere is in contact with fast moving air at temperature T_2. Owing to the high rate of heat transfer, the surface remains at the same constant temperature.
3. Surface of the cavity is kept at constant temperature T_1. A simple method to achieve this would be to continuously inject saturated steam at temperature T_1 into the cavity using a very fine probe, which is assumed to have negligible interference in the physical situation. If temperature T_2 is below T_1 condensation will occur. Another fine probe may be used to withdraw the condensed water from the cavity. As the heat transfer coefficients during condensation are usually very high, the temperature at the cavity surface would remain very nearly constant.
4. Shell material has constant thermal conductivity and density. This also constitutes an approximation, which would become critical for larger differences between T_2 and T_1.

In view of the above, only heat transfer is observed through the shell, and from Equation A.36 it is possible to write

$$\frac{1}{r^2} \frac{d}{dr} \left(r^2 \frac{dT}{dr} \right) = 0 \tag{4.74}$$

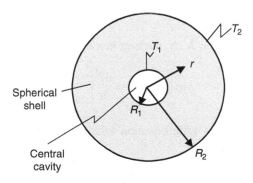

FIGURE 4.8 Sphere with internal cavity and surfaces at different temperatures.

In the case of the shell, the radial coordinate can never become zero, therefore

$$\frac{\mathrm{d}}{\mathrm{d}r}\left(r^2\frac{\mathrm{d}T}{\mathrm{d}r}\right) = 0 \tag{4.75}$$

and

$$r^2\frac{\mathrm{d}T}{\mathrm{d}r} = C_1 \tag{4.76}$$

This yields

$$T = -\frac{C_1}{r} + C_2 \tag{4.77}$$

Despite being a second-order differential problem, it could be reduced to two first-order differential equations. These equations lead to two constants, which can be found using the following boundary conditions:

$$T(R_1) = T_1 \tag{4.78}$$

$$T(R_2) = T_2 \tag{4.79}$$

Using condition given by Equation 4.77, one gets

$$C_1 = -\frac{T_1 - T_2}{\frac{1}{R_1} - \frac{1}{R_2}} \tag{4.80}$$

and finally

$$\frac{T - T_1}{T_2 - T_1} = \frac{\frac{1}{R_1} - \frac{1}{r}}{\frac{1}{R_1} - \frac{1}{R_2}} \tag{4.81}$$

According to Equation A.25, the heat flux at the external surface is given by

$$q_r|_{r=R_2} = -\lambda\frac{\mathrm{d}T}{\mathrm{d}r}\bigg|_{r=R_2} \tag{4.82}$$

Using Equations 4.76 and 4.80, Equation 4.82 can be written as

$$q_r|_{r=R_2} = \lambda\frac{1}{R_2^2}\frac{T_1 - T_2}{\frac{1}{R_1} - \frac{1}{R_2}} \tag{4.83}$$

Finally, the rate of heat exchange with environment is

$$\dot{Q}_r|_{r=R_2} = \pi\lambda \frac{T_1 - T_2}{\dfrac{1}{R_1} - \dfrac{1}{R_2}} \tag{4.84}$$

As always, there is a flux of energy given by the ratio between the potential represented by the numerator $(T_1 - T_2)$ and the resistance represented by the denominator $\frac{1}{\pi\lambda}\left(\frac{1}{R_1} - \frac{1}{R_2}\right)$. Obviously, the resistance decreases for thinner shells (R_1 approaching R_2) or for higher conductivities of the shell material.

4.7 ABSORPTION WITHOUT REACTION

Several industrial processes employ selected absorption for separation of a given component from a mixture. Among these, there is gas–liquid absorption, where a mixture of gases is put in contact with a specific liquid capable of diluting only one among the components in the gaseous mixture. An example is hydrogen chloride, which can be extracted from a mixture with air when in contact with water. This is so because the acid is much more soluble in water than nitrogen and oxygen; therefore, after some time the concentration of HCl in the air decreases significantly. Of course, the concentration of acid in water may find a limit or saturation. However, during most of the process, a flow of HCl is established from the gas to the water.

The usual interest resides in estimating the rate of transfer of absorbed species from one phase to the other. To simplify and generalize the problem, let us imagine a gas mixture with a component A soluble in a liquid B, as illustrated by Figure 4.9. It is desired to determine the concentration profile of component A in the liquid phase as well as the rate of its absorption by liquid B.

The following assumptions are set here:

1. Steady-state regime.
2. Among all possible components of gas mixture, only A is soluble in the liquid B and these two chemical species do not react.

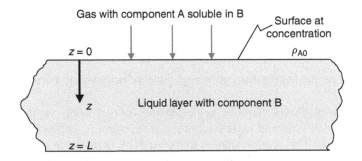

FIGURE 4.9 Diffusion through a deep fluid.

3. Liquid B is still and its rate of evaporation into the gas is negligible. Of course, this is an approximation because all liquids would evaporate into the gas phase. A velocity field would be observed in the liquid and in the gas.
4. Area between liquid and gas is indefinite.
5. Despite the absorption of A, diffusivity, density, and any other property of the liquid phase remain constant. This would become even more critical if A is too soluble in B, i.e., if large concentrations of A in B may be observed.
6. Temperature and pressure remain constant in the gas and fluid. This is also an approximation because even in the case of very low evaporation rate of the liquid, the absorption process usually leads to temperature variations.
7. Concentration of A is known at depth L of the liquid. For now, this concentration is assumed equal to zero, another critical assumption because at steady-state regime concentration of A tends to be uniform within the fluid. The only possibility for such a situation would be if at the bottom of the vase containing B another substance immediately reacts with A. This is seen in a process where a catalyst surface provides the disappearance of A.

During such a process, a concentration ρ_{A0} of diffusing component A is established at the liquid surface. If temperature, pressure, and concentration of A in the gas phase remain unchanged, ρ_{A0} is also constant. This situation characterizes the saturation value for mixture A in B. An approximated example would be the case of air and water, with oxygen as component A and water as component B. At the surface of a lake, the concentration of oxygen in the water reaches a saturation level where the concentration of oxygen can no longer increase. However, this does not mean that oxygen can no longer be transferred from air to the water. As long as the oxygen concentration beneath the surface is smaller than the saturation value at the surface, the gradient of concentration ensures that more oxygen would migrate from air into water.

There are two possibilities of treatment for the present problem. One is very approximated, by assuming negligible velocity field in the liquid, and the other more realistic, where the assumption is not made. However, the mass flow of component A into resting B would inevitably lead to a velocity field. To understand this, consider Equation A.55, which shows the relation between the sum of mass fluxes and velocity, or

$$N_{Az} + N_{Bz} = \rho v_z \qquad (4.85)$$

Therefore, the approximation of negligible velocity would occur only if

1. Both mass fluxes could be neglected and approximately valid only for very small values of mass fractions of component A, i.e., low solubility of A in B. This is equivalent to low diffusivity or mass transfer rate of A into B, which due to Equation 4.86 leads to low velocity field.
2. Mass fluxes are exactly equal but in the opposing directions.

In the present case, only flux of B is zero. Therefore, a velocity field should exist.

As no momentum or energy transport is involved, let us depart from Equation A.37. Because of assumption 1, no time variations may occur. In addition, as the liquid area is indefinite, all variations would take place only in the vertical direction z. Therefore, the above leads to

$$\frac{dN_{Az}}{dz} = 0 \qquad (4.86)$$

A relationship for mass flux is given by Equation A.53 and on just one direction is written as

$$N_{Az} = -D_{AB}\rho \frac{dw_A}{dz} + w_A(N_A + N_B) \qquad (4.87)$$

As component B does not move, the above equation becomes

$$N_{Az} = -D_{AB}\rho \frac{1}{1 - w_A} \frac{dw_A}{dz} \qquad (4.88)$$

Thus, Equation 4.87 would be written as

$$\frac{d}{dz}\left(\frac{1}{1 - w_A} \frac{dw_A}{dz}\right) = 0 \qquad (4.89)$$

Alternatively, we may depart from Equation A.40, which after the assumptions is reduced to

$$\frac{d^2 w_A}{dz^2} = 0 \qquad (4.90)$$

However, arriving at the above relation required neglecting the velocity in the vertical (z) direction. The circumstances of such strong approximation have been discussed already.

Equations 4.90 and 4.91 are second-order equations that can be easily reduced to first order.

The boundary conditions are

$$w_A(0) = \frac{\rho_{A0}}{\rho} = w_{A0} \qquad (4.91)$$

$$w_A(L) = 0 \qquad (4.92)$$

The solutions are as follows:

1. Using Equation 4.90

$$\left(\frac{1}{1 - w_A}\right)\left(\frac{dw_A}{dz}\right) = C_1 \tag{4.93}$$

and

$$-\ln(1 - w_A) = C_1 z + C_2 \tag{4.94}$$

Applying the two boundary conditions, one gets

$$w_A = 1 - (1 - w_{A0})^{1 - \frac{z}{L}} \tag{4.95}$$

2. Using Equation 4.91 the following is obtained:

$$w_A = C_1 z + C_2 \tag{4.96}$$

After the boundary conditions, one arrives at

$$w_A = w_{A0}\left(1 - \frac{z}{L}\right) \tag{4.97}$$

Despite the differences, solutions in Equations 4.95 and 4.97 provide similar concentrations depending on the value of saturation concentration w_{A0}. This is so because more solubility leads to higher saturation concentration and therefore higher rates of mass transfers. In turn, these induce higher velocity fields in the solution. The profiles provided by the solution in Equation 4.95 are shown in Figure 4.10 and by Equation 4.97 in Figure 4.11.

From the figures, it is easy to see they are similar for small values of w_{A0}. However, great errors can be made if one applies Equation 4.97 for cases of highly soluble A or high saturation concentrations (w_{A0}). Numerical comparisons between the two solutions can be easily provided through the use of flux rate of absorbed component A, given by Equation 4.89. The solution given by Equation 4.95 leads to the following flux at the liquid surface:

$$N_{Az}\big|_{z=0} = \frac{\rho D_{AB}}{L} \ln\left(\frac{1}{1 - w_{A0}}\right) \tag{4.98}$$

On the other hand, using solution given by Equation 4.97 one arrives at

$$N_{Az}\big|_{z=0} = -\rho D_{AB} \frac{dw_A}{dz}\bigg|_{z=0} = \frac{\rho D_{AB}}{L} w_{A0} \tag{4.99}$$

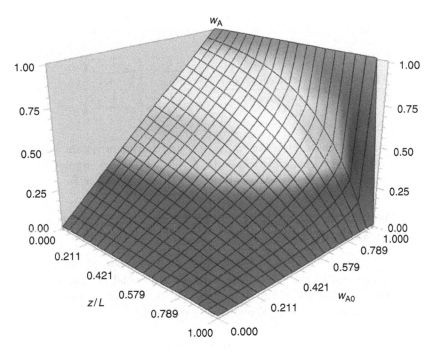

FIGURE 4.10 Concentration profiles provided by Equation 4.95.

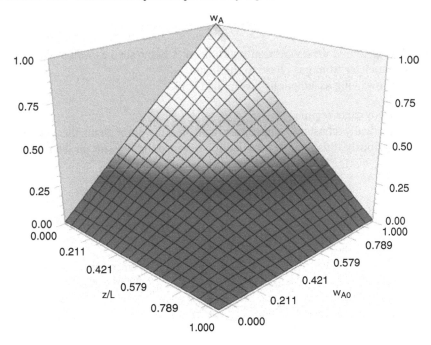

FIGURE 4.11 Concentration profiles provided by Equation 4.97.

The relative deviation between the fluxes provided by the two solutions can be written as

$$\frac{|\text{Difference between solutions}|}{|\text{Sum of two solutions}|} = \frac{1 - \dfrac{1}{w_{A0}} \ln\left(\dfrac{1}{1 - w_{A0}}\right)}{1 + \dfrac{1}{w_{A0}} \ln\left(\dfrac{1}{1 - w_{A0}}\right)} \qquad (4.100)$$

Just to show a few values, for $w_{A0} = 0.1$, the deviation would be 2.61%. For $w_{A0} = 0.01$, the deviation would be 0.25%. This demonstrates how the assumption of no velocity is valid only for low solubility of A in B.

4.8 DIFFUSION THROUGH A SPHERICAL SHELL WITH ZERO-ORDER REACTION

Consider the spherical shell as shown in Figure 4.12. Gaseous species A is kept in the central cavity whereas pure liquid B is kept in the shell, which is sealed by rigid membranes permeable only to molecules of A. Such membranes are called "molecular sieves," with several industrial applications.

Gas A is continuously injected into the cavity just to keep the pressure constant or to replace the amount diffused through the shell. While diffusing, species A reacts with B following an irreversible and zero-order reaction:

$$A + B \rightarrow 2C$$

It is desired to find the concentration profile of A in the shell as well as the rate of its mass transfer from gas phase to the liquid.

The following assumptions are applied here:

- Steady-state regime.
- Practically constant temperature and pressure throughout the system. Obviously, this is an approximation because the pressure in the central

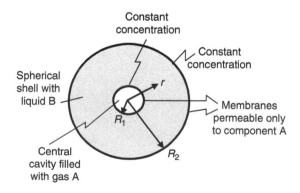

FIGURE 4.12 Diffusion through a liquid in a spherical shell.

cavity should be higher than that of the exterior to allow the flux through the shell.

- Negligible variations of physical properties for any composition of the liquid mixture in the shell. This is the strongest assumption and should be seen as an approximation valid for cases where the concentration of B remains high throughout the shell. This can happen in various circumstances such as
  1. Reactant B is in great excess related to stoichiometric reaction
  2. Reaction with equilibrium leading to low concentrations of product C combined with low solubility of A into B
- Concentration at the external surface of the interior membrane $(r = R_1)$ equals to $\tilde{\rho}_{A1}$ and at the internal surface of the external membrane $(r = R_2)$ equals to $\tilde{\rho}_{A2}$. Those values are known and are constants.
- Liquid components B and C are static, i.e., no velocity fields related to these species are present. This does not assume that no velocity field is present. The very fact that component A is passing through the liquid shell involves a velocity and this is not neglected.
- Zero-order reaction with constant coefficient (k).

As no major velocity field or temperature variations are present, only mass transfer equations are necessary and sufficient to solve the problem. Using the molar version of Equation A.39, one gets

$$\frac{1}{r^2} \frac{d}{dr} \left(r^2 \tilde{N}_{Ar} \right) = \tilde{R}_{M,A} \tag{4.101}$$

The molar version of the equation is usually preferred when chemical transformations are involved because reaction rates are easily represented using this basis. Of course, if a zero-order reaction is present, no particular advantage is gained with one form or the other.

As mentioned, a zero-order reaction takes place, therefore

$$\tilde{R}_{M,A} = -\tilde{\rho}k \tag{4.102}$$

where the SI unit of k is s^{-1}.

The second-order differential equation (Equation 4.101) can be easily integrated. However, as the boundary conditions are related to concentrations, it is convenient to apply the form linking the flux and concentration and is given by Equation A.54. Since the fluid B and the new generated component C are still, their mass fluxes are zero, therefore

$$\tilde{N}_{Ar} = -\tilde{\rho}D_{AB} \frac{1}{1 - x_A} \frac{dx_A}{dr} \tag{4.103}$$

After the first integration, Equation 4.101 becomes

$$\frac{1}{1 - x_A} \frac{dx_A}{dr} = \frac{a}{3} r + \frac{C_1}{r^2} \tag{4.104}$$

and

$$-\ln(1 - x_A) = \frac{a}{6} r^2 - \frac{C_1}{r} + C_2 \tag{4.105}$$

where

$$a = \frac{k}{D_{AB}} \tag{4.106}$$

Since the total concentration is constant, the boundary conditions for this problem are

$$x_A(R_1) = \frac{\tilde{\rho}_{A1}}{\tilde{\rho}} = x_{A1} \tag{4.107}$$

$$x_A(R_2) = \frac{\tilde{\rho}_{A2}}{\tilde{\rho}} = x_{A2} \tag{4.108}$$

After application of these conditions the following are obtained:

$$C_1 = \frac{\ln\left(\dfrac{1 - x_{A1}}{1 - x_{A2}}\right) - \dfrac{a}{6}(R_2^2 - R_1^2)}{\dfrac{1}{R_1} - \dfrac{1}{R_2}} \tag{4.109}$$

and

$$\frac{1 - x_A}{1 - x_{A1}} = \left(\frac{1 - x_{A2}}{1 - x_{A1}}\right)^{\frac{(1/R_1)-(1/r)}{(1/R_1)-(1/R_2)}} \exp\left\{\frac{a}{6}\left[-(r^2 - R_1^2) + (R_2^2 - R_1^2)\frac{(1/R_1)-(1/r)}{(1/R_1)-(1/R_2)}\right]\right\} \tag{4.110}$$

Figure 4.13 illustrates profiles of mole fraction of A against the radius and parameter a for the particular case where $R_2 = 0.4$ m, $R_1 = 0.1$ m, $x_{A1} = 0.5$, and $x_{A2} = 0.1$. It is possible to notice the lower concentration of component A inside the shell for faster reactions or higher values of parameter a.

4.9 ABSORPTION WITH HOMOGENEOUS REACTION

Consider a similar example as presented in Section 4.7. However, diffusing components A and B react under a first-order and irreversible reaction, or

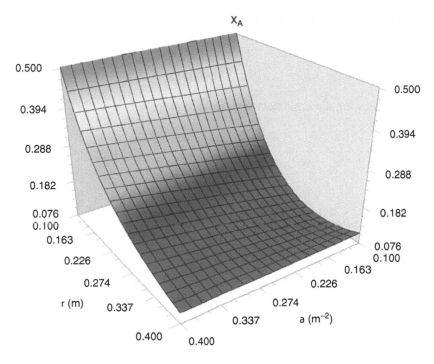

FIGURE 4.13 Molar fraction of component A through a spherical shell against reaction rate parameter (a) in the particular case of $R_2 = 0.4$, $R_1 = 0.1$, $x_{A1} = 0.5$, and $x_{A2} = 0.1$.

$$\tilde{R}_{M,A} = -k\tilde{\rho}_A \tag{4.111}$$

Since the molar basis is used, the rate of A consumption would be given, for instance, in kmol (of A) m^{-3} s^{-1}, and the unit of kinetic coefficient k would be s^{-1}.

The same assumptions set in Section 4.7 should be used.

4.9.1 CASE OF LOW SOLUBILITY

In the present simplified approach, low solubility of A in the liquid is assumed. Therefore, no velocity field is negligible. The reader must refer to the discussion made before regarding this approximation.

If the velocity induced by the mass transfer of A into liquid B is neglected and after the other applying the assumptions listed in Section 4.7, the molar form of Equation A.40 can be written as

$$\frac{d^2\tilde{\rho}_A}{dz^2} = \frac{k}{D_{AB}}\tilde{\rho}_A \tag{4.112}$$

The boundary conditions would be

$$\tilde{\rho}_A(0) = \tilde{\rho}_{A0} \tag{4.113}$$

$$\tilde{\rho}_A(L) = 0 \tag{4.114}$$

The following dimensionless variables are proposed:

$$\phi = \frac{\tilde{\rho}_A}{\tilde{\rho}_{A0}} = \frac{x_A}{x_{A0}} \tag{4.115}$$

$$\zeta = \frac{z}{L} \tag{4.116}$$

With this, Equation 4.111 becomes

$$\frac{d^2\phi}{d\zeta^2} = a^2\phi \tag{4.117}$$

where

$$a = L\sqrt{\frac{k}{D_{AB}}} \tag{4.118}$$

The boundary conditions are now written as

$$\phi(0) = 1 \tag{4.119}$$

$$\phi(1) = 0 \tag{4.120}$$

As seen earlier, Equation 4.117 is homogeneous, linear, second-order, and with constant coefficients. According to Appendix B, Equation 4.117 has the following general solution:

$$\phi = C_1 \sinh(a\zeta) + C_2 \cosh(a\zeta) \tag{4.121}$$

After the application of boundary conditions, it becomes

$$\phi = \cosh(a\zeta) - \frac{\sinh(a\zeta)}{\tanh(a)} \tag{4.122}$$

This equation is represented by Figure 4.14 within the range of characteristic parameter a between 0.5 and 10. It is interesting to notice the linear dependence of concentration ϕ against liquid depth ζ for low values of parameter a, or slow reaction rates. Of course, the other lines are also approximations valid only for low solubility of A in B.

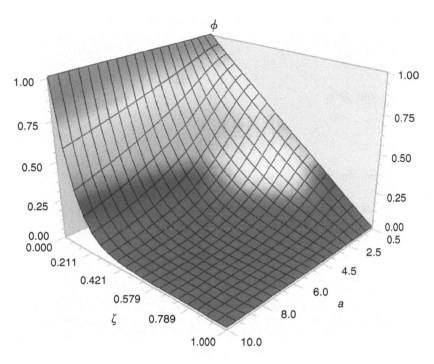

FIGURE 4.14 Concentration profiles of component A into the reacting liquid B; case of low solubility.

The mass flux at the liquid surface can now be computed by

$$\tilde{N}_A\big|_{z=0} = -D_{AB}\frac{\tilde{\rho}}{1-x_A(0)}\frac{dx_A}{dz}\bigg|_{z=0} \tag{4.123}$$

Further work to arrive at a final form is left as an exercise.

4.9.2 Case of High Solubility

As commented, Equation 4.112 is just an approximation valid for constant overall concentration $\tilde{\rho}$ and constant diffusivity. If component A is appreciably soluble in B, this concentration would not remain constant and the velocity field induced by the mass transfer would not be negligible. In this case, the molar form of Equation A.37 should be employed, which after the application of assumptions listed at Section 4.7 combined with Equation 4.111 becomes

$$\frac{\partial N_{Az}}{\partial z} = -k\tilde{\rho}_A$$

Since no flux of species B is observed, Equation A.54 leads to

$$\tilde{N}_{Az} = -D_{AB}\tilde{\rho}\,\frac{1}{1-x_A}\frac{dx_A}{dz} \tag{4.124}$$

Therefore, one gets

$$\frac{d}{dz}\left(\frac{1}{1-x_A}\frac{dx_A}{dz}\right) = \frac{k}{D_{AB}}x_A \tag{4.125}$$

In the above, the global density was assumed constant. Comments on this are made below.

Using the same dimensionless variables as set by Equations 4.115, 4.116, and 4.118, it is possible to write

$$\frac{d}{d\zeta}\left(\frac{1}{1-x_{A0}\phi}\frac{d\phi}{d\zeta}\right) = a^2\phi \tag{4.126}$$

Before solving this equation, it is interesting to remember that the saturation concentration at the liquid–gas interface (x_{A0}) is an indication of how soluble component A is in B, or A is in a mixture of B and C. This is why Equation 4.117 approaches Equation 4.126 for low values of x_{A0}.

The previous equation can be put in the following form:

$$(1 - x_{A0}\phi)\phi'' - x_{A0}\phi'^2 - a^2(1 - x_{A0}\phi)^2\phi = 0 \tag{4.127}$$

while the boundary conditions continue to be given by Equations 4.119 and 4.120.

4.9.3　SOLUTION BY MWR

Equation 4.127 is a nonlinear, second-order differential equation, and an analytical exact solution is very difficult at the least. Therefore, let us apply an approximated method, such as MWR.

As mentioned in Appendix E, an approximate solution should be proposed, and a possible form is

$$\bar{\phi}_n(x) = \sum_{j=1}^{n} C_j\varphi_j + \varphi_0 \tag{4.128}$$

where functions φ_0 and φ_j are trial functions. Preferably, they should satisfy the boundary conditions given by Equations 4.119 and 4.120. Having this in mind, an alternative would be

$$\bar{\phi}_n(x) = (1 - \zeta) + \sum_{j=1}^{n} C_j(1 - \zeta)^{j+1} \tag{4.129}$$

The first approximation becomes

$$\bar{\phi}_1(x) = (1 - \zeta) + C_1(1 - \zeta)^2 \qquad (4.130)$$

The corresponding residual is obtained by using Equation 4.130 in Equation 4.127, and is given by

$$\Lambda_1 = 2C_1\left[1 - x_{A0}(1 - \zeta) - C_1 x_{A0}(1 - \zeta)^2\right] - x_{A0}\left[-1 - 2C_1(1 - \zeta)\right]$$
$$- a^2\left[1 - x_{A0}(1 - \zeta) - C_1 x_{A0}(1 - \zeta)^2\right]^2\left[(1 - \zeta) + C_1(1 - \zeta)^2\right] \qquad (4.131)$$

Now, several submethods can be used. For the sake of simplicity, let us just apply the collocation method and choose $\zeta = 0.5$ as the collocation point. For a particular case of $x_{A0} = 0.5$ and $a = 5$, the solution is $C_1 = -1.99683$.

This first approximation is plotted in Figure 4.15 along with the solutions for low solubility as given by Equation 4.122 for various values of characteristic parameter a.

It is interesting to notice that the first approximation given by Equation 4.131 is

- Already able to represent the concentration profile not too far from the real trend
- Closer to the solution given by Equation 4.122 representing the case of low solubility when the value of parameter a is around 3, therefore far from 5, which was the value used to determine constant C_1 at Equation 4.130.

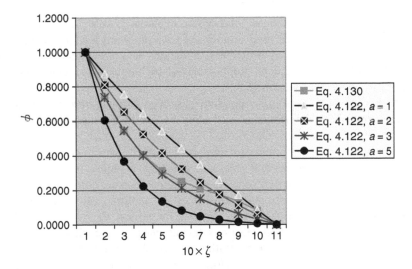

FIGURE 4.15 Concentration profiles of component A into the reacting liquid B; first approximation by MWR method compared with the case of low solubility against various values of parameter a.

This shows that the solution of the problem assuming low solubility may lead to important errors

The reader is invited to derive the second approximation using MWR methods.

4.9.4 VARIABLE GLOBAL DENSITY OR CONCENTRATION

Actually, in all cases of mass transfer, variations on the global density (mass or molar) occur. The molar density or concentration should vary because the involved species does not have the same density or molecular mass. In such cases, the global concentration can be evaluated at each point using the correlation

$$\tilde{\rho} = \tilde{\rho}_A + \tilde{\rho}_B + \tilde{\rho}_C \qquad (4.132)$$

Therefore, another equation for the concentration of B would be necessary. This is given by the stoichiometry of the reaction. For instance, if the reaction $A + B \rightarrow 2C$ could be considered irreversible, the following holds:

$$\frac{1}{2}\frac{d\tilde{\rho}_C}{dt} = -\frac{d\tilde{\rho}_A}{dt} = -\frac{d\tilde{\rho}_B}{dt} \qquad (4.133)$$

This more complete problem would, therefore, involve a system of differential equations, which is beyond the scope of this text.

4.10 REACTING PARTICLE

The field of reactions within a solid porous material is vast with innumerable applications in industry. Several solutions of such problems lead to second-order equations including Bessel differential equations and are therefore excellent examples to be described here. Applications of such equations include combustion of carbonaceous particles [6,8,9].

A porous particle of pure graphite has been thrown into a combustion chamber and one is interested in estimating its rate of consumption at a given position of that chamber. Obviously, the situation is much complex in real situations where coal or biomass are used as fuels. First, these fuels are composed of several chemical species and phases of moisture, volatiles, fixed carbon, and ash. One should be aware that volatiles and fixed carbon are composed of a large amount of organic and inorganic substances. In addition, ash is a mixture of several oxides; many of them may catalyze or poison the various pyrolysis, combustion, and gasification reactions taking place in the chamber. Second, drying and devolatilization occur within a very short time after the injection of a particle in a combustor, therefore involving fast reactions and physical processes. The complexity of pyrolysis or devolatilization of particles is considerable, and therefore beyond the scope of this text. Third, once the solid fuel particle starts burning, an ash layer is formed at the outer shell or layer of that particle. If conditions allow, this layer stays around the unreacted nucleus of fuel; otherwise it detaches from the particle. The previous model is called an unreacted-core model and the other is known as an exposed-core model [6].

As the particle is exposed to an atmosphere with oxygen, the gas as well as other substances in the chamber atmosphere diffuses into the particle through the pores. During this process, the oxygen reacts with carbon leading to the formation of carbon monoxide, which in turn combines with oxygen, leading to the formation of carbon dioxide. To simplify, let us add the two steps and write the complete oxidation as

$$C + O_2 \rightarrow CO_2$$

The general problem is to find the concentration profile of oxygen inside the particle and from there to deduce the rate at which oxygen is consumed by the particle. In turn, this also provides the consumption rate of solid fuel during the combustion.

The present treatment can be used for other cases of reaction involving gases and particles as well. These include situations where the reacting solid species impregnates an inert porous material such as ceramics.

On the other hand, the shape of any particle could be approximated by one among the three basic forms: plate, cylindrical, or spherical, as shown in Figure 4.16, and it is easy to adapt the present treatment to forms different from spherical forms. Moreover, the present model can be applied to particles with irregular shapes, which are better approximations for most cases of the solid fuels. This is accomplished through the use of the sphericity concept, as commented ahead.

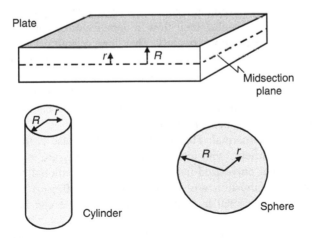

FIGURE 4.16 Basic shapes of particles.

During this treatment, the basic forms have been simplified by assuming the following:

- Plane or flat particle with thickness ($2R$) much smaller than any other dimension. Consequently, the particle would be defined by its total area and thickness. Mass and energy transfers would occur in the direction offering the lower resistance or in the direction (r) perpendicular to the main flat surfaces.
- Cylinder with length much higher than diameter ($2R$). Again, critical transfers would occur in the direction offering the lower resistance or in the radial direction (r).
- Sphere with diameter $2R$ and, of course, the main transfers occur in the radial direction.

To precisely define the simplified situations, let us list the other assumptions made here:

1. Steady-state regime. Consider a continuous flow of fuel particles injected into a combustor. During their travel, the carbon particles react with oxygen, and therefore the temperature, density, and concentration of chemical components inside them vary according to their position in the equipment. However, average characteristics of particles passing at a given position in a reactor remain constant. In other words, we are interested in the oxygen concentration profile inside the particle (or profile in the radial direction) and its combustion rate at a given instant or position in the combustor chamber.
2. Injected solid particles are homogeneous in all directions, except of course the one at which the main mass transfer occurs. For instance, variations on concentrations occur only in the radial (r) direction.
3. The velocity field inside the particles is negligible. As gases are entering and leaving the particles, their flow in opposite directions tend to cancel each other, at least approximately. Therefore, overall mass transfer does not create a substantial velocity component in the radial or any other direction. Even if an overall mass flow of gas from inside toward the particle surface exists, the convective term will be assumed negligible compared with source/sink term (or production/consumption due to chemical reactions) or with the diffusion term.
4. Particles are isothermal. This assumption can be also criticized, mainly when large particles with low thermal conductivity are involved. However, the usual pulverized particles used in conventional suspension of pneumatic combustion are so small (bellow 100 μm) that the Biot number is much smaller than 0.1 and the particle can be considered isothermal.
5. Velocity field of gas around the particle is such that the concentration of oxygen or reacting gas A is constant and known at the particle surface.

This is a strong simplification and made here to allow first-kind boundary conditions. A more realistic approach would set third-kind boundary conditions at the surface. This originates from the equality between the convective and diffusive mass transfer at the surface [6,9].

6. Heat and mass transfer are assumed independent of transport phenomena. Hence, Soret's effect is neglected*. This hypothesis is a consequence of the isothermal approximation for the particle.

7. Particle pores do not suffer severe blockage by related process.

8. As described before, it is assumed that only the reaction of carbon and oxygen will occur. Of course, this is not the case in real combustion where the fuel is composed of various other species than carbon (such as H, N, O, S, etc.) and the surrounding atmosphere would contain several additional gases to oxygen, as water, carbon dioxide, etc. These gases also react with carbon as well as with other components of the fuel. Additionally, and as has been commented, solid fuels such as coal and biomass go through drying and pyrolysis [6,8,9].

9. Indicated reaction will be assumed as first order and irreversible. This is not too far from reality in the case of oxidation of carbon.

Of course, the mass transfer for each basic shape would be easier treated if the respective differential mass balance were applied, i.e., Equation A.40 for planar or flat shape, Equation A.41 for cylindrical shape, and A.42 for spherical one. Nonetheless, it will be shown that if the treatment starts with the spherical or using Equation A.42, a generalization that would allow treatment for all other shapes would be obtained. This equation in molar basis is

$$
\frac{\partial \tilde{\rho}_A}{\partial t} + v_r \frac{\partial \tilde{\rho}_A}{\partial r} + \frac{v_\theta}{r} \frac{\partial \tilde{\rho}_A}{\partial \theta} + \frac{v_\phi}{r \sin \theta} \frac{\partial \tilde{\rho}_A}{\partial \phi}
$$
$$
= D_{AB} \left[\frac{1}{r^2} \frac{\partial}{\partial r} \left(r^2 \frac{\partial \tilde{\rho}_A}{\partial r} \right) + \frac{1}{r^2 \sin \theta} \frac{\partial}{\partial \theta} \left(\sin \theta \frac{\partial \tilde{\rho}_A}{\partial \theta} \right) + \frac{1}{r^2 \sin^2 \theta} \frac{\partial^2 \tilde{\rho}_A}{\partial \phi^2} \right] + \tilde{R}_{M,A}
$$

(4.134)

Here component A might be understood as oxygen and component B as the solid material of the sphere (porous carbon). Actually, the diffusion coefficient D_{AB} should be replaced by another coefficient that reflects the real diffusivity of gas in a porous medium. This is known as effective diffusivity. More details can be found elsewhere [5,6,9]. From now on, D_{AB} will be just written as D_A.

The use of molar basis is justified because reaction rates are easily expressed in this basis, therefore simplifying the mathematical treatment.

Assumption 1 allows elimination of the first term on the left side of Equation 4.134, while assumption 3 permits to neglect all convective terms represented in

---

* Soret effect is caused by interference of temperature gradients into mass transfer process [1].

the left side. Simplification of assumption 2 eliminates the two last terms inside the brackets of the right side and it is possible to write

$$D_A r^{-2} \frac{d}{dr}\left(r^2 \frac{d\tilde{\rho}_A}{dr}\right) + \tilde{R}_{M,A} = 0 \qquad (4.135)$$

Here the source (or sink) term on the right is the total rate of production (or consumption) of component A due to one or various competing reactions. For the present, it is assumed that rate of consumption of component A can be written in the following form:

$$\tilde{R}_{M,A} = -k\tilde{\rho}_A \qquad (4.136)$$

where k is the reaction rate coefficient with unit s^{-1}.

In order to transform Equation 4.135 into a dimensionless form, the following change of variables are proposed:

$$x = \frac{r}{R} \qquad (4.137)$$

$$y = \frac{\tilde{\rho}_A}{\tilde{\rho}_{A,s}} \qquad (4.138)$$

In the above, R is the radius (or the dimension at the direction in which the main transfers occur) of the particle and $\tilde{\rho}_{A,s}$ is the concentration of gas A at the particle surface. Both values are assumed constant and known at the instant in which we are interested on determining the concentration profile of A in the particle and its combustion or reaction rate.

After that, Equation 4.135 can be rewritten as

$$\nabla^2 y = \Phi^2 y \qquad (4.139)$$

The Laplacian operator can be generalized to include other coordinate systems (rectangular and cylindrical) and written as

$$\nabla^2 = x^{-a} \frac{d}{dx}\left(x^a \frac{d}{dx}\right) \qquad (4.140)$$

Parameter a may assume the following possible values: 0 for plane geometry, 1 for cylindrical, and 2 for spherical. In the present case, it equals 2.

The Thiele coefficient is given by

$$\Phi = r_p \left[\frac{k}{D_A}\right]^{1/2} \qquad (4.141)$$

4.10.1 SOLUTION FOR SPHERICAL PARTICLES

In the case of spherical shape, Equation 4.139 can be written as

$$x^2y'' + 2xy' - \Phi^2 x^2 y = 0 \tag{4.142}$$

From Appendix C, this is a modifier Bessel differential equation with solutions given by

$$x^{-1/2} I_{1/2}(x\Phi) \tag{4.143}$$

and

$$x^{-1/2} I_{-1/2}(x\Phi) \tag{4.144}$$

Notice that an alternative for this last solution would be $x^{-1/2} K_{1/2}(x\Phi)$.

Also from Appendix C, relations between these and hyperbolic functions allow writing the general solution in the form

$$y = C_1 \frac{\sinh(x\Phi)}{x} + C_2 \frac{\cosh(x\Phi)}{x} \tag{4.145}$$

Assumption 5 leads to one boundary condition as

$$y(1) = 1 \tag{4.146}$$

The other conditions come from the fact that the concentration at the particle center must acquire finite values. Since

$$\lim_{x \to 0} \frac{\sinh(x\Phi)}{x} = \Phi \tag{4.147}$$

and

$$\lim_{x \to 0} \frac{\cosh(x\Phi)}{x} = \infty \tag{4.148}$$

the coefficient C_2 must be identical to zero. Therefore

$$y = C_1 \frac{\sinh(x\Phi)}{x} \tag{4.149}$$

Using condition given by Equation 4.146, it is possible to write

$$y = \frac{1}{x} \frac{\sinh(x\Phi)}{\sinh(\Phi)} \tag{4.150}$$

This provides the concentration profile of reacting gas A inside the porous particle, which would be an approximation for the case of oxygen inside a porous carbon particle.

Despite the above very simplified approach to the problem, let us obtain the rate of consumption of A. This can be accomplished by realizing that the diffusion rate of oxygen (or gaseous component A) into the particle is usually a much slower process than combustion. Therefore, it is reasonable to assume that, once the oxygen molecule penetrates the particle, it will be consumed by the reaction with the solid phase. In other words, the rate of mass transfer of A at the particle surface would provide the total reaction rate or rate of gas consumption. Such a rate is given by Fick's law applied at the particle surface, or

$$\tilde{r}_A = D_A \frac{d\tilde{\rho}_A}{dr}\bigg|_{r=r_p} = \frac{D_A}{r_p} \tilde{\rho}_{A,s} \frac{dy}{dx}\bigg|_{x=1} \tag{4.151}$$

Using Equation 4.151, the above becomes

$$\tilde{r}_A = \frac{D_A}{r_p} \tilde{\rho}_{A,s} B \tag{4.152}$$

where

$$B = \frac{\Phi}{\tanh \Phi} - 1 \tag{4.153}$$

Figure 4.17 illustrates the behavior of parameter B against Thiele modulus (Equation 4.141).

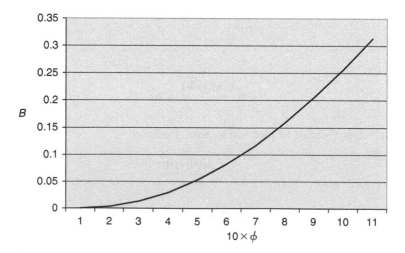

FIGURE 4.17 Graph showing dependence of parameter B on Thiele modulus.

Of course, the reaction rate or consumption of oxygen should increase with the reaction constant or Thiele modulus.

4.10.1.1   Solution through Power Series

To illustrate the application of power series method or its more elaborate version, the Frobenius method, the solution of Equation 4.142 will be repeated.

The method is applicable for cases of linear differential equations. Of course, as the solution of that equation is known, the use of Frobenius method is unnecessary. Nonetheless, the very definition of Bessel functions appeared after the application of power series method is known today as Bessel differential equations.

Frobenius method is described in Appendix C and starts by rewriting that equation in the form similar to Equation C.22, or

$$y'' + \frac{2}{x}y' - \Phi^2 y = 0 \tag{4.154}$$

Therefore

$$b(x) = 2 \tag{4.155}$$

and

$$c(x) = -\Phi^2 x^2 \tag{4.156}$$

As analytical functions of x with center in zero, functions $b(x)$ and $c(x)$ allow at least one solution of Equation 4.155 in the form similar to Equation C.23, or

$$y(x) = x^r \sum_{m=0}^{\infty} a_m x^m \tag{4.157}$$

Using Equations 4.155 and 4.156 and the series in the forms given by Equations C.25 and C.26, it is possible to see that

$$b_0 = 2 \tag{4.158}$$

and

$$c_2 = -\Phi^2 \tag{4.159}$$

The other terms of $b(x)$ and $c(x)$ series are zero. Therefore, the indicial equation (C.29) becomes

$$r^2 + r = 0 \tag{4.160}$$

with roots $r_1 = 0$ and $r_2 = -1$. Hence, this is a case of distinct roots, but differing by an integer. One solution is

$$y_1(x) = \sum_{m=0}^{\infty} A_m x^m \tag{4.161}$$

The other solution is given by

$$y_2(x) = C y_1(x) \ln x + x^{-1} \sum_{m=0}^{\infty} B_m x^m \tag{4.162}$$

Let us start by collecting the coefficients A_m belonging to the first solution. Using Equation 4.143 we arrive at the following:

$$\sum_{m=0}^{\infty} m(m-1)A_m x^m + 2 \sum_{m=0}^{\infty} mA_m x^m - \Phi^2 \sum_{m=0}^{\infty} A_m x^{m+2} = 0 \tag{4.163}$$

- No relationship is obtained for terms with zero powers of x.
- From terms with x^1 the following can be written:

$$A_1 = 0 \tag{4.164}$$

- From terms with x^2 the following can be written:

$$6A_2 - \Phi^2 A_0 = 0 \tag{4.165}$$

- From terms with x^3 the following can be written:

$$12A_3 - \Phi^2 A_1 = 0 \tag{4.166}$$

However, due to Equation 4.164, $A_3 = 0$, as well as all others with odd indexes.

- From terms with x^4 the following can be written:

$$20A_4 - \Phi^2 A_2 = 0 \tag{4.167}$$

Using Equation 4.165, it is possible to write

$$A_4 = \Phi^4 \frac{A_0}{5!} \tag{4.168}$$

One may verify that in general

$$A_{2j} = \Phi^{2j} \frac{A_0}{(2j+1)!}, \quad A_{2j-1} = 0, \quad j = 1, 2, 3 \ldots \tag{4.169}$$

Thus, the first solution becomes

$$y_1(x) = A_0 \sum_{m=0}^{\infty} \Phi^{2m} \frac{x^{2m}}{(2m+1)!} \tag{4.170}$$

From Equation C.5, the above is recognized as

$$y_1(x) = A_0 \frac{\sinh(\Phi x)}{x} \tag{4.171}$$

which, according to Equation 4.145, reproduces one possible solution.

The second possible solution, as given by Equation 4.162, can be substituted into Equation 4.142. If the first term on the left of Equation 4.162 is used in Equation 4.142, one should verify that constant C must be equal to zero. Therefore, we are left with the second term. Using this in Equation 4.142, the following is obtained:

$$\sum_{m-0}^{\infty} (m-1)(m-2)B_m x^{m-1} + 2\sum_{m=0}^{\infty} (m-1)B_m x^{m-1} - \Phi^2 \sum_{m=0}^{\infty} B_m x^{m+1} = 0$$

$$\tag{4.172}$$

After collecting the terms related to the same powers of x, their general forms become

$$B_{2j} = \Phi^{2j} \frac{B_0}{(2j)!}, \quad \text{and} \quad B_{2j+1} = \Phi^{2j} \frac{B_1}{(2j+1)!}, \quad j = 1, 2, 3 \dots \tag{4.173}$$

Comparing the above with the series given by Equations C.5 and C.6, one concludes that the even terms may be written as $B_0 \cosh(\Phi x)$ and the odd terms as $B_1 \sinh(\Phi x)$.

Therefore, the second solution would be

$$y_1(x) = B_0 \frac{\cosh(\Phi x)}{x} + B_1 \frac{\sinh(\Phi x)}{x} \tag{4.174}$$

Of course, the hyperbolic sinus term has already been found as the first solution and therefore repetitive. The general final result is the sum of y_1 and y_2, or

$$y(x) = C_2 \frac{\cosh(\Phi x)}{x} + C_1 \frac{\sinh(\Phi x)}{x} \tag{4.175}$$

which is exactly the same as given by Equation 4.145. Equation 4.150 is obtained after application of boundary conditions.

4.10.1.2   Solution Using Laplace Transform

Equation 4.142 can also be solved using Laplace transform. For such, the properties listed in Appendix D should be applied.

That equation can be written as

$$xy'' + 2y' - \Phi^2 xy = 0 \qquad (4.176)$$

Using properties given in Equation D.10 and in Equations D.33 through D.36 (all summarized at Table D.1), it is possible to transform the above equation and arrive at

$$-s^2 Y'(s) - 2sY(s) + y(0) + 2sY(s) - 2y(0) + \Phi^2 Y'(s) = 0 \qquad (4.177)$$

or

$$(\Phi^2 - s^2)Y'(s) = y(0) \qquad (4.178)$$

This is a separable equation and the integration gives

$$Y(s) = y(0)\frac{1}{2\Phi} \ln\left(\frac{\Phi + s}{\Phi - s}\right) + C_3 \qquad (4.179)$$

As the sought function $y(x)$ should be continuous, property given by Equation D.1a ensures that constant C_3 must be equal to zero. Using the Tables in Appendix D, the inverse of $Y(s)$ above would be

$$y(x) = y(0)\frac{1}{2\Phi x}\left(e^{\Phi x} - e^{-\Phi x}\right) = C_1 \frac{\sinh(\Phi x)}{x} \qquad (4.180)$$

As $y(0)$ is unknown, its combination with Thiele modulus has been replaced by a constant (C_1). Therefore, the solution is exactly as given by Equation 4.149.

4.10.1.3   Solution Using Method of Weighted Residuals

Another example of MWR application is shown here.

According to Appendix E, a possible approximate solution might be given by

$$\bar{y}_n(x) = x + \sum_{j=1}^{n} C_j\left(1 - x^{j+1}\right) \qquad (4.181)$$

It should be noticed that such approximation complies with the boundary condition given by Equation 4.146.

The first approximation is

$$\bar{y}_1 = x + C_1(1 - x^2) \qquad (4.182)$$

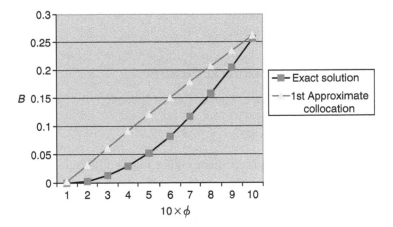

FIGURE 4.18 Graph comparing exact and approximate solutions for the dependence of parameter B on Thiele modulus.

The corresponding residual is obtained by combining the above and Equation 4.142 to obtain*

$$\Lambda_1 = 2 \quad 6C_1 x \quad \Phi^2 C_1 x - \Phi^2 x^2 + \Phi^2 C_1 x^3 \qquad (4.183)$$

One might apply any MWR method, among them collocation. If the point $x = 2/3$ is chosen as collocation one, parameter C_1 becomes

$$C_1 = \frac{54 - 12\Phi^2}{108 + 10\Phi^2} \qquad (4.184)$$

In a general situation, the second and other approximations should be obtained and the decision regarding satisfactory solution would emerge from comparisons between the deviations between consecutive approximations. However, in the present case, it is possible to verify that even this first approximation may be acceptable for crude calculation by comparing the result with the exact one, given by Equation 4.150. In particular, it would be easier here to compare the rate of species A consumption from exact and approximated solution.

Using Equation 4.182—with C_1 given by Equation 4.184—and Equation 4.151, the first approximation for the rate can be written as Equation 4.152, where parameter B is given by

$$B_1 = 1 - \frac{108 - 24\Phi^2}{108 + 10\Phi^2} \qquad (4.185)$$

Figure 4.18 illustrates the value of B_1 against Thiele modulus. This can be compared with the exact solution shown in Figure 4.17.

---

* To simplify, here Equation 4.142 was divided by x, therefore the solution is not valid at $x = 0$.

The reader is invited to obtain the same as well as further levels of approximations by other MWR methods.

4.10.2 SOLUTION FOR CYLINDRICAL PARTICLES

In the case of cylindrical shape, Equation 4.139 becomes

$$xy'' + y' - \Phi^2 xy = 0 \tag{4.186}$$

As seen in Appendix C, this is another modifier Bessel differential equation with solutions given by

$$I_{1/2}(x\Phi) \tag{4.187}$$

and

$$K_{1/2}(x\Phi) \tag{4.188}$$

Using the alternative forms for the above equations as shown in Appendix C, it is possible to write the general solution as

$$y = C_1 \frac{\sinh(x\Phi)}{\sqrt{x}} + C_2 \frac{\cosh(x\Phi)}{\sqrt{x}} \tag{4.189}$$

The boundary conditions to be applied are the same as in the case of spherical particle. In addition, the same conclusions related to the limits of functions of general solution for positions approaching the center ($x \to 0$) are obtained. Therefore, the final solution is

$$y = \frac{1}{\sqrt{x}} \frac{\sinh(x\Phi)}{\sinh(\Phi)} \tag{4.190}$$

This provides the concentration profile of reacting gas A inside the porous particle.

The consumption rate of species A due to gas–solid reaction can be obtained similar to that for the case of spheres.

4.10.3 SOLUTION FOR FLAT PARTICLES

In the case of flat or planar shape, Equation 4.139 is simplified to

$$y'' - \Phi^2 y = 0 \tag{4.191}$$

As shown in Appendix C, this is a linear second-order differential equation with general solution given by

$$y = C_1 \sinh(x\Phi) + C_2 \cosh(x\Phi) \qquad (4.192)$$

Like before, the boundary conditions are applied and the final solution is

$$y = \frac{1}{\sqrt{x}} \frac{\sinh(x\Phi)}{\sinh(\Phi)} \qquad (4.193)$$

Again, the consumption rate of species A due to gas–solid reaction can be obtained in the same way as for the spherical shape.

4.10.4 GENERAL IRREGULAR SHAPE

After proper series of grinding processes, coal particles reach a desired range of size distribution. Other solid fuels as biomasses are fed to combustors after cutting. In most of those cases, the particle shapes may not even approximate any of the previous studied cases of sphere, cylinder, or plate. For such situations, the use of sphericity concept provides reasonable approximation for computations of fuel consumption. It is defined by

$$\phi = \frac{A_{sp}}{A_p} \qquad (4.194)$$

Here A_{sp} is the surface area of a spherical particle with the same volume of the real studied particle and A_p is the surface area of the real particle. Therefore, the sphericity is always a number between 0 and 1. Its usual values are 0.7 for general coal particle after leaving the mill and 0.2 for wood chips.

To compute the fuel consumption rate for the present case, one should just apply the formulas deduced in the case of spherical particle and replace the diameter by

$$d_p = \bar{d}_p \phi^{1/3} \qquad (4.195)$$

where the diameter indicated on the right side is the average value determined by screen analysis of a fuel sample [6].

4.11 HEAT TRANSFER THROUGH A REACTING PLATE

Let us consider a plate with a material such that it undergoes an exothermic reaction when heated. It is in contact with air or any fluid, as shown in Figure 4.19. One is interested in obtaining the temperature profile in the plate as well as the rate of heat transfer through it.

To clearly set the problem, let us list the assumptions:

1. Wall is very wide and long, but with relatively small thickness. In other words, it is considered infinite in all directions except coordinate x.

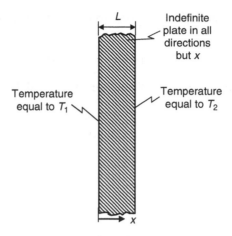

FIGURE 4.19 A reacting semi-infinite solid wall with faces at different temperatures.

2. Steady-state regime is established, i.e., the temperatures will be only a function of the position x.
3. No velocity field exists inside the solid wall.
4. All physical properties of the plate material are constants. Of course, if transformations in the material occur due to chemical reactions, this assumption will be difficult to ensure. However, for the sake of illustrating methods to solve the mathematical problem, it will be assumed that these transformations are not too severe. Improvements to the solution might consider the thermal conductivity and other properties as functions of temperature.
5. Temperatures of the plate surfaces are kept at constant values, as shown in Figure 4.19. Such a situation may be forced by several means, one among them is high heat transfer by convection with fluids in contact with each face. For this, the fluids may be flowing at such velocity that the surface temperatures are very similar to the values found at positions far from the plate in each respective fluid. This usually constitutes a strong approximation because the temperature at the plate surfaces must be different from those found at the respective fluids. Actually, a more realistic approach for such a situation should include heat transfer by convection between the fluids and the plate. This is discussed in Chapter 6. Other methods would also allow the assumption of constant temperatures at each plate main surface, one among them being the condensation of a vapor at the surface. The heat transfer coefficients found in such phase changes are usually high, therefore leading to constant temperatures at the condensing surface.
6. Plate goes through an exothermic reaction and the rate of energy generation per unit of plate volume (W m^{-3}) is given by

$$R_Q = cT \qquad (4.196)$$

where c is a positive real constant. Of course, this is a simplification and only used here for the sake of showing applications of analytical mathematical methods.

After the above assumptions, Equation A.34 would lead to

$$\lambda \frac{d^2 T}{dx^2} + cT = 0 \tag{4.197}$$

The boundary conditions are

$$T(0) = T_1 \tag{4.198}$$

and

$$T(L) = T_2 \tag{4.199}$$

Therefore, they are first-kind boundary conditions.

The solution of homogeneous linear second-order differential equation with constant coefficients is described in Section B.4.1. From this, the general solution is

$$T(x) = C_1 \sin\left(x\sqrt{\frac{c}{\lambda}}\right) + C_2 \cos\left(x\sqrt{\frac{c}{\lambda}}\right) \tag{4.200}$$

Applying boundary conditions leads to

$$T(x) = \frac{T_2 - T_1 \cos\left(L\sqrt{c/\lambda}\right)}{\sin\left(L\sqrt{c/\lambda}\right)} \sin\left(x\sqrt{c/\lambda}\right) + T_1 \cos\left(x\sqrt{c/\lambda}\right) \tag{4.201}$$

This can be written in dimensionless form, or

$$\theta = \frac{T - T_1}{T_2 - T_1} = \frac{T_2 \frac{\sin(\varphi\zeta)}{\sin(\varphi)} - T_1\left[1 - \frac{\sin[\varphi(1-\zeta)]}{\sin(\varphi)}\right]}{T_2 - T_1} \tag{4.202}$$

where

$$\varphi = L\sqrt{\frac{c}{\lambda}} \text{ and } \zeta = \frac{x}{L} \tag{4.203a,b}$$

The dimensionless temperature profiles through the plate for each parameter φ are shown by Figure 4.20. The values for T_1 and T_2 were chosen as 300 and 600 K, respectively.

It is easy to recognize that a linear profile is approached for cases of small reaction rates, drawn when c or φ are close to zero. The distortion from this results into higher peaks of temperature inside the plate, caused by the exothermic

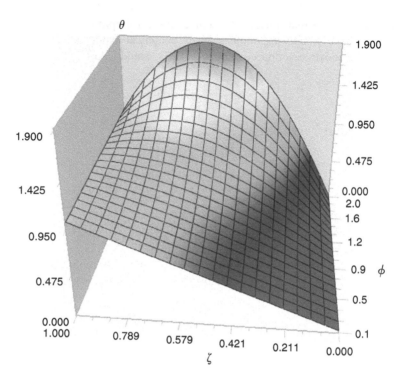

FIGURE 4.20 Dimensionless temperature profiles in the reacting plate.

reaction. To maintain the temperatures at the plate surfaces at constant values, faster reactions would require higher heat transfer rates between the surfaces and the environment. These rates of heat transfer are easily obtained by

$$q_x(0) = -\lambda \left. \frac{dT}{dx} \right|_{x=0} \quad \text{and} \quad q_x(L) = -\lambda \left. \frac{dT}{dx} \right|_{x=L} \qquad (4.204a,b)$$

The deductions of formulas for computations of heat transfer rates are left as exercises to the reader.

EXERCISES

1. Using Equations 4.7, 4.8, and A.45 show that the sum of fluxes A and B at any point of the rod (Figure 4.1) is equal to zero.
2. Solve the differential Equation 4.20 to arrive at Equation 4.21.
3. Find the first approximation of the problem posed in Section 4.3 using the method of sub-domain. Likewise, as presented by Figures 4.2 through 4.5, plot a graph of Ω against x (between 0 and 1) with parameter γ (between 0.1 and 0.6).

4. Repeat the previous problem using the method of least squares.
5. Complete the application of Galerkin method to obtain the second approximation for the problem proposed in Section 4.3.
6. Solve the problem proposed in Section 4.3 in the case of diffusivity given by

$$D_{AB} = D_0 + a(\rho_A - \rho_{A0})^n$$

where D_0, a, and n are constants. Choose any MWR method and determine the number of approximations necessary to obtain a 5% deviation between successive trials.
7. Solve the problem of temperature profile in the pipe wall presented in Section 4.4 when its conductivity depends on the temperature as given by

$$\lambda = \lambda_0 + a(T - T_0)$$

8. Try to solve Equation 4.65 by Laplace transform.
9. Using Equation 4.71, obtain the heat transfer between the wall and fluid as well as between the external surface and surrounding air.
10. Solve the same problem presented in Section 4.5 but in the case where insulation covers the electric heating blanket. Assume the following information:
   - Temperature of outside surface of the insulation is known.
   - Temperature at the interface between the electric blanket and the insulation are equal.
   - Diameter of pipe, thickness of electric heating blanket, and insulation are given.
   - All physical properties such as electric heating coefficient (a), thermal conductivities of blanket, and insulation material are provided.
   Is this information sufficient to solve the problem?
11. With the results of the previous exercises, compare the rate of heat transfers to the pipe and environment with and without insulation around the electric heating blanket.
12. Let us consider a sphere made of a porous solid catalyst substance with a central cavity, as shown by Figure 4.21. Gas with component A is injected into the cavity just to keep the pressure constant. The component A diffuses through the shell. An irreversible and zero-order reaction (A → B) takes place between component A and the catalyst material. Assuming
   - Constant temperature and practically constant pressure in the whole system.
   - Component A behaves as an ideal gas.
   - Concentration at the interior surface ($r = R_1$) equals to $\tilde{\rho}_{A1}$ and at the external surface ($r = R_2$) equals to $\tilde{\rho}_{A2}$.
   - Component B is retained by the catalyst, and therefore does not flow as fluid through the porous material.

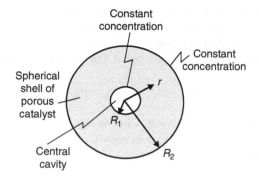

FIGURE 4.21 Sphere with internal cavity and surfaces at different concentrations.

- Effective diffusivity (D_A) of A through the porous substance, as well as all other properties are kept constant.
- Mass flux of component A in the radial direction through the porous substance can be approximately given by $N_{Ar} = -D_A \dfrac{d\tilde{\rho}_A}{dr}$.

Determine

1. The concentration profile of component A in the shell
2. The mass flux of species A from the external surface to the surrounding environment

Answers: $\dfrac{\tilde{\rho}_{A1} - \tilde{\rho}_A}{\tilde{\rho}_{A1} - \tilde{\rho}_{A2}} = a \dfrac{r^2 - R_1^2}{\tilde{\rho}_{A1} - \tilde{\rho}_{A2}} + \left(1 - a \dfrac{R_2^2 - R_1^2}{\tilde{\rho}_{A1} - \tilde{\rho}_{A2}}\right) \dfrac{(1/r) - (1/R_1)}{(1/R_2) - (1/R_1)}$

and

$$\tilde{N}_A\bigg|_{r=R_2} = 2aD_A R_2 + \frac{D_A}{R_2^2} \frac{(\tilde{\rho}_{A1} - \tilde{\rho}_{A2}) - a(R_2^2 - R_1^2)}{(1/R_1) - (1/R_2)}$$

where $a = \dfrac{k}{6D_A}$

13. From the results presented in Section 4.8, obtain the molar flux of component A at the internal $(r = R_1)$ and external $(r = R_2)$ surface of the shell. Interpret the difference between those two results.
14. Repeat example shown in Section 4.8 for the case where a first-order irreversible chemical reaction takes place between components A and B.
15. Using the same approach as followed in Section 4.9.2, find the second approximation for concentration profile of A. Apply the assumed form given by Equation 4.128 by collocation MWR method. Plot a graph similar to that shown in Figure 4.15 to compare with solutions found for low solubility.

16. Repeat the previous problem using Galerkin's MWR method.
17. Solve the problem presented in Section 4.9 for the case of a second-order chemical reaction.
18. From Equation 4.117 arrive at solutions given by Equations 4.121 and 4.122.
19. From Equation 4.142 arrive at solutions given by Equations 4.120 and 4.121. Hint: use a new variable $z = \Phi x$.
20. From Equation 4.186 arrive at solutions given by Equations 4.187 and 4.188, as well as to the form given by Equation 4.189.
21. Arrive at the same solutions as that of Equation 4.190 by solving Equation 4.186 using the Frobenius method.
22. Arrive at the solution given by Equation 4.189 using the Laplace transform.
23. Use Galerkin's MWR method to find the first and second approximations of solutions to Equation 4.142. Deduce the consumption rates using these approximations and compare with the exact solution given by Equation 4.152.
24. In Section 4.10, arrive at the solution for the case of flat plate through the use of Frobenius method and Laplace transform.
25. Using a similar mathematical treatment as shown at Section 4.10 for the case of spheres, obtain the consumption rate of species A for the cases of cylindrical and flat particles.
26. Using Equation 4.204, deduce the formulas for fluxes of heat transfers at the plate surfaces.
27. Repeat the treatment made at Section 4.11 for the case where the rate of energy release in the plate because of exothermic reaction is given by $R_Q = cT^2$.
28. Solve the following differential equations by reducing them to the Bessel form using the indicated hints for variable changes:
    (a) $x^2y'' + xy' + (a^2x^2 - b^2)y = 0$ $(ax = z)$
    (b) $xy'' + y' + \frac{1}{4}y = 0$ $(\sqrt{x} = z)$
    (c) $x^2y'' + xy' + (4x^4 - \frac{1}{4})y = 0$ $(x^2 = z)$
    (d) $4x^2y'' + 4xy' + (x - a^2)y = 0$ $(\sqrt{x} = z)$
    (e) $y'' + a^2x^4y = 0$ $(y = w\sqrt{x}, \frac{1}{3}ax^3 = z)$
    (f) $x^2y'' + (1 - 2a)xy' + a^2(x^{2a} + 1 - a^2)y = 0$ $(y = x^aw, x^a = z)$
    (g) $xy'' + (1 + 2a)y' + xy = 0$ $(y = x^{-a}w)$
    (h) $x^2y'' + \frac{1}{4}(x + \frac{3}{4})y = 0$ $(y = w\sqrt{x}, \sqrt{x} = z)$
    (i) $y'' + a^2x^2y = 0$ $(y = w\sqrt{x}, \frac{1}{2}ax^2 = z)$
    (j) $y'' + a^2xy = 0$ $(y = w\sqrt{x}, \frac{2}{3}ax^{3/2} = z)$
    (k) $x^2y'' - 3xy' + 4(x^4 - 3)y = 0$ $(y = wx^2, x^2 = z)$
    (l) $y'' + xy = 0$ $(y = w\sqrt{x}, \frac{2}{3}x^{3/2} = z)$
29. Solve the following differential boundary-value problems by Laplace transforms (leave the other condition as a constant):
    (a) $y''(x) + xy'(x) - 2y(x) = 1$, $y(0) = 0$
    (b) $xy''(x) + (x - 1)y'(x) + y(x) = 0$, $y(0) = 0$
    (c) $xy''(x) + (2x + 3)y'(x) + (x + 3)$, $y(0) = 3$
    (d) $x^2y''(x) - 2y(x) = 2x$, $y(2) = 2$

REFERENCES

1. Bird, R.B., Stewart, W.E., and Lightfoot, E.N., *Transport Phenomena*, John Wiley, New York, 1960.
2. Slattery, J.C., *Momentum, Energy, and Mass Transfer in Continua*, Robert E. Kriefer, Huntington, New York, 1978.
3. Luikov, A.V., *Heat and Mass Transfer*, Mir, Moscow, 1980.
4. Incropera, F.P. and DeWitt, D.P., *Fundamentals of Heat and Mass Transfer*, John Wiley, New York, 1996.
5. Walker, Jr., P.L., Rusinko, Jr., F., and Austin, L.G., Gas Reactions of Carbon. In *Advances in Catalysis*, Academic Press, New York, 1959, vol. XI.
6. de Souza-Santos, M.L., *Solid Fuels Combustion and Gasification: Modeling, Simulation and Equipment Operation*, Marcel Dekker, New York, 2004.
7. Villadsen, J. and Michelsen, M.L., *Solution of Differential Equation Models by Polynomial Approximation*, Prentice Hall, New Jersey, 1978.
8. Smith, K.L., Smoot, L.D., Fletcher, T.H., and Pugmire, R.J., *The Structure and Reaction Processes of Coal*, Plenum Press, New York, 1994.
9. Smoot, L.D. and Pratt, D.T., *Pulverized-Coal Combustion and Gasification*, Plenum Press, New York, 1979.

5 Problems 122; One Variable, 2nd Order, 2nd Kind Boundary Condition

5.1 INTRODUCTION

This chapter presents methods to solve problems with one independent variable involving second-order differential equation and second-kind boundary condition. Therefore, only ordinary differential equations are involved. Mathematically, this class of cases can be summarized as $f\left(\phi, \omega, \frac{d\phi}{d\omega}, \frac{d^2\phi}{d\omega^2}\right)$, second-kind boundary condition.

5.2 FLOW ON AN INCLINED PLATE

Despite being very simplified, this example is important as it shows a solution for the important problem of calculating the drag force between surfaces and flowing fluids around them.

Consider a thin film of fluid flowing on an inclined plate, as shown in Figure 5.1.

One is interested in obtaining the velocity and tension profiles as well as the formula to estimate the drag force between fluid and plate.

For the sake of simplicity, the following are assumed:

1. Steady-state regime.
2. Liquid fluid is Newtonian with constant physical properties; among them are the density and viscosity. This is an approximation because the very stress throughout the fluid causes energy dissipation. According to the first law of thermodynamics such an effect can be correlated to work on the fluid. If not properly dissipated as heat, the fluid would experience increases in temperature, which would lead to changes in its viscosity and density. However, for relatively low velocities and thin fluid layers, the assumption of constant properties is a very reasonable approximation.

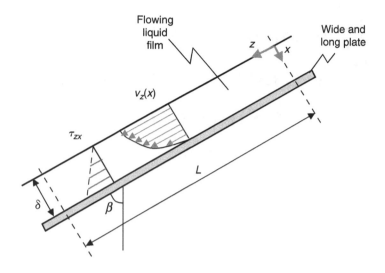

FIGURE 5.1 Flowing film on an inclined surface.

3. Laminar flow. This is also an approximation and valid for a very smooth solid surface and within a relatively short length of plate. Of course, there is no such thing as perfectly smooth surface. Therefore, every fluid would experience perturbations on the laminar flow due to imperfections. These perturbations are more or less cumulative and, after a certain length, the flow would reach a turbulent behavior.
4. Plate is very wide and long and the studied region is far from the borders, therefore possible interference of them is negligible.
5. Flow is already developed at position $z = 0$.
6. Since the fluid density is assumed constant, which is the same to say that the fluid is incompressible, its thickness (δ) is constant, at least within the plate length where the present solution is approximately valid.
7. Shear stress at the interface between the fluid and air above it is negligible. Owing to very low viscosity of air when compared with usual fluids, this is also a very reasonable approximation.

In such problems, the routine is to apply the total continuity and then the momentum or motion equation at each direction. Obviously, rectangular coordinates are the best choice here and the basic equation of mass continuity (Equation A.1) should be the first to be applied. According to assumptions 3, 4, and 6, there are no velocity components in direction x or y. With the aid of other assumptions, one is allowed to write

$$\frac{\partial v_z}{\partial z} = 0 \qquad (5.1)$$

Once assumption 2 is observed, the momentum conservation in the x-coordinate can be written in the form of Equation A.7. After the application for this simple situation, Equation A.7 becomes

$$\frac{\partial p}{\partial x} = \rho g_x = \rho g \sin \beta \tag{5.2}$$

The momentum at the y-coordinate, or Equation A.8, does not provide any information, while the z-coordinate, or Equation A.9, leads to

$$\frac{\partial p}{\partial z} = \mu \left(\frac{d^2 v_z}{dx^2} \right) + \rho g \cos \beta \tag{5.3}$$

It is important to notice that as velocity v_z depends only on the x-coordinate, its partial derivative on this direction could be replaced by a total derivative. Moreover, pressure does not vary in the z direction because the film is exposed to ambiance, which is at constant atmospheric value. Therefore, the above equation becomes

$$\frac{d^2 v_z}{dx^2} = -\frac{\rho g}{\mu} \cos \beta \tag{5.4}$$

The following are the boundary conditions for the present case:

$$v_z(\delta) = 0 \tag{5.5}$$

$$\left. \frac{dv_z}{dx} \right|_{x=0} = 0 \tag{5.6}$$

Of course, condition given by Equation 5.5 is imposed because the plate is stationary regarding the chosen coordinates. Equation 5.6 is the consequence of assumption 7 and is represented by a second-kind boundary condition.

Equation 5.4 is a second-order separable differential equation and its first integration yields

$$\frac{dv_z}{dx} = -\frac{\rho g}{\mu} x \cos \beta + C_1 \tag{5.7}$$

The second integration provides

$$v_z = -\frac{\rho g}{\mu} \cos \beta \frac{x^2}{2} + C_1 x + C_2 \tag{5.8}$$

Boundary conditions given by Equations 5.5 and 5.6 can be applied to arrive at

$$C_1 \delta + C_2 = \frac{\rho g}{\mu} \cos \beta \frac{\delta^2}{2} \tag{5.9}$$

$$C_1 = 0 \tag{5.10}$$

Therefore,

$$v_z = \frac{\rho g}{2\mu} \cos\beta \left(\delta^2 - x^2\right) \tag{5.11}$$

As expected, a parabolic velocity profile with decreasing velocity for higher viscosity is obtained. From this, the profile of shear stress in the fluid is deduced by

$$\tau_{zx} = -\mu \frac{dv_z}{dx}\bigg|_{x=\delta} = \rho g \delta \cos\beta \tag{5.12}$$

This shows a linear distribution of stress in the fluid.

From the above, the drag force imposed on the plate due to the shear stress between fluid and surface can be easily calculated by

$$F = A\tau_{zx}(x=\delta) = A\rho g\delta \cos\beta \tag{5.13}$$

where A is the wetted area of the surface.

It is interesting to notice that the shear stress, and consequently the drag force between fluid and plate, does not depend on the fluid viscosity. This happens here due to the inverse linear dependence of velocity on the fluid viscosity, as given by Equation 5.11.

Finally, one should observe that Equation 5.2 provides the variation of static pressure for points inside the fluid, and of course due to the weight of fluid above the considered position x.

5.3 FLOW IN AN INCLINED TUBE

Instead of the flat plate of the last section, the flow now occurs inside a cylindrical tube inclined by an angle θ in relation to the horizontal line, as shown in Figure 5.2.

In addition, a cross section at $z=0$ is shown in Figure 5.3 of a cross section.

Notwithstanding the similarity, the present problem includes the effect of pressure gradient, which did not exist in the last problem.

Again, it is desired to determine the following:

1. Velocity profile of the fluid
2. Shear stress profile in the fluid
3. Drag force between fluid and tube

The following conditions or assumptions are used:

1. Steady-state regime.
2. Liquid fluid is Newtonian with constant physical properties; among them are the density and viscosity. Considerations regarding this assumption can be found at the similar point of the last section.

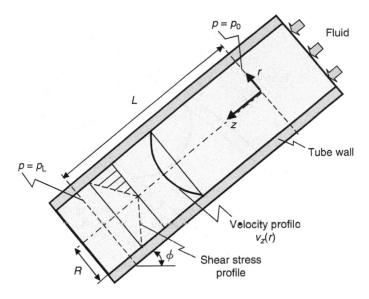

FIGURE 5.2 Circular inclined tube.

3. Laminar flow. This is also an approximation and valid for very smooth tube internal surface and within a relatively short tube length.
4. Flow is already developed at position $z = 0$. The analysis is valid for the region where the flow is completely developed, therefore unaffected by border effects. The situation near the entrance border is illustrated in Figure 5.4. At the entrance, attrition forces the fluid near the wall to decelerate. Therefore, a region near the wall would present lower velocities whereas the region near the tube centerline is still at the same

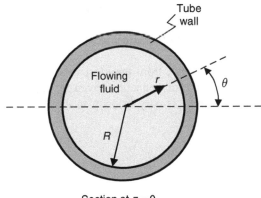

Section at $z = 0$

FIGURE 5.3 Cross section of Figure 5.2 at $z = 0$.

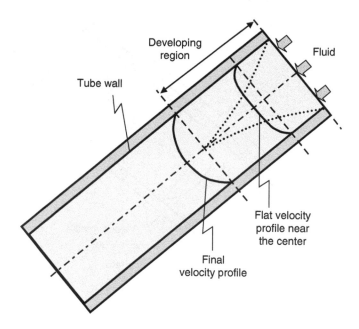

FIGURE 5.4 Developing flow region in a tube.

velocity of the original entering flow. In other words, the velocity profile for regions near the centerline of the tube is flat or unaffected by the presence of the internal tube walls. This situation remains for a short length after the entrance and soon the velocity profile reaches a stable shape. The present study refers just to the developed region.

5. No swirl on the fluid is imposed or observed.
6. Pressure measured at position $z = 0$ is equal to p_0. Therefore, pressure assumed is uniformly applied at each section z, thus independent of radial or angular coordinates.
7. Pressure measured at position is equal to p_L.

As always, one should depart from the most fundamental equations, which in this case is the mass continuity equation given by Equation A.2. From the above assumption, and as there is only velocity in the z direction, this equation yields

$$\frac{\partial v_z}{\partial z} = 0 \tag{5.14}$$

Since there is no movement in the r or θ directions and pressure depends only on z, the momentum conservation equations at these coordinates lead to trivial identity.

Equation A.15, or momentum conservation in the z direction, can be simplified to

$$\mu \frac{1}{r} \frac{d}{dr} \left(r \frac{dv_z}{dr} \right) = -\frac{\partial p}{\partial z} + \rho g \sin \phi \qquad (5.15)$$

According to information given by conditions 5–7, it is possible to write

$$\frac{\partial p}{\partial z} = \frac{p_L - p_0}{L} \qquad (5.16)$$

Therefore, Equation 5.15 gives

$$\frac{d}{dr} \left(r \frac{dv_z}{dr} \right) = ar \qquad (5.17)$$

Here

$$a = \frac{p_0 - p_L}{\mu L} + \frac{\rho g \sin \phi}{\mu} \qquad (5.18)$$

Despite being a second-order differential equation, Equation 5.18 is a separable one and can be integrated by a sequence of two first-order equations. The first integration leads to

$$\frac{dv_z}{dr} = \frac{a}{2} r + \frac{C_1}{r} \qquad (5.19)$$

and the second integration to

$$v_z = \frac{a}{4} r^2 + C_1 \ln r + C_2 \qquad (5.20)$$

The following are the boundary conditions:

$$v_z(R) = 0 \qquad (5.21)$$

$$\left. \frac{dv_z}{dr} \right|_{r=0} = 0 \qquad (5.22)$$

This last condition (Equation 5.22) is the consequence of symmetry of velocity field as well as due to zero shear stress at the centerline of the flow. As seen, it is a second-kind boundary condition.

Since the velocity is finite at $r = 0$, $C_1 = 0$. Using Equation 5.21, C_2 is obtained and the velocity is given by

$$v_z = -\frac{a}{4} \left(R^2 - r^2 \right) = \frac{1}{4\mu} \left(\frac{p_0 - p_L}{L} + \rho g \sin \phi \right) \left(R^2 - r^2 \right) \qquad (5.23)$$

The parabolic velocity profile of a Newtonian incompressible laminar flow is well known. Again, as expected, the velocity decreases for fluids with larger viscosities.

Using Equation 5.19, the shear stress profile is deduced as

$$\tau_{rz} = -\mu \frac{dv_z}{dr} = \mu \frac{a}{2} r = \frac{r}{2} \left(\frac{p_L - p_0}{L} + \rho g \sin \phi \right) \tag{5.24}$$

Hence, the shear stress has a linear profile, as illustrated in Figure 5.2. Notice that again, the shear stress, and consequently the drag force between fluid and tube walls, does not depend on the fluid viscosity.

Equation 5.24 allows obtaining the drag force of the fluid on the tube wall, or vice versa. This is left as an exercise.

To understand the nature and role of pressure, let us consider the situation of a vertical tube ($\phi = 90°$). Looking at Equation 5.23 and imagining a situation where the velocity were zero, or static condition, the following would result:

$$p_L = p_0 + \rho g L \tag{5.25}$$

which is the simple relation of pressures in a static column of incompressible fluid. However, the movement of the fluid imposes a pressure loss. In the cases of horizontal tube, the pressure would always decrease (or $p_L < p_0$) due to the attrition between tube and fluid.

5.4 RECTANGULAR FIN

Fins are used to enhance the heat transfer between surfaces and a fluid and can be found at the top of cylinders of engines, air conditioning, radiators, as well as several other equipment.

The basic effect provided by the fin is to increase the area for heat transfer between the surface and surrounding fluid, as shown in Figure 5.5.

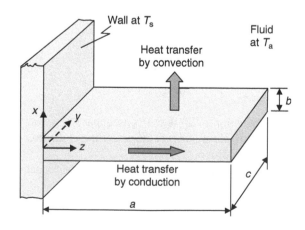

FIGURE 5.5 Rectangular fin.

The rectangular fin is welded to a vertical wall at T_s and is in contact with a wall and ambient air or any fluid at temperature T_a. The objective here is to find the temperature profile in the fin as well as the total rate of heat transfer between the fin and the surrounding fluid. Moreover, a comparison between this rate and the rate achievable without the fin would allow evaluating the fin efficiency.

To arrive at a reasonable approximated solution, let the following be the assumptions:

1. Steady-state regime.
2. In the fin, heat is transferred by conduction only in the z direction. Of course, this is an approximation because heat is transferred also in the x and y directions. Nonetheless, usually fins are very thin and wide, or the dimensions of b and c are small and large, respectively, when compared with a. Thus, the rates of heat transferred between the lateral borders (including the tip of the fin at $z = a$) and the air or fluid are negligible when compared with the rates transferred between the horizontal surfaces (up and down). The consequence of this assumption is that temperature would be only a function of the z-coordinate.
3. Thermal conductivity of the fin material is constant, or does not depend on the temperature.
4. Coefficient (α) for the heat transfer by convection between fin surfaces and fluid is constant. This is a reasonable assumption since steady state is achieved.
5. No phase change is observed in the fin, therefore, no velocity field is possible in it.

Under these assumptions, Equation A.34 is simplified to

$$\lambda \frac{d^2 T}{dz^2} + R_Q = 0 \tag{5.26}$$

Heat transfer between the fin and surrounding air occur by convection, and the total area for such exchange is given by

$$A = 2a(b + c) \tag{5.27}$$

The volume of the fin is

$$V = abc \tag{5.28}$$

Therefore, the rate of energy gain per unit of fin volume is equal to

$$R_Q = -\frac{\alpha 2a(b + c)}{abc}(T - T_a) \tag{5.29}$$

Thus, Equation 5.26 becomes

$$\frac{d^2 T}{dz^2} = \frac{2\alpha(c+b)}{bc\lambda}(T - T_a) \tag{5.30}$$

The following are the boundary conditions for this problem:

$$T(0) = T_s \tag{5.31}$$

and

$$\left.\frac{dT}{dz}\right|_{z=a} = 0 \tag{5.32}$$

Equation 5.32 is a second-kind condition and is due to the approximation of negligible heat transfer between the tip of the fin and the surrounding fluid.

Equation 5.30 is a nonhomogeneous linear ordinary second-order differential equation with constant coefficients. The method for solution of these equations is described in Appendix B. However, this equation can be easily transformed into a homogeneous equation through the introduction of the following dimensionless variables:

$$\Psi = \frac{T - T_a}{T_s - T_a} \tag{5.33}$$

and

$$\gamma = \frac{z}{a} \tag{5.34}$$

With the introduction of the above variables, Equation 5.30 becomes

$$\frac{d^2 \Psi}{d\gamma^2} = \beta^2 \Psi \tag{5.35}$$

where

$$\beta = a\sqrt{\frac{2\alpha(c+b)}{bc\lambda}} \tag{5.36}$$

The boundary conditions given by Equations 5.31 and 5.32 are respectively written as

$$\Psi(0) = 1 \tag{5.37}$$

and

$$\left.\frac{d\Psi}{d\gamma}\right|_{\gamma=1} = 0 \qquad (5.38)$$

This example illustrates, once more, the advantages of using dimensionless variables.

Following the method presented in Appendix B, the solution will be assumed as

$$\Psi = \exp(C\gamma) \qquad (5.39)$$

Here C is a constant, and from Equation 5.35, one would arrive at

$$C^2 = \beta^2 \qquad (5.40)$$

Consequently, there are two roots for the characteristic equation and the general solution can be written as

$$\Psi = C_1 \exp(\beta\gamma) + C_2 \exp(-\beta\gamma) \qquad (5.41)$$

Alternatively, this may be put in the following form:

$$\Psi = C_3 \sinh(\beta\gamma) + C_4 \cosh(\beta\gamma) \qquad (5.42)$$

The boundary conditions given by Equations 5.37 and 5.38 can be applied to reach at

$$\Psi = \cosh(\beta\gamma) - \tanh(\beta)\sinh(\beta\gamma) \qquad (5.43)$$

With the above temperature profile, it is possible, for instance, to compute the total rate of heat transfer from the fin by

$$\dot{Q}_{\text{Fin}} = \iint_{A=\text{fin surface}} \alpha(T - T_a)\,dA = 2(b + c)a\alpha(T_s - T_a)\int_0^1 \psi\,d\gamma$$

$$= 2(b + c)a\alpha(T_s - T_a)\frac{\tanh(\beta)}{\beta} \qquad (5.44)$$

Using Equations 5.36 and 5.43 the following is obtained:

$$\dot{Q}_{\text{Fin}} = [2bc(b + c)\lambda\alpha]^{1/2}\,(T_s - T_a)\tanh\left\{a\left[\frac{2\alpha(b + c)}{bc\lambda}\right]^{1/2}\right\} \qquad (5.45)$$

As for large values of the argument, the hyperbolic tangent is limited to 1. Thus, the rate of heat transfer from the fin increases when any of the following terms increases:

- Difference of temperature between the surface and the air
- Area of the fin
- Heat transfer coefficients (α or λ)

On the other hand, if no fin were attached to the surface, the rate of heat transfer at the contact area between fin and vertical wall would be

$$\dot{Q}_{\text{no fin}} = bc\alpha(T_s - T_a) \tag{5.46}$$

The efficiency of a fin can be defined by

$$\varepsilon = 1 - \frac{\dot{Q}_{\text{no fin}}}{\dot{Q}_{\text{fin}}} = 1 - \frac{1}{\left[\dfrac{2(b+c)\lambda}{bc\alpha}\right]^{1/2} \tanh\left\{a\left[\dfrac{2\alpha(b+c)}{bc\lambda}\right]^{1/2}\right\}} \tag{5.47}$$

Figure 5.6 illustrates the dependence of fin efficiency against the thermal conductivity of the fin (λ) and the heat transfer coefficient (α) between fin and

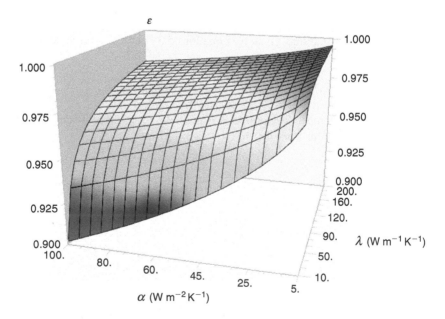

FIGURE 5.6 Efficiency of a rectangular fin against heat transfer coefficients for $a = 0.3$ m, $b = 2$ mm, $c = 0.2$ m.

surrounding fluid. The dimensions adopted for the fin here are $a = 0.3$ m, $b = 2$ mm, and $c = 0.2$ m.

The following are clearly observed:

- Efficiency increases with the thermal conductivity (λ) and this is one among the most important factor.
- Efficiency decreases with α. In other words, for cases where the air is blowing with high velocity and therefore higher values of α, the presence of fin becomes less important.
- In addition, Equation 5.47 shows that efficiency increases for larger fin perimeter $[2(b + c)]$ and small cross-sectional area (bc). Therefore, thin fins are more effective than thick ones regarding heat exchange with the ambiance. Of course, the limitation would rest on the mechanical resistance of the fin against eventual impacts.

To illustrate the effects of dimensions, Figures 5.7 and 5.8 are variations of Figure 5.6. In Figure 5.7, the following dimensions were adopted: $a = 0.05$ m, $b = 2$ mm, and $c = 0.05$ m, therefore much smaller fin perimeter than for the case shown in Figure 5.6. One should notice the relatively higher values for the efficiency obtained now when compared with the previous case. In Figure 5.8, the following dimensions were adopted: $a = 0.05$ m, $b = 10$ mm, and $c = 0.05$ m, therefore thicker fin than for the case shown in the two previous figures. One should verify another decrease in efficiency when compared with the previous cases.

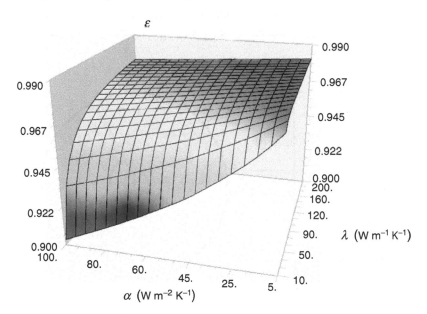

FIGURE 5.7 Efficiency of a rectangular fin against heat transfer coefficients for $a = 0.05$ m, $b = 2$ mm, $c = 0.05$ m.

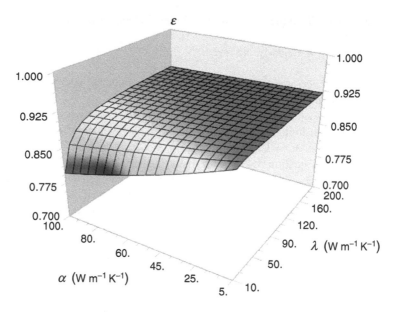

FIGURE 5.8 Efficiency of a rectangular fin against heat transfer coefficients for $a = 0.05$ m, $b = 10$ mm, $c = 0.05$ m.

5.5   CIRCULAR FIN

Likewise the case of rectangular fin, the circular fin is in contact with air or any fluid at temperature T_a. The fin is welded to a tube with surface temperature T_s, as shown in Figure 5.9.

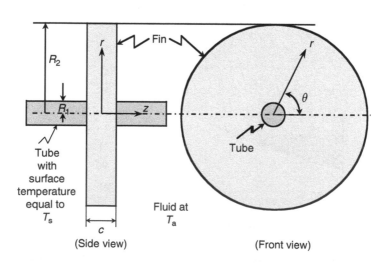

FIGURE 5.9 Side and front views of a circular fin attached to a central tube.

The objective is the same as before, or to find the temperature profile in the fin as well as the rate of heat transfer between the fin and the fluid. A comparison between this rate and the rate achievable without a fin would allow evaluating its efficiency.

To arrive at a reasonable solution, let us assume the following:

1. Steady-state regime.
2. In the fin, heat is transferred by conduction only in the radial direction. Therefore, the temperature is just a function of the r-coordinate. Hence, the temperature profiles are flat in axial coordinate z. Of course, this is an approximation because heat is transferred also in the z direction. Nonetheless, it is assumed here that the fin is very thin when compared with its radius (R_2). Thus, the rates of heat transfer between the border ($r = R_2$) and the air are negligible when compared with the rate transferred from the surfaces with large areas, or sides of the disk, to the surrounding fluid.
3. Fin is homogeneous, or there is no variation of flux in the angular θ-coordinate.
4. Thermal conductivity of the fin material is constant, or does not depend on the temperature.
5. Constant coefficient α for the heat transfer between fin surfaces and fluid.

Departing from Equation A.35 and using the above assumption, it is possible to arrive at

$$0 = \lambda \left[\frac{1}{r} \frac{d}{dr} \left(r \frac{dT}{dr} \right) \right] + R_Q \tag{5.48}$$

The rate of heat transfer from the fin by its volume is given by

$$R_Q = -\frac{\alpha(T - T_a)2\pi r^2}{\pi r^2 c} \tag{5.49}$$

Thus, Equation 5.48 becomes

$$r^2 \frac{d^2 T}{dr^2} + r \frac{dT}{dr} - (T - T_a)\frac{\alpha r^2}{\lambda c} = 0 \tag{5.50}$$

The following are the boundary conditions:

$$T(R_1) = T_s \tag{5.51}$$

and

$$\left. \frac{dT}{dr} \right|_{r=R_2} = 0 \tag{5.52}$$

The boundary condition given by Equation 5.52 is a second-kind equation, which is similar to the equation derived in the case of rectangular fins.

Again, the application of dimensionless variables is convenient. Of course, there are several possibilities and the following is one among them:

$$\Psi = \frac{T - T_a}{T_s - T_a} \tag{5.53}$$

and

$$\zeta = \frac{r}{R_1} \tag{5.54}$$

Finally,

$$\zeta^2 \frac{d^2\Psi}{d\zeta^2} + \zeta \frac{d\Psi}{d\zeta} - b^2\zeta^2\Psi = 0 \tag{5.55}$$

where

$$b = R_1\sqrt{\frac{\alpha}{\lambda c}} \tag{5.56}$$

Equation 5.55 is a modified Bessel differential equation. Its solution can be found through the following transformation of variables:

$$\chi = b\zeta \tag{5.57}$$

Applying Equation 5.57 to Equation 5.55 results in

$$\chi^2 \frac{d^2\Psi}{d\chi^2} + \chi \frac{d\Psi}{d\chi} - \chi^2\Psi = 0 \tag{5.58}$$

Its general solution is shown in Appendix C, which when applied here provides the following:

$$\Psi = C_1 I_0(\chi) + C_2 K_0(\chi) \tag{5.59}$$

Here I_0 is the modified zero-order, first-kind Bessel function and K_0 is the modified zero-order, second-kind Bessel function.

The boundary conditions given by Equations 5.51 and 5.52 become

$$\Psi(b) = 1 \tag{5.60}$$

and

$$\frac{d\Psi}{d\chi}\bigg|_{\chi=a} = 0 \tag{5.61}$$

Here the parameter a is given by

$$a = b\frac{R_2}{R_1} \tag{5.62}$$

Using Equation 5.60 in Equation 5.59 leads to

$$1 = C_1 I_0(b) + C_2 K_0(b) \tag{5.63}$$

The derivative of Equation 5.59 is (see Appendix C)

$$\frac{d\Psi}{d\chi} = C_1 I_0'(\chi) + C_2 K_0'(\chi) \tag{5.64}$$

On the other hand,

$$I_0'(\chi) = I_1(\chi) \tag{5.65}$$

$$K_0'(\chi) = -K_1(\chi) \tag{5.66}$$

Therefore, the condition given by Equation 5.61 yields

$$0 = C_1 I_1(a) - C_2 K_1(a) \tag{5.67}$$

Finally,

$$\Psi = \frac{K_1(a)I_0(\chi) + I_1(a)K_0(\chi)}{K_1(a)I_0(b) + I_1(a)K_0(b)} \tag{5.68}$$

or

$$\Psi = \frac{K_1(bR_2/R_1)I_0(b\zeta) + I_1(bR_2/R_1)K_0(b\zeta)}{K_1(bR_2/R_1)I_0(b) + I_1(bR_2/R_1)K_0(b)} \tag{5.69}$$

For the case of R_2/R_1 equals to 5, Figure 5.10 illustrates the dimensionless temperature (Ψ) as function of dimensionless position in the fin (ζ) according to parameter b.

One should notice the zero derivatives at the outer border of the fin, or $\zeta = 5$. In addition, the temperatures in the fin are higher for lower values of b. From the definition of b in Equation 5.56, this is easy to understand because higher conductivities tend to even the temperature throughout the fin.

As shown in the previous section, it would be simple to obtain the rate of heat transfer between the fin and air as well as its efficiency. These are proposed as exercises.

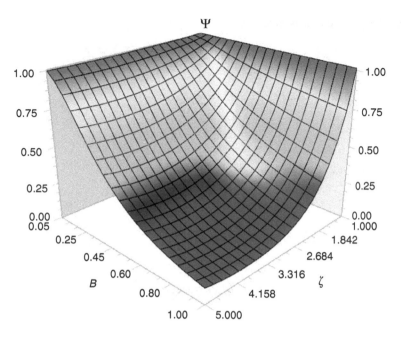

FIGURE 5.10 Dimensionless temperature (Ψ) as function of dimensionless position in the circular fin (ζ) according to parameter b for $R_2/R_1 = 5$.

5.5.1 SOLUTION BY LAPLACE TRANSFORM

The Laplace transform can also be used to solve problems of linear differential equations with nonconstant coefficients, such as the Bessel ones.

Equation 5.58 can be written as

$$\chi\frac{d^2\Psi}{d\chi^2} + \frac{d\Psi}{d\chi} - \chi\Psi = 0 \tag{5.70}$$

Let the transform be

$$\Omega(s) = L\{\Psi(\chi)\} \tag{5.71}$$

Applying Equation D.34 (see also Table D.1) to Equation 5.58 provides

$$L\left\{\chi\frac{d^2\Psi}{d\chi^2}\right\} = \frac{d^2}{ds^2}\left[L\left\{\frac{d^2\Psi}{d\chi^2}\right\}\right] = -\frac{d}{ds}\left[s^2\Omega(s) - s\Omega(0) - \Omega'(0)\right]$$

$$= -2s\Omega(s) - s^2\frac{d\Omega}{ds} + \Psi(0)$$

The second term would lead to

$$L\left\{\frac{d\Psi}{d\chi}\right\} = s\Omega(s) - \Psi(0) \tag{5.72}$$

And the third yields

$$L\{\chi\Psi(\chi)\} = -\frac{d\Omega}{ds} \tag{5.73}$$

Therefore, the transform of entire Equation 5.70 leads to

$$(1 - s^2)\frac{d\Omega}{ds} - s\Omega(s) = 0 \tag{5.74}$$

Equation 5.74 is a separable equation and the solution is

$$\Omega(s) = C(s^2 - 1)^{-1/2} \tag{5.75}$$

The inversion of Equation 5.75 is found in Table D.5 and

$$\Psi(\chi) = CI_0(\chi) \tag{5.76}$$

The above reflects a particular solution among all possible derived from a general solution given by Equation 5.59. As the pair of Bessel functions is always the general solution, one would expect that a general solution would be given exactly as Equation 5.59. The previous sequence follows from this point.

It should be noticed that if the Bessel equation (Equation 5.58) could not be simplified to Equation 5.70, the application of Laplace transform would lead to another differential equation with the same order of difficulty. In addition, it is possible to eliminate parameter $\Psi(0)$ in the process leading to Equation 5.74. Actually, according to the definition of variable χ by Equation 5.57, it cannot attain value zero. Therefore, since $\Psi(0)$ is unknown, a difficulty would appear if it had remained in the final differential equation (Equation 5.74).

5.6 FILM CONDENSATION

The classical problem of heat transfers during the condensation of a vapor on a vertical surface is worth mentioning because it involves coupling between this and momentum transfer. In addition, it is an excellent example to illustrate the resolution of a second-order differential equation.

Let the vertical plate be at a surface temperature T_s in contact with saturated steam (or any vapor) at temperature T_{sat}, as illustrated in Figure 5.11.

If these conditions persist, a liquid film would be formed and flow would be gaining mass due to additional condensation.

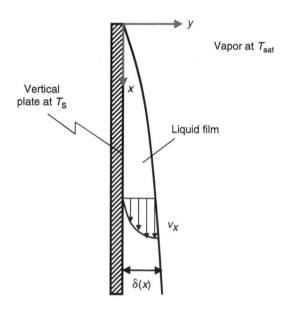

FIGURE 5.11 Scheme of a film formed by the condensation of a vapor on a vertical surface.

It is desired to obtain the rate of vapor condensation as well as the temperature and velocity profiles in the liquid film.

Despite the simplifications listed below, the equations obtained through the treatment ahead are useful in design of industrial condensers.

The following are the basic assumptions adopted here:

1. Steady-state regime. Therefore, the surface of the vertical plate in contact with the liquid film is kept at constant temperature T_s as well as the vapor at the saturation temperature T_{sat}. The vapor pressure and any other condition also remain constant. As a consequence, the shape of the film does not change.
2. Plate is broad enough to avoid border effects. In other words, the plate is very wide in the orthogonal direction z to plane x–y. Thus, border effects on velocity and temperature profiles would be neglected, which also implies in no velocity component in the z direction.
3. Plate is very smooth and laminar flow is observed in the film within the length (x) at which the present treatment is valid. Of course, due to several factors, after a certain extent in the vertical direction, small ripples appear on the film surface. This process is cumulative leading to fully turbulent flow.
4. Shear stress between liquid and vapor at the film surface is negligible. Actually, the force balance dictates equal stress for the liquid and the vapor at that surface. However, as the viscosity of the vapor is usually much smaller than the viscosity of the liquid, therefore, justifying the present assumption.

5. Proprieties (density, viscosity, and thermal conductivity) of the liquid are considered constant throughout the whole film. As a gradient of temperature is involved, this hypothesis could be criticized. However, this difference usually stays within the range that allows the approximation, at least as a first attack to the problem.
6. Velocity component in the y direction is much smaller than the component in the x direction. In the present approach, the magnitude of v_y will be approximated to zero. Despite the apparent crudeness, this assumption is found in the first works in the area of heat transfer [1] and was dropped in subsequent treatments [2–15].
7. Liquid is Newtonian.
8. Any energy source due to viscous dissipation is negligible. This is very likely due to the relatively small velocity gradients involved in the process.

5.6.1 MOMENTUM CONSERVATION

The overall mass conservation is given by Equation A.1, which after the above assumption is simplified to

$$\frac{\partial v_x}{\partial x} = 0 \tag{5.77}$$

Therefore, the vertical velocity is solely a function of the y-coordinate, or $v_x = v_x(y)$.

For a Newtonian and constant density fluid, the momentum conservation in the x direction is given by Equation A.7 and using the above hypothesis it becomes

$$0 = -\frac{\partial p}{\partial x} + \mu \frac{\partial^2 v_x}{\partial y^2} + \rho g \tag{5.78}$$

The motion equation for the horizontal direction is given by Equation A.8. If the momentum (or velocity) in this direction is neglected when compared to the momentum transfer in the vertical direction and with the help of Equation 5.77, it is possible to write

$$\frac{\partial p}{\partial y} = 0 \tag{5.79}$$

Consequently, pressure is a function only of the x direction. In addition, at a given vertical position, the pressure values at the liquid should be the same as that in the vapor. Since owing to assumption 4 there is no movement in the vapor phase, the pressure variation is given just by static influence, or

$$\frac{dp}{dx} = \rho_v g \tag{5.80}$$

Forced by the equality of pressure in both phases, this is the same variation observed in the liquid phase. Thus, Equation 5.78 becomes

$$\frac{d^2v_x}{dy^2} = -\frac{g}{\mu}(\rho - \rho_v) \tag{5.81}$$

Here, the properties without subscripts should be understood as related to the liquid phase.

The momentum conservation equations for other directions lead to trivial equalities and bring no additional information.

Equation 5.81 is an ordinary second-order differential equation with constant coefficients. It is easily solvable by the methods shown in Appendix B.

The boundary conditions for the present problem are

$$v_x(0) = 0 \tag{5.82}$$

$$\left.\frac{dv_x}{dy}\right|_{y=\delta} = 0 \tag{5.83}$$

The first condition (Equation 5.82) is obvious and the second (Equation 5.83) is due to assumption 4. Notice that δ is a function of position x. It is also interesting to comment that Equation 5.83 is an approximation and a consequence of Newton's second law or momentum conservation, which imposes equal shear stress at each side the liquid–vapor interface, or

$$\tau = -\mu \left.\frac{dv_x}{dz}\right|_{y=\delta^-} = -\mu_v \left.\frac{dv_{x,v}}{dz}\right|_{y=\delta^+} \tag{5.84}*$$

Remembering that the viscosity of the vapor is usually much smaller than the viscosity of the liquid, the derivative on the left-hand side (or at the liquid side of the interface) should be much smaller than the derivative on the right-hand side (or at the vapor side of the interface). In other words, the contact with the vapor leads to no considerable drag force to the liquid layers near the interface.

Equation 5.81 under the boundary conditions given by Equations 5.82 and 5.83 leads to

$$v_x(y) = \frac{g(\rho - \rho_v)\delta^2}{\mu}\left[\frac{y}{\delta} - \frac{1}{2}\left(\frac{y}{\delta}\right)^2\right] \tag{5.85}$$

---

* The notation $y=\delta^-$ symbolizes a limit position y tending to δ by smaller values, or from the left to the right in Figure 5.11. The contrary when positive sign is used.

The above equation allows obtaining the mass flow of condensate at any position x by

$$F = b \int_0^{\delta(x)} \rho v_x dy = \frac{gb\rho(\rho - \rho_v)\delta^3}{3\mu} \tag{5.86}$$

where b is the width of the plate.

5.6.2 Energy Conservation

The energy balance within the liquid layer can be obtained using Equation A.34, which under assumptions 1 and 8 becomes

$$\rho C_p \left(v_x \frac{\partial T}{\partial x} + v_y \frac{\partial T}{\partial y} \right) = \lambda \left(\frac{\partial^2 T}{\partial x^2} + \frac{\partial^2 T}{\partial y^2} \right) + R_Q \tag{5.87}$$

Equation 5.87 can be further simplified because of the following:

1. Derivative of temperature in the horizontal (y) direction is much higher than in the vertical direction (x).
2. Velocity in the vertical direction is much higher than in the horizontal direction.
3. Second derivative of temperature in the vertical (x) direction is much smaller than the similar in the horizontal (y) direction.
4. Because of the small thickness of the film and relatively large difference between the temperature of the plate and condensing vapor, the heat transfer by conduction in the y direction is greater or even considerably greater than heat transfer by convection.

In view of the above, it is reasonable to neglect the entire left-hand side of Equation 5.87 and the first term of the right-hand side to give

$$\lambda \left(\frac{d^2 T}{dy^2} \right) = -R_Q \tag{5.88}$$

In the present case, the rate of energy generation per unit of volume is given by the rate of vapor condensation times the energy involved in that phase change per unit of volume, or

$$R_Q = \frac{h_{fg}}{b\delta} \frac{dF}{dx} \tag{5.89}$$

Using Equation 5.86, Equation 5.89 becomes

$$R_Q = \frac{h_{fg} g \rho(\rho - \rho_v)}{\mu} \delta \frac{d\delta}{dx} \tag{5.90}$$

Therefore

$$\frac{d^2T}{dy^2} = -\frac{h_{fg}g\rho(\rho - \rho_v)}{\mu\lambda}\,\delta\frac{d\delta}{dx} = f(x) \tag{5.91}$$

It should be noticed that the right-hand side is solely a function of x variable. Therefore, this equation can be easily integrated.

The conditions for a particular solution are

$$T(y = 0) = T_s \tag{5.92}$$

and

$$T(y = \delta) = T_{sat} \tag{5.93}$$

Using Equations 5.92 and 5.93, it is possible to arrive at

$$T - T_s = (T_{sat} - T_s)\frac{y}{\delta} + f(x)\frac{\delta^2}{2}\left(\frac{y^2}{\delta^2} - \frac{y}{\delta}\right) \tag{5.94}$$

Keeping in mind the small thickness of the film, it is possible [1] to make an interesting approximation by assuming a linear profile and equating the heat flux through the film with the energy delivered by the condensation, or

$$q = -\lambda\frac{dT}{dy} = -\lambda\frac{T_s - T_{sat}}{\delta} = \frac{h_{fg}}{b}\frac{dF}{dx} \tag{5.95}$$

Applying Equation 5.95 to Equation 5.86, it is possible to obtain the description of film thickness:

$$\delta(x) = \left[\frac{4\lambda\mu(T_{sat} - T_s)x}{g\rho(\rho - \rho_v)h_{fg}}\right]^{1/4} \tag{5.96}$$

An overall convection heat transfer coefficient between the plate and vapor phase can be defined as

$$q = \alpha(T_{sat} - T_s) \tag{5.97}$$

Using the above equations, it is possible to arrive at

$$\alpha = \left[\frac{g\rho(\rho - \rho_v)\lambda^3 h_{fg}}{4\mu(T_{sat} - T_s)x}\right]^{1/4} \tag{5.98}$$

5.6.3 COMMENTS

As seen, the above treatment allows the computation of heat transfer rates between a plate and the condensing vapor, including the rate of condensation. However, such treatment cannot be applied to long plates due to departing from laminar conditions. Further and more rigorous studies can be found in the literature [2–15].

Another interesting aspect of this problem concerns the dependence of film properties on temperature. Similar to the achieved in Section 5.2, the parabolic velocity profile obtained at Equation 5.85 was expected because the momentum transfer problem is the same as the set for an isothermal film flowing on an inclined plate. Nonetheless, here the temperature in the film varies and if a more realistic approach were to be followed, the physical properties of the film would be allowed to vary. This would lead to departure from the parabolic velocity profile and therefore modification on parameters related to the heat transfer between film and plate. Kruzhilin [2], Voskresenkiy [3], Bromley [4], and Labunt-sov [7] were among the first to verify that variations of film physical properties with temperature had nonnegligible influence on the rate of condensation. Depending on the relationships for specific heat and viscosity, the solution of coupled momentum and energy transfers might become awkward to solve by analytical methods.

Another somewhat strong assumption made during the solution was to neglect the film acceleration or momentum convection. If included, the terms $v_x \frac{\partial v_y}{\partial x}$ and $v_y \frac{\partial v_y}{\partial y}$ were considered, the momentum equation to solve would be

$$v_x \frac{\partial v_y}{\partial x} + v_y \frac{\partial v_y}{\partial y} = v \frac{\partial^2 v_y}{\partial y^2} + g \qquad (5.99)$$

Additionally, the convective heat transfer terms would be added, and the energy equation would become

$$v_x \frac{\partial T}{\partial x} + v_y \frac{\partial T}{\partial y} = \frac{\lambda}{\rho C_p} \frac{\partial^2 T}{\partial y^2} \qquad (5.100)$$

Finally, for surfaces different from the flat vertical one, the gravitational term of Equation 5.99 would be a function describing its dependence on variable y. Treatments of laminar condensation on surfaces of several forms can be found elsewhere [9–15].

5.7   HEAT TRANSFER THROUGH A REACTING PLATE

This situation is similar to situation shown in Section 4.11, however with different boundary conditions, as illustrated in Figure 5.12. Again, the plate material goes through an exothermic reaction when heated but only one surface is kept at constant temperature, while the other exchanges heat with the environment at a

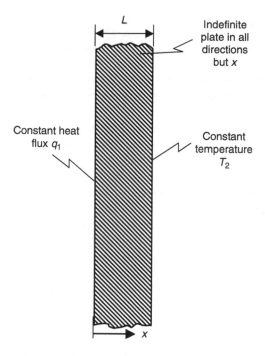

L

Indefinite
plate in all
directions
but x

Constant heat
flux q_1

Constant
temperature
T_2

x

FIGURE 5.12 A reacting semi-infinite solid wall with a face receiving a constant heat flux and another at constant temperature.

given and constant heat flux. It is desired to obtain the temperature profile in the plate as well as the rate of heat transfer through it.

To clearly set the problem, let us list the assumptions:

1. Wall is very wide and long, but with relatively small thickness. In other words, it is considered infinite in all the directions except the x-coordinate.
2. Steady-state regime has been established, i.e., the temperatures will be only a function of the position x.
3. No velocity field exists inside the solid wall.
4. All physical properties of the plate material are constants. Comments on this can be found in Section 4.11.
5. Temperature at one face is known, while the heat flux at the other surface is kept at constant value, as shown in Figure 5.12. Such might be obtained through several means, among them by setting electrical resistances in contact with the plate. Varying current or resistance values can control the energy delivered to the surface.
6. Plate goes through an exothermic reaction. The rate of energy generation per unit of plate volume (W m^{-3}) depending on the temperature is given by

$$R_Q = cT \qquad (5.101)$$

where c is a positive real constant.

With variable conductivity, Equation A.34 would lead to

$$\lambda \frac{d^2 T}{dx^2} + cT = 0 \tag{5.102}$$

The boundary conditions are

$$q_x(0) = -\lambda \left. \frac{dT}{dx} \right|_{x=0} = q_1 \tag{5.103}$$

and

$$T(L) = T_2 \tag{5.104}$$

Therefore, Equation 5.103 is a second-kind boundary condition.

The solution of homogeneous linear second-order differential equation with constant coefficients is described in Appendix B, and given by

$$T(x) - C_1 \sin\left(x\sqrt{\frac{c}{\lambda}}\right) + C_2 \cos\left(x\sqrt{\frac{c}{\lambda}}\right) \tag{5.105}$$

Applying the boundary conditions, the temperature profile can be written as

$$\theta = \frac{\cos(\varphi\zeta)}{\cos(\varphi)} + \frac{\gamma}{\varphi}[\tan(\varphi)\cos(\varphi\zeta) - \sin(\varphi\zeta)] \tag{5.106}$$

Here

$$\theta = \frac{T}{T_2}, \quad \varphi = L\sqrt{\frac{c}{\lambda}}, \quad \zeta = \frac{x}{L}, \quad \text{and} \quad \gamma = \frac{q_1 L}{T_2 \lambda} \tag{5.107a,b,c,d}$$

The dependence of dimensionless temperature profiles throughout the plate on parameter φ are exemplified in Figure 5.13, for the case of γ equal to 10.

Of course, higher values of temperatures (or its corresponding dimensionless value θ) are obtained for larger values of parameter φ, which reflects values of reaction rate factor (c). The constant derivative of temperature at $x=0$ (or $\zeta=0$) is the consequence of boundary condition given by Equation 5.103.

The rate of heat transfer at $x=L$ (or $\zeta=1$) can be obtained by

$$q_x(L) = -\lambda \left. \frac{dT}{dx} \right|_{x=L} \tag{5.108}$$

The deduction of the formula to allow calculations is left as an exercise.

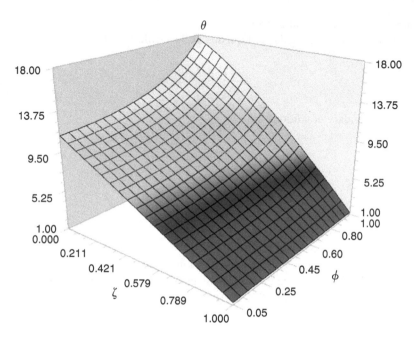

FIGURE 5.13 Dimensionless temperature profile in a reacting plate with one face at constant heat flux and the other face at constant temperature.

EXERCISES

1. Obtain the rate of heat transfer between the circular fin and air. Use the temperature profile as obtained in Section 5.5 and follow the procedure shown in Section 5.4.
2. Obtain the expression to evaluate the efficiency of a circular fin.
3. Work the details to arrive at Equation 5.85.
4. Use the result for velocity profile to perform the integration shown in Equation 5.86 to arrive at the last term on the right-hand side.
5. Demonstrate Equations 5.96 and 5.98.
6. Using the results in Section 5.6, compute the rate of vapor condensation if a plate maintained at 280 K is kept in steam at 1 MPa. Assume the plate to be 0.1 m wide and 1 m long.
7. Solve the following differential boundary-value problems by Laplace transform:
   (a) $y''(x) + xy'(x) - 2y(x) = 1$, $y(0) = y'(0) = 0$
   (b) $xy''(x) + (x - 1)y'(x) + y(x) = 0$, $y(0) = 0$, $y'(0) = 1$
   (c) $xy''(x) + (2x + 3)y'(x) + (x + 3)$, $y(0) = 3e^{-x}$, $y'(0) = 2$
   (d) $x^2 y''(x) - 2y(x) = 2x$, $y(2) = 2$, $y'(0) = 0$
8. Deduce Equation 5.105.
9. Work the details to arrive at Equation 5.106.
10. Using Equation 5.106 into Equation 5.108, obtain the expression to calculate the heat flux at $x = L$.

11. Solve a similar problem as set in Section 5.7, however, under the following boundary conditions:

$$q_x(0) = -\lambda \frac{dT}{dx}\bigg|_{x=0} = q_1$$

and

$$q_x(L) = -\lambda \frac{dT}{dx}\bigg|_{x=L} = q_2$$

In other words, assuming the flux at both faces of the plate (shown in Figure 5.12) as known constants.

Answer:

$$T(x) = \frac{\gamma_2 \cos(\varphi\zeta) - \gamma_1 \cos[\varphi(1 - \zeta)]}{\varphi \sin(\varphi)}$$

where

$$\varphi = L\sqrt{\frac{c}{\lambda}},$$

$$\zeta = \frac{x}{L},$$

$$\gamma_1 = \frac{q_1 L}{\lambda}, \quad \text{and}$$

$$\gamma_2 = \frac{q_2 L}{\lambda}.$$

REFERENCES

1. Nusselt, W., Die Oberflaechenkondensation des Wasserdampfes, Z. *Ver. Deutscher Ing.*, 60, 541, 1916.
2. Kruzhilin, D.A., On the precision of the Nusselt theory concerning the heat transfer during condensation (original paper in Russian), *Zhurn. Tekhn. Fisizi*, 7, 2011, 1937.
3. Voskresenkiy, K.D., Calculation of heat transfer in film condensation allowing for the temperature dependence of the physical properties of the condensate, *U.S.S.R. Acad. Sci., OTK*, 1948.
4. Bromley, L.A., Effect of heat capacity of condensate, *Ind. Eng. Chem.*, 44, 2966, 1952.
5. Hermann, R., Heat transfer by free convection from horizontal cylinders in diatomic gases, *NACA TM*, 1366, 1954.
6. Rosehnow, W.M., Heat transfer and temperature distribution in laminar-film condensation, *Trans. ASME*, 78, 1645, 1956.
7. Labuntsov, D.A., The effect of the temperature dependence of the physical properties of the condensate during laminar film condensation, *Teploenergetika*, 4, 49, 1957.
8. Sparrow, E.M. and Gregg, J.L., A boundary-layer treatment of laminar film condensation, *J. Heat Transfer, Trans. ASME*, C81, 13, 1959.

9. Sparrow, E.M. and Gregg, J.L., A theory of rotating condensation, *J. Heat Transfer, Trans. ASME*, C81, 113, 1959.

10. Sparrow, E.M. and Gregg, J.L., Laminar condensation heat transfer on a horizontal cylinder, *J. Heat Transfer, Trans. ASME*, C81, 291, 1959.

11. Dhir, V.K. and Lienhard, J., Laminar film condensation on plane and axisymmetric bodies in nonuniform gravity, *J. Heat Transfer, Trans. ASME*, C93, 97, 1971.

12. Dhir, V.K. and Lienhard, J., Similar solutions for film condensation with variable gravity or body shape, *J. Heat Transfer, Trans. ASME*, C95, 483, 1973.

13. Dhir, V.K. and Lienhard, J., Laminar film condensation on nonisothermal arbitrary-heat-flux surfaces and fins, *J. Heat Transfer, Trans. ASME*, C96, 197, 1974.

14. Dhir, V.K. and Lienhard, J., Laminar film condensation on submerged isothermal bodies, *J. Heat Transfer, Trans. ASME*, C96, 555, 1974.

15. de Souza-Santos, M.L., Explicit forms for the calculation of heat and momentum transfer coefficients for vapour condensation on surfaces of various forms, *Can. J. Chem. Eng.*, 68, 29, 1990.

6 Problems 123; One Variable, 2nd Order, 3rd Kind Boundary Condition

6.1 INTRODUCTION

This chapter presents methods to solve problems with one independent variable involving second-order differential equation and third-kind boundary condition. Mathematically, this class of cases can be summarized as $f\left(\phi, \omega, \frac{d\phi}{d\omega}, \frac{d^2\phi}{d\omega^2}\right)$, a third-kind boundary condition.

6.2 HEAT TRANSFER BETWEEN A PLATE AND FLUIDS

Figure 6.1 illustrates the situation where two fluids—at temperatures T_1 and T_2, respectively—are separated by a plate. However, the temperatures T_{w1} and T_{w2} at the wall surfaces are not known. The heat flux between the two fluids is to be calculated. Let us simplify the problem by assuming the following:

1. Wall can be considered infinite in all directions but x. In other words, its thickness is much smaller than its width and length.
2. Steady-state regime has been established; therefore, no condition changes with time.
3. No phase change occurs in the solid. Thus, no macroscopic velocity field is involved inside the wall.
4. All other physical properties of the plate material remain approximately constant.
5. The convection heat transfer coefficients (α_1 and α_2) between each fluid and the wall are constants. These coefficients can be calculated by appropriate semiempirical equations.

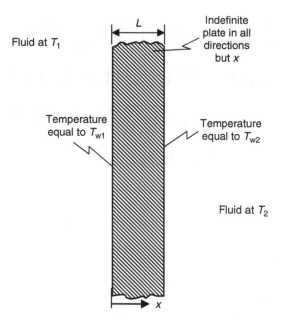

FIGURE 6.1 A semi-infinite solid wall separating two fluids at different temperatures.

The energy balance given by Equation A.34, and owing to the above assumptions, becomes

$$\frac{d^2T}{dx^2} = 0 \qquad (6.1)$$

The boundary conditions are

$$\alpha_1(T_1 - T_{w1}) = -\lambda \left.\frac{dT}{dx}\right|_{x=0} \qquad (6.2)$$

and

$$\alpha_2(T_{w2} - T_2) = -\lambda \left.\frac{dT}{dx}\right|_{x=L} \qquad (6.3)$$

The above conditions indicate that at the fluid–solid interfaces, heat flux by convection equals the respective due to conduction. These are typical third-kind boundary conditions.

According to the explanation given in Appendix B, the solution of Equation 6.1 is easily obtained as

$$T = C_1 x + C_2 \tag{6.4}$$

Therefore, Equations 6.2 and 6.3, respectively, become

$$\alpha_1 (T_1 - C_2) = -\lambda C_1 \tag{6.5}$$

$$\alpha_2 (C_1 L + C_2 - T_2) = -\lambda C_1 \tag{6.6}$$

From Equations 6.5 and 6.6, it is possible to obtain the indicated constants, and this is left as an exercise.

The main interest on this problem is the computation of heat flux between the two fluids. From Equation 6.4 one sees that in the present case the flux is constant and is given by

$$q_x = -\lambda \frac{dT}{dx} = -\lambda C_1 \tag{6.7}$$

Using the value for C_1 obtained earlier, it is possible to write

$$q_x = \frac{T_1 - T_2}{\dfrac{1}{\alpha_1} + \dfrac{L}{\lambda} + \dfrac{1}{\alpha_2}} \tag{6.8}$$

The numerator represents the potential, and the denominator the three resistances in series for the heat transfer: two due to convection and the center one due to conduction.

6.3   HEAT TRANSFER IN A SPHERICAL SHELL

Consider a spherical shell, as shown in Figure 6.2. The surface of central cavity is kept at temperature T_1 while the external one exchanges heat by convection with the environment. One is interested in determining the temperature profile in the shell material as well as the rate of heat transfer between the external surface and surrounding environment.

Despite some similarity with the problem shown in Section 4.6, the change in boundary condition leads to a very different solution.

For the sake of clarity, the following are the main characteristics or assumptions:

1. Steady-state regime.
2. Solid shell with no phase change, therefore there is no velocity field in the shell.
3. All properties of the shell material are constants.

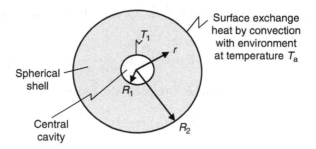

FIGURE 6.2 Sphere with internal cavity and external surface exchanging heat with environment.

4. All properties of the external fluid as well as the heat transfer coefficient by convection remain constant.
5. Far from the shell surface, the environment temperature remains constant and equals to T_a.

Therefore, only heat transfer is observed in the process. Given the above assumptions, Equation A.36 can be simplified to provide

$$\frac{1}{r^2}\frac{d}{dr}\left(r^2\frac{dT}{dr}\right) = 0 \tag{6.9}$$

In the shell, the radius is never equal to zero, thus

$$\frac{d}{dr}\left(r^2\frac{dT}{dr}\right) = 0 \tag{6.10}$$

Despite being a second-order differential equation, Equation 6.10 can be easily reduced to separable first-order equations, with the following solutions:

$$r^2\frac{dT}{dr} = C_1 \tag{6.11}$$

This yields

$$T = -\frac{C_1}{r} + C_2 \tag{6.12}$$

The boundary conditions are

$$T(R_1) = T_1 \tag{6.13}$$

and

$$-\lambda \frac{dT}{dr}\bigg|_{r=R_2} = \alpha(T_2 - T_a) \tag{6.14}$$

At this point, the solution departs from the one presented in Section 4.6. Using Equations 6.11 and 6.14, one gets

$$C_1 = \frac{T_a - T_1}{\dfrac{1}{R_1} - \dfrac{1}{R_2} + \dfrac{\lambda}{\alpha R_2^2}} \tag{6.15}$$

After Equation 6.13, it is possible to write

$$\frac{T - T_1}{T_a - T_1} = \frac{\dfrac{1}{R_1} - \dfrac{1}{r}}{\dfrac{1}{R_1} - \dfrac{1}{R_2} + \dfrac{\lambda}{\alpha R_2^2}} \tag{6.16}$$

The heat flux at the external surface is given by

$$q_r\big|_{r=R_2} = -\lambda \frac{dT}{dr}\bigg|_{r=R_2} \tag{6.17}$$

Using Equations 6.11 and 6.15, the heat flux becomes

$$q_r\big|_{r=R_2} = \frac{T_1 - T_a}{\dfrac{R_2^2}{\lambda R_1} - \dfrac{R_2}{\lambda} + \dfrac{1}{\alpha}} \tag{6.18}$$

The rate of heat exchange with environment is

$$\dot{Q}_r\big|_{r=R_2} = \pi \frac{T_1 - T_a}{\dfrac{1}{\lambda}\left(\dfrac{1}{R_1} - \dfrac{1}{R_2}\right) + \dfrac{1}{R_2^2 \alpha}} \tag{6.19}$$

It is important to notice the following:

- Driving force for the heat transfer is the difference of temperatures between the center and ambiance.
- Resistance for heat transfer is composed of two terms that appear in the denominator. The first resistance decreases for higher thermal conductivity as well as for smaller thickness of the shell. The second decreases for higher convective transfer coefficient as well as for larger superficial area.
- Introduction of another resistance due to convection is easily verified by comparing Equation 6.16 with Equation 4.81 and Equation 6.19 with Equation 4.84.

6.4   REACTING PARTICLE

The simplified treatment shown in Section 4.10 is assumed as a constant concentration of reacting gas at the particle surface. However, this is not the case in a real situation. Here, a more realistic treatment for the heterogeneous chemical reactions is presented. This can be used for modeling of gas–solid reactions found in industrial process such as catalytic conversions, combustion and gasification processes, etc. It is based on the following two limiting cases that are assumed as possible models:

1. Unreacted-core model is schematically shown in Figure 6.3a. Here, a shell of inert or spent material surrounds the core where reactions occur. In the case of combustion, this spent material is known as ash, which is basically a mixture of inert oxides. Actually, a better name for this case is unexposed-core model. This model can include cases where no reaction takes place in the core and just absorption or desorption occurs.
2. Exposed-core model is schematically shown in Figure 6.3b. Here, the shell of spent material cannot stand the stress suffered by the particle surface and, once formed, disintegrates into fine particles. Therefore, the core is always exposed to the gas environment. Another name for this model is segregation model because the reacting core is continuously separated from the spent material.

Relations to allow the computation of overall consumption of fuel and reacting gas are shown below for both cases. Before this, the following are the basic hypotheses assumed in the present treatments:

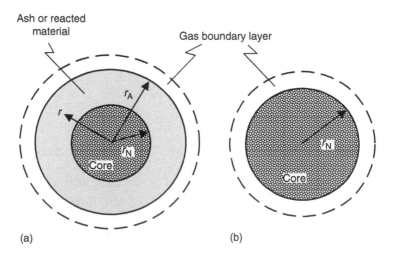

(a)   (b)

FIGURE 6.3 Unreacted-core (a) and exposed-core (b) models for the gas–solid particle reactions.

1. Particle is isothermal. This is a common assumption in the area of combustion and gasification of particles. However, this is only approximately valid in cases of relatively small particle sizes with high thermal conductivity. The conditions for this approximation are related to the Biot number and have already been made in Section 1.2.
2. At a given phase interface (core shell or shell gas), the concentrations of a chemical component at each side of the interface are assumed to be equal. In fact, the concentrations are near the equilibrium at each interface.
3. Heat and mass transfer do not interfere with each other. This hypothesis is a consequence of the isothermal approximation for the particle. Some improvements of this can be found elsewhere [1].
4. Each reaction can be treated independently. This assumption should become clear with the treatment ahead.

6.4.1 UNREACTED-CORE MODEL

As idealized by unreacted-core model (Figure 6.3a), the reactions occur only in an internal core with radius r_N. The spent material (ash in cases of combustion process) constitutes the external shell coating the core. This is a good approximation of the reality for cases in which the mechanical resistance of the formed shell is enough to withstand attrition suffered by the particle. Therefore, the equivalent diameter $(2 \times r_A)$ of a particle, at any point of the process, is the same as its original value.

For a spherical particle, other geometric shapes are treated ahead, the mass continuity differential equation for component j at any layer can be written as (Equation A.39)

$$D_j r^{-2} \frac{d}{dr}\left(r^2 \frac{d\tilde{\rho}_j}{dr}\right) = -\tilde{R}_j \tag{6.20}$$

The rate of production of component j is given by

$$\tilde{R}_j = \sum_{i=1}^{M} v_{ij}\tilde{r}_i \tag{6.21}$$

where v_{ij} indicates the reaction stoichiometry coefficient for each competing reaction.

It should be noticed that the coefficient D_j is the diffusivity of component j into the phase in which the process takes place. If this phase is the particle core, or nucleus, the parameter is called the effective diffusivity of j in that porous structure, and it is represented here by $D_{j,N}$. A similar notation is used for the diffusivity of j in the shell of inert porous solid that covers the core, namely, $D_{j,A}$. $D_{j,G}$ is the diffusivity of component A in the boundary layer of gas mixture surrounding the particle. Usually, the values for effective diffusivity can be correlated to the gas–gas diffusivity $D_{j,G}$.

For the present, it is assumed that all reaction rates can be written in the following form:

$$\tilde{r}_i = -k_i \left(\tilde{\rho}_j - \tilde{\rho}_{j,eq} \right)^n \tag{6.22}$$

where k_i is the reaction rate coefficient.

To put Equation 6.20 in a dimensionless form, the following change of variables is used:

$$x = \frac{r}{r_A} \tag{6.23}$$

$$y = \frac{\tilde{\rho}_j - \tilde{\rho}_{j,eq}}{\tilde{\rho}_{j,\infty} - \tilde{\rho}_{j,eq}} \tag{6.24}$$

The treatment for a single reaction and on the basis of each consumed mole of reacting gas component j by reaction i ($v_{ij} = -1$), the Equation 6.20 can be rewritten as

$$\nabla^2 y = \Phi^2 y^n \tag{6.25}$$

The Laplacian operator is generalized as

$$\nabla^2 = x^{-p} \frac{d}{dx} \left(x^p \frac{d}{dx} \right) \tag{6.26}$$

Like before, the coefficient p takes the following possible values: 0 for plane geometry, 1 for cylindrical, and 2 for spherical. In the present case, $p = 2$. It should be noticed that Equation 6.25 was written for a single reaction and consequence of approximation announced by assumption d.

The Thiele coefficient is given by

$$\Phi = r_A \left[\frac{k_i \left(\tilde{\rho}_{j,\infty} - \tilde{\rho}_{j,eq} \right)^{n-1}}{D_{j,N}} \right]^{1/2} \tag{6.27}$$

The above treatment is valid for most of the combustion and gasification reactions. For combustion processes, the reaction order n varies between 0 and 2. However, the main reactions that control most of the processes (carbon–oxygen and carbon–water) follow a first-order behavior.

The objective now is to obtain the rate of mass transfer between the particle and the external gas layer, which is possible after solving Equation 6.25. Such a solution should be accomplished in two steps: initially, for the external shell layer and then for the core. This is necessary because of the interface boundary conditions imposed by the problem, as described below.

For the external shell, no reaction is observed, therefore Equation 6.25 becomes

$$\frac{dy}{dx} = A_1 x^{-2}, \quad a \le x \le 1 \tag{6.28}$$

Equation 6.28 is a separable first-order differential equation and therefore

$$y = -A_1 x^{-1} + B_1, \quad a \le x \le 1 \tag{6.29}$$

where

$$a = \frac{r_N}{r_A} \tag{6.29a}$$

The surface concentration is given by

$$y(1) = \frac{\tilde{\rho}_{j,s} - \tilde{\rho}_{j,eq}}{\tilde{\rho}_{j,\infty} - \tilde{\rho}_{j,eq}} = -A_1 + B_1 \tag{6.30}$$

The mass transfer between the particle surface and the surrounding gaseous atmosphere gives one boundary condition, which is written as

$$D_{j,A} \frac{d\tilde{\rho}_j}{dr}\bigg|_{r=r_{A(\ )}} = \beta_G(\tilde{\rho}_{j,\infty} - \tilde{\rho}_{j,s}) \tag{6.31}$$

The mass transfer coefficient β_G is given as a function of the transport parameters [1–3]. The signal at the coefficient with the radius value indicates whether the derivative should be computed as limits from below (−) or above (+) the specified value.

Equation 6.31 can be written as

$$y'(1) = N_{Sh}[1 - y(1)]\frac{D_{j,G}}{D_{j,A}} = C_1[1 - y(1)] \tag{6.32}$$

The Sherwood number and associated parameter are given by

$$N_{Sh} = \frac{\beta_G r_A}{D_{j,G}} \tag{6.33}$$

$$C_1 = N_{Sh}\frac{D_{j,G}}{D_{j,A}} \tag{6.34}$$

Using Equations 6.28, 6.20, and 6.32, it is possible to write

$$B_1 = 1 - \frac{1 - C_1}{C_1} A_1 \tag{6.35}$$

Hence, one extra relationship is necessary to compute the values of A_1 and B_1. This would be possible if the problem is solved for the core as well. Since reaction is involved, Equation 6.25 becomes

$$xy'' + 2y' - \Phi^2 xy = 0 \tag{6.36}$$

The following new variables are introduced in order to put the above in the form of a Bessel equation:

$$y = x^{-1/2} u \tag{6.37}$$

and

$$z = \Phi x \tag{6.38}$$

After this, Equation 6.36 can be written as

$$z^2 \frac{d^2 u}{dz^2} + z \frac{du}{dz} - u\left(z^2 - \frac{1}{4}\right) = 0 \tag{6.39}$$

As described in Appendix C, Equation 6.39 is a modified Bessel differential equation whose solutions are

$$u = I_{1/2}(z) \tag{6.40}$$

and

$$u = I_{-1/2}(z) \tag{6.41}$$

In relation to variable y, they are

$$x^{-1/2} I_{1/2}(x\Phi) \tag{6.42}$$

and

$$x^{-1/2} I_{-1/2}(x\Phi) \tag{6.43}$$

These can be transformed into an equivalent form to give the following general solution:

$$y = A_2 \frac{\sinh(x\Phi)}{x} + B_2 \frac{\cosh(x\Phi)}{x}, \quad 0 \leq x \leq a \tag{6.44}$$

Of course, at the particle center, as well as anywhere, the concentration must acquire finite values. In addition

$$\lim_{x \to 0} \frac{\sinh(x\Phi)}{x} = \Phi \tag{6.45}$$

and

$$\lim_{x \to 0} \frac{\cosh(x\Phi)}{x} = \infty \tag{6.46}$$

Consequently, the coefficient B_2 in Equation 6.44 must be identical to zero, and

$$y = A_2 \frac{\sinh(x\Phi)}{x}, \quad 0 \leq x \leq a \tag{6.47}$$

Owing to assumption 2, the concentrations at each side of the interface between the core and the shell are equal. Therefore, using Equation 6.29 and the above relation one arrives at

$$y(a) = A_2 \frac{\sinh(a\Phi)}{a} = -\frac{A_1}{a} + B_1 \tag{6.48}$$

Moreover, at the interface, the mass flux must be conserved, or

$$D_{j,A} y'(a+) = D_{j,N} y'(a-) \tag{6.49}$$

From Equations 6.29 and 6.47 the following equation results:

$$D_{j,A} A_1 = D_{j,N} A_2 [a\Phi \cosh(a\Phi) - \sinh(a\Phi)] \tag{6.50}$$

Combining Equations 6.35 and 6.48 the relation between A_1 and A_2 is obtained:

$$\frac{A_2}{a} = \frac{1 - A_1 \left(\frac{1}{a} + \frac{1 - C_1}{C_1} \right)}{\sinh(a\Phi)} \tag{6.51}$$

Finally, Equations 6.50 and 6.51 lead to

$$\frac{A_1}{a} = \frac{1}{1 + C_2 + C_3} \tag{6.52}$$

where

$$C_2 = \frac{D_{j,A}}{D_{j,N}[a\Phi \coth(a\Phi)]} \tag{6.53}$$

and

$$C_3 = a\frac{1 - C_1}{C_1} \tag{6.54}$$

The consumption rate of reacting gas j by reaction i is given by the rate of mass transfer through the external surface of the particle, or

$$D_{j,A}\frac{d\tilde{\rho}_j}{dr}\bigg|_{r=r_{A-}} = \frac{D_{j,A}}{r_A}\left(\tilde{\rho}_{j,\infty} - \tilde{\rho}_{j,eq}\right)\frac{dy}{dx}\bigg|_{x=1-} \tag{6.55}$$

Employing the above with Equation 6.28, it is possible to deduce the consumption rate of component j as follows:

$$\frac{1}{V_P}\frac{d\tilde{m}_j}{dt} = n_j = -\frac{D_{j,A}}{r_A}A_1(\tilde{\rho}_{j,\infty} - \tilde{\rho}_{j,eq}) \tag{6.56}$$

Using Equation 6.52 now, one gets

$$|n_j| = \frac{2}{d_{PI}}\frac{\tilde{\rho}_{j,\infty} - \tilde{\rho}_{j,eq}}{\sum\limits_{k=1}^{3} U_{U,k}} \tag{6.57}$$

The three resistances indicated at the denominator are given by

$$U_{U,1} = \frac{1}{N_{Sh}D_{j,G}} \tag{6.58}$$

$$U_{U,2} = \frac{1 - a}{aD_{j,A}} \tag{6.59}$$

$$U_{U,3} = \frac{1}{aD_{j,N}[a\Phi \coth(a\Phi) - 1]} \tag{6.60}$$

6.4.1.1   Comments

The rate of consumption (or production) of the reacting gas is given by a simple formula. In it, the role played by the three resistances to the mass transfer combined with the kinetics are explicitly shown as follows:

- $U_{U,1}$ for the gas boundary layer
- $U_{U,2}$ for the shell, which surrounds the core
- $U_{U,3}$ for the core

It is interesting to notice that such parameters are inversely proportional to the first power of the diffusivities of gases through the respective layers.

The relative importance between these three factors determines the ruling or controlling mechanism of the process related to each reaction. There is no predetermined rule for this and the controlling factor would depend on the combined conditions to which the particle is subjected to at each point in the reactor. An example can be given for particles leaving the furnace, where almost all solid fuels have already been consumed. Therefore, relatively thick layers of ash surround the cores of the particles, and it is likely that at this condition the ruling resistance would be offered by the shell. The picture can be different for the particles entering the furnace. At these points, the layer of ash coating is relatively thin and the gas layer or the diffusion through the core could constitute the main resistances.

6.4.2   EXPOSED-CORE MODEL

In this case, it is assumed that no layer of spent material withstand, among other effects, the fast variations of temperature combined with attrition with other particles. Therefore, as soon as it is formed, the layer breaks into small particles, which break free from the original particle (Figure 6.3b).

Adopting the same notation as before, the governing differential equation for the core is given by Equation 6.25 and the solution by Equation 6.47, or

$$y = A_3 \frac{\sinh(x\Phi)}{x}, \quad 0 \le x \le a \tag{6.61}$$

The finite value condition at the particle center has already been used. Similar to Equations 6.31 and 6.32 at the gas–solid interface, it is possible to write

$$y'(a) = C_4[1 - y(a)] \tag{6.62}$$

where

$$C_4 = N_{Sh} \frac{D_{j,G}}{D_{j,N}} \tag{6.63}$$

Using Equation 6.62, this last condition leads to

$$A_3 = \frac{C_5}{1 + C_6} \tag{6.64}$$

where

$$C_5 = \frac{a}{\sinh(a\Phi)} \tag{6.65}$$

and

$$C_6 = \frac{a\Phi \coth(a\Phi) - 1}{a\Phi} \tag{6.66}$$

The consumption or production rate of the reacting gas by the particle is given in a similar way as before or

$$D_{j,\mathrm{N}} \frac{d\tilde{\rho}_j}{dr}\bigg|_{r=r_N^-} = \frac{D_{j,\mathrm{N}}}{r_\mathrm{A}} \left(\tilde{\rho}_{j,\infty} - \tilde{\rho}_{j,\mathrm{eq}} \right) \frac{dy}{dx}\bigg|_{x=a^-} \tag{6.67}$$

Using Equations 6.61, 6.64, and 6.67, it is possible to write

$$|n_j| = \frac{2}{d_{\mathrm{PI}}} \frac{\tilde{\rho}_{j,\infty} - \tilde{\rho}_{j,\,\mathrm{eq}}}{\displaystyle\sum_{k=1}^{3} U_{\mathrm{X},k}} \tag{6.68}$$

Here the three resistances are given by

$$U_{\mathrm{X},1} = \frac{1}{N_{\mathrm{Sh}} D_{j,\mathrm{G}}} = U_{\mathrm{U},1} \tag{6.69}$$

$$U_{\mathrm{X},2} = 0 \tag{6.70}$$

$$U_{\mathrm{X},3} = \frac{a}{D_{j,\mathrm{N}}[a\Phi \coth(a\Phi) - 1]} = a^2 U_{\mathrm{U},3} \tag{6.71}$$

The above formulas are similar to the case of unreacted-core model; however, and of course no reference to the shell diffusivity is made. Other considerations on the above treatment and application are beyond the scope of this chapter, but can be found elsewhere [1].

6.5 HEAT TRANSFER BETWEEN A REACTING PLATE AND A FLUID

The situation is shown in Figure 6.4.

It differs from the situation discussed in Section 6.2, because one face is kept at a constant temperature (T_2) and the plate material goes through an exothermic reaction when heated. It is desired to obtain the temperature profile in the plate as well as the rate of heat transfer between the fluids.

To clearly set the problem, let us list the assumptions:

1. Wall is infinite in all directions but x.
2. Steady-state regime has been established, i.e., the temperatures will be only a function of the position x.
3. No velocity field exists inside the solid wall.
4. All physical properties of the plate material are considered constants. Of course, if transformations in the material due to chemical reactions occur, this assumption will be difficult to maintain. However, for the sake of illustrating methods for the solution of the mathematical problem, it will be assumed that these transformations are not too severe.
5. Convection heat transfer coefficient (α_1) between the fluid at temperature T_1 and the wall surface does not vary significantly. Such coefficients can be calculated by appropriate semiempirical equations.

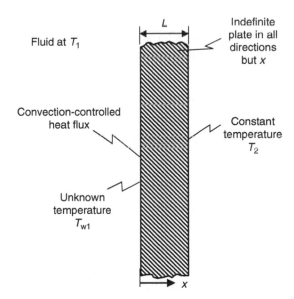

FIGURE 6.4 A reacting semi-infinite solid wall where a face is exchanging heat with fluid and another face at constant temperature.

6. Plate goes through an exothermic reaction. The rate of energy generation per unit of plate volume ($W \, m^{-3}$) is given by

$$R_Q = cT \tag{6.72}$$

where c is a positive real constant. Obviously, this is a simplification of dependence of energy released by any chemical reaction as a function of temperature. More elaborate and realistic situations would probably require numerical solutions, which is out of the scope of the present book.

Using the above assumptions and Equation A.34, it is possible to write

$$\lambda \frac{d^2 T}{dx^2} + cT = 0 \tag{6.73}$$

Following are the boundary conditions:

$$\alpha_1 (T_1 - T_{w1}) = -\lambda \frac{dT}{dx}\bigg|_{x=0} \tag{6.74}$$

and

$$T(L) = T_2 \tag{6.75}$$

As seen, the boundary condition given by Equation 6.74 is a third-kind boundary condition.

The solution of a homogeneous linear second-order differential equation with constant coefficients is described in Appendix B. After this, it is possible to write

$$T(x) = C_1 \sin\left(x\sqrt{\frac{c}{\lambda}}\right) + C_2 \cos\left(x\sqrt{\frac{c}{\lambda}}\right) \tag{6.76}$$

Applying the boundary conditions, one would arrive at the following dimensionless form:

$$\theta = \frac{N_{Bi} \sin(\varphi\zeta) + \varphi\cos(\varphi\zeta)}{N_{Bi} \sin(\varphi) + \varphi\cos(\varphi)} \tag{6.77}$$

where

$$\theta = \frac{T + T_1}{T_2 + T_1}, \quad \varphi = L\sqrt{\frac{c}{\lambda}}, \quad \zeta = \frac{x}{L}, \quad \text{and} \quad N_{Bi} = \frac{\alpha_1 L}{\lambda} \tag{6.78a,b,c,d}$$

The solution is illustrated by Figure 6.5 for the case when N_{Bi} is equal to 10 and by Figure 6.6 in the case when N_{Bi} is equal to 100.

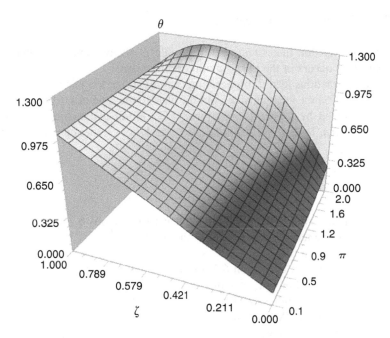

FIGURE 6.5 Dimensionless temperature profile in a reacting plate with one face exchanging heat with a fluid and the other face at constant temperature; case for $N_{Bi} = 10$.

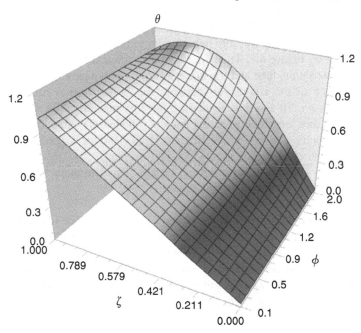

FIGURE 6.6 Dimensionless temperature profile in a reacting plate with one face exchanging heat with a fluid and the other face at constant temperature; case for $N_{Bi} = 100$.

The following is worth mentioning:

- Obviously, the linear temperature profile is always obtained when no reaction occurs in the plate (φ approaching zero).
- Faster reactions lead to higher temperatures inside the plate.
- Increase on the heat transfer from the plate to the fluid at $x = 0$ (or $\zeta = 0$) is reflected by higher Biot numbers. Therefore, maintaining all other conditions, higher N_{Bi} leads to lower peaks of temperature inside the plate.

EXERCISES

1. Solve Equation 6.1 under conditions given by Equations 6.2 and 6.3 in order to arrive at the temperature profile in the plate.
2. Deduce Equation 6.8.
3. Repeat problem set in Section 6.2 for the case where the thermal conductivity of the plate is a function of temperature given by $\lambda = a + bT + cT^2$. Here parameters a, b, and c are known constants. If necessary, use method of weighted residuals (MWR) methods.
4. Consider the fin as shown in Figure 5.5. Derive the solution for the temperature profile in the fin in the case where the heat transfer by convection at its tip ($z = a$) cannot be assumed negligible.
5. Repeat Problem 4 for the case of circular fin, i.e., considering heat transfer by convection at the outer border of the fin.
6. Consider an insulated pipe and heat transfer to fluid inside and to ambiance, as illustrated by Figure 6.7. The heat transfer coefficient by convection between the fluid inside the tube and the internal surface of the pipe is given as α_i. The one relative to the convection between the most external surface and the ambiance is α_o. It is asked to deduce the formula for the heat transfer rate per unit of tube length between the internal fluid and the ambiance as function of the geometry, convection coefficients, T_f, T_∞, and the conductivities of pipe wall (λ_p) and of insulation (λ_i).

    Assume constant conductivities. In addition, determine the temperature profiles in the pipe and in the insulation.
7. Deduce Equation 6.77.
8. Repeat the problem presented in Section 6.5 for the case where the rate of energy generation due to chemical reaction is given by $R_Q = cT^2$. As a suggestion, apply approximate or MWR methods.
9. Repeat the problem presented in Section 6.5 for the case of variable thermal conductivity of the wall. Assume that it is given by a linear function of temperature, or $\lambda = a + bT$. Again, as a suggestion, apply approximate or MWR methods.

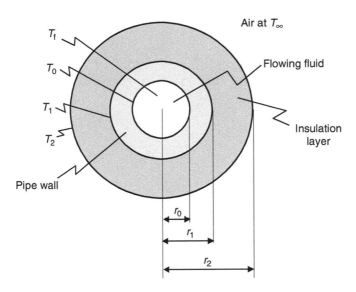

FIGURE 6.7 Pipe with insulation.

REFERENCES

1. de Souza-Santos, M.L., *Solid Fuels Combustion and Gasification: Modeling, Simulation and Equipment Operation*, Marcel Dekker, New York, 2004.
2. Bird, R.B., Stewart, W.E., and Lightfoot, E.N., *Transport Phenomena*, John Wiley, New York, 1960.
3. Slattery, J.C., *Momentum, Energy, and Mass Transfer in Continua*, Robert E. Kriefer, Huntington, New York, 1978.

7 Problems 211; Two Variables, 1st Order, 1st Kind Boundary Condition

7.1 INTRODUCTION

This chapter presents methods to solve problems with two independent variables involving first-order differential equation and first-kind boundary condition. Therefore, the problems fall in the category of partial differential ones. Mathematically, this class of cases can be summarized as $f\left(\phi, \omega_1, \omega_2, \frac{\partial \phi}{\partial \omega_1}, \frac{\partial \phi}{\partial \omega_2}\right)$, first-kind boundary condition.

7.2 PRESSURE IN FLUID UNDER ROTATIONAL MOVEMENT

In Section 1.8, the velocity profile of the fluid between the vertical drums was determined. Now, one is interested in determining its pressure profile.

The situation is shown in Figure 7.1 and the assumptions set in Section 1.8 are still valid.

As seen, Equation A.13, describing the momentum continuity in the radial direction provided

$$\frac{\partial p}{\partial r} = \rho \frac{v_\theta^2}{r} \qquad (7.1)$$

While, Equation A.15 representing the momentum in vertical direction gave

$$\frac{\partial p}{\partial z} = \rho g_z = -\rho g \qquad (7.2)$$

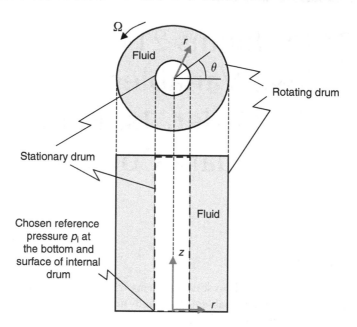

FIGURE 7.1 Scheme of concentric drums to impose a rotating field in a fluid between them.

7.2.1 SEPARABLE EQUATIONS

Since all terms on the right-hand side of Equation 7.1 are functions of only r or constants, the integration is immediate to give

$$p = \rho \int_{r=r_i}^{r=r} \frac{v_\theta^2}{r} \, dr + a(z) \tag{7.3}$$

It should be noticed that the arbitrary choice of one end of the integration is possible and this was taken at position $r = r_i$. Of course, any other choice (for instance $r = r_o$) would not change the final result. In addition, when dealing with a partial differential equation, the integration constants should, in principle, be assumed as a function of the variable or variables not involved in the integration. This is the case of the function $a(z)$ in Equation 7.3, which appears after the integration related to the radial direction.

A similar procedure can be applied to Equation 7.2, where the right-hand side is only a function of variable z. Its integration leads to

$$p = -\rho g z + b(r) \tag{7.4}$$

Here, the lower limit of integration was set at $z = 0$.

As seen, both differential equations, i.e., Equations 7.1 and 7.2, were separable.

The pressure is now given either by Equation 7.3 or 7.4, therefore

$$\rho \int_{r=r_i}^{r=r} \frac{v_\theta^2}{r} dr + a(z) = -\rho gz + b(r) \tag{7.5}$$

Equation 7.5 can be written as

$$\rho \int_{r=r_i}^{r=r} \frac{v_\theta^2}{r} dr - b(r) = -\rho gz - a(z) \tag{7.6}$$

Now, one should notice that the left-hand side contains only functions of r and the right only functions of z. As these variables are independent, this would be possible if and only if both sides are equal to a constant, or

$$\rho \int_{r=r_i}^{r=r} \frac{v_\theta^2}{r} dr - b(r) = C \tag{7.7}$$

and

$$-\rho gz - a(z) = C \tag{7.8}$$

Combining Equations 7.8 and 7.3 results into

$$p = \rho \int_{r=r_i}^{r=r} \frac{v_\theta^2}{r} dr - \rho gz - C \tag{7.9}$$

The same would be achieved using Equations 7.7 and 7.4.

The constant at Equation 7.9 can be found if the pressure at a given position of the drum interior is known. For instance, assume that at $z = 0$ (bottom of the drum) and $r = r_i$ (surface of the internal drum) the pressure is p_i. The above equation allows writing

$$C = -p_i \tag{7.10}$$

Finally,

$$p = p_i + \rho \int_{r=r_i}^{r=r} \frac{v_\theta^2}{r} dr - \rho gz \tag{7.11}$$

The velocity profile is given by Equation 1.105

$$v_\theta = \Omega \frac{r_0^2}{r} \frac{r^2 - r_i^2}{r_0^2 - r_i^2}$$

Using the above equation into Equation 7.11, the pressure profile is provided as follows:

$$p = p_i - \rho g z + \frac{\rho \Omega^2}{2} \frac{r_0^4}{r_0^2 - r_i^2} \left[\frac{r^2 - r_i^2}{r_0^2 - r_i^2} \left(1 + \frac{r_i^2}{r^2} \right) - 4 \frac{r_i^2}{r_0^2 - r_i^2} \ln\left(\frac{r}{r_i}\right) \right] \quad (7.12)$$

The pressure profile against the height (z) in the drum and the radius (r) is presented in Figure 7.2. In this case, the following values have been assumed: $\rho = 1000 \text{ kg m}^{-3}$, $g = 9.81 \text{ m s}^{-2}$, $\Omega = 10 \text{ s}^{-1}$, $r_i = 0.1 \text{ m}$, $r_0 = 1.0 \text{ m}$, $p_i = 0.2 \text{ MPa}$. One should observe the following:

- Dependence of pressure on the vertical direction due to the static term leading to a linear decrease from the bottom at $z = 0$ to the top at $z = 1$ m.

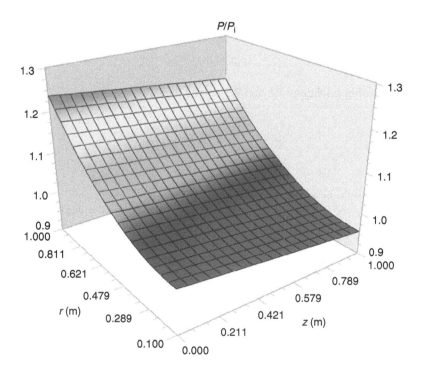

FIGURE 7.2 Pressure profile against the height (z) and radius (r) in the space between the drums.

- Increase of pressure from the inner drum ($r = r_i$) to the outer one ($r = r_o$). This difference on pressures is one of the most important aspects of the so-called centrifuges.*

In the present particular case, the result arrived at Equation 7.11 could be achieved by another route.

Since the total derivative of a function (p) on its independent variables (r and z) is given by

$$dp = \frac{\partial p}{\partial r} dr + \frac{\partial p}{\partial z} dz \qquad (7.13)$$

Applying Equations 7.1 and 7.2 into Equation 7.13 leads to

$$dp = \rho \frac{v_\theta^2}{r} dr - \rho g \, dz \qquad (7.14)$$

This can now be integrated using inferior limits $z = 0$ and $r = r_i$ to arrive at Equation 7.11. The above was only possible because the explicit forms for partial derivatives at Equations 7.1 and 7.2 were available.

7.3   HEATING A FLOWING LIQUID

Several heating systems are designed to heat a fluid while passing through a tube. Among these devices, many employ a series of gas burners where the flames are directed on the tube surface. Others use electrical resistances either wrapped around the tube or set in a grid at the tube entrance. At this point, consider a simple scheme, as illustrated in Figure 7.3.

The liquid flows in a very long tube and passes through electrical resistances in the form of a grid located at $z = 0$.

For a first approximation, let us assume the following simplifications:

1. Plug-flow regime or the velocity profile is approximately flat. Again, this is approximately valid when convective momentum transfer in the main flow direction is much higher than viscous influences.
2. Grid does not interfere in the flow. In fact, grids are often employed to provide a flat velocity profile for the flow.
3. Tube is very long and all possible border (inlet or outlet) effects are negligible, at least in the studied region.
4. Tube is perfectly insulated.

---

* One should be careful with the concept of centrifuge. Actually, the fluid experiences a centripetal force applied by the wall of the external drum. This is the force responsible for the fact that the fluid of any body under circular movement does not follow the inertial movement, which would lead it to escape the circular path following a tangential one.

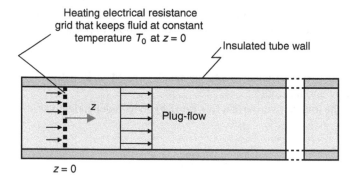

FIGURE 7.3 Flowing liquid heated by a grid of electrical resistances.

5. Liquid is Newtonian and its properties are approximately constant. Of course, this might be a strong approximation if large variations of temperature are present. This assumption would be invalid in cases of gases.
6. No energy dissipation due to friction is observed.
7. Before the heating system is turned on, the fluid is at uniform temperature T_b.
8. Once the heating starts working, the fluid leaving the grid (or at $z=0$) is at temperature T_0. This implies an instantaneous jump of temperature, which is physically impossible. Despite this, such situations can be approximated in practice by a high heat transfer between the grid and the flowing fluid. On the other hand, sharp step functions are very useful in studies of process and control performance.
9. Rate of heat transfer by convection is much higher than the rate by conduction. This is reasonable for cases where the velocity of the fluid is relatively high.

The basic equations of momentum transfer are not necessary here because the fluid keeps a constant velocity. In other words, the only velocity component (v_z) does not vary in axial (z), radial (r), or angular (θ) coordinates.

Owing to assumption 5, Equation A.35 can be applied to describe the energy conservation. Assumption 4 allows discarding all terms related to temperature variation in the radial direction. Also, terms related to temperature variations in the angular direction are zero due to the imposed uniform heating at the tube cross section. Consequently, it is simplified to

$$\rho C_p \left(\frac{\partial T}{\partial t} + v_z \frac{\partial T}{\partial z} \right) = \lambda \frac{\partial^2 T}{\partial z^2} \qquad (7.15)$$

It is important to notice that the terms of Equation A.35 involving the viscosity are negligible owing to assumption 6. Assumption 9 allows another simplification and Equation 7.15 becomes

$$\frac{\partial T}{\partial t} + v \frac{\partial T}{\partial z} = 0 \tag{7.16}$$

Here the axial velocity component (v_z) is written just as v.

The boundary conditions for the problem are

$$T(0, z) = T_b, \quad z \geq 0 \tag{7.17}$$

and

$$T(t, 0) = T_0, \quad t > 0 \tag{7.18}$$

As already discussed and justified, it is convenient to change the variables into dimensionless ones. Here, just the dependent one will be changed, and the following is proposed:

$$\psi = \frac{T - T_b}{T_0 - T_b} \tag{7.19}$$

Of course, it is always possible to find dimensionless variables related to the independent variables, such as time and position.

Applying Equation 7.19 one gets

$$\frac{\partial T}{\partial t} = (T_0 - T_b) \frac{\partial \psi}{\partial t} \tag{7.20}$$

and

$$\frac{\partial T}{\partial z} = (T_0 - T_b) \frac{\partial \psi}{\partial z} \tag{7.21}$$

Therefore, Equation 7.16 becomes

$$\frac{\partial \psi}{\partial t} + v \frac{\partial \psi}{\partial z} = 0 \tag{7.22}$$

The boundary conditions given by Equations 7.17 and 7.18 are written as

$$\psi(0, z) = 0, \quad z \geq 0 \tag{7.23}$$

$$\psi(t, 0) = 1, \quad t > 0 \tag{7.24}$$

7.3.1 SOLUTION BY LAPLACE TRANSFORM

As seen in Appendix D, the partial differential equation can be transformed into an ordinary one by the correct application of Laplace transform.

The first task is to choose the independent variable to be transformed. Of course, the final solution should be the same regardless of the choice. However, the amount of work required to arrive at the solution usually depends on this decision. For instance, it is very convenient to apply the transform to the variable with known initial value. This is so because the Laplace transform of a derivative involves the value of the function at zero (see Equation D.10). Therefore, either time (t) or position coordinate (z) could be used in the transform. Nonetheless, time seems even more attractive because, according to the condition given by Equation 7.23, no term due would be added to the resulting differential equation. With this in mind, let us consider the following transform:

$$\Psi(s,z) = L\{\psi(t,z)\} \tag{7.25}$$

According to Appendix D, the transform of Equation 7.22 is

$$s\Psi - \psi(0,z) + v\frac{d\Psi}{dz} = 0 \tag{7.26}$$

Using Equation 7.23, the equation above becomes

$$\frac{d\Psi}{dz} = -\frac{s}{v}\Psi \tag{7.27}$$

Equation 7.27 is a separable ordinary differential and on integrating, one gets

$$\ln\Psi = -\frac{s}{v}z + \ln C_1 \tag{7.28}$$

or

$$\Psi = C_1 \exp\left(-\frac{s}{v}z\right) \tag{7.29}$$

The Laplace transform of the remaining boundary condition given by Equation 7.24 is

$$\Psi(s,0) = \frac{1}{s} \tag{7.30}$$

After the application of Equation 7.30, Equation 7.29 becomes

$$\Psi = \frac{1}{s}\exp\left(-\frac{s}{v}z\right) \tag{7.31}$$

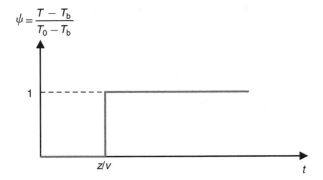

FIGURE 7.4 Dimensionless temperature profile against time at a given axial position.

Since the exponential term indicates a t-shift (see Appendix D), the inverse of such a function is

$$\psi(t,z) = u\left(t - \frac{z}{v}\right) \tag{7.32}$$

This was the expected result and is illustrated in Figure 7.4.

As plug-flow regime was assumed, a wave front of hot fluid advances through the tube at a velocity v. Therefore, the front will reach a position z in the tube only after the period equals z/v.

7.3.2 Comments

Just for the sake of discussion, let us apply the Laplace transform at the space variable (z) instead of time. In this case, the following is written:

$$\Psi(t,s) = L\{\psi(t,z)\} \tag{7.33}$$

The transform of Equation 7.22 would be

$$\frac{d\Psi}{dt} + vs\Psi - v\psi(t,0) = 0 \tag{7.34}$$

or

$$\frac{d\Psi}{dt} = -vs\Psi + v \tag{7.35}$$

The transformed boundary condition given by Equation 7.23 would be

$$\Psi(0,s) = 0 \tag{7.36}$$

After that application of parameter variation (see Appendix B) the solution of Equation 7.35 is written as

$$\psi = \frac{1}{s} + C_2 e^{-svt} \tag{7.37}$$

After the application of the condition given by Equation 7.36, Equation 7.37 becomes

$$\psi = \frac{1}{s}(1 - e^{-svt}) \tag{7.38}$$

By inverting the above, one obtains

$$\psi = 1 - u(z - vt) \tag{7.39}$$

Despite being in a different form than Equation 7.32, the two solutions are completely equivalent. For instance, it is easy to see from Figure 7.5 that for positions smaller than $z = vt$, the function ψ is equal to 1, or $T = T_0$. Ahead of position $z = vt$, or the wave front, the function ψ is equal to zero, or $T = T_b$.

The combination of Figures 7.4 and 7.5, for a particular case of $v = 2$ m s^{-1}, is presented by Figure 7.6.

Then, it is possible to see that for a given position z, the temperature would jump to the final value after a period, which would increase for positions farther from the tube entrance ($z = 0$). It is also interesting to notice that even if a reasonably sharp step-increase of temperature could be reproduced in practice, the flat wave-front would dissipate as it travels along the tube length. This dissipation would be slower (or take greater lengths of tube) or faster according to thermal conductivity of the fluid. This can be easily understood if one looks at Equation 7.15. The term at the right-hand side represents the dissipation term and its influence would be to "dissolve" the wave front or smooth the temperature variations. This can be seen as a degeneration of the sharp increase of temperature into a more realistic and smooth variation, similar to that shown in Figure 7.7.

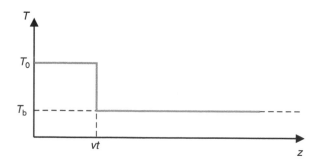

FIGURE 7.5 Temperature profile throughout the tube.

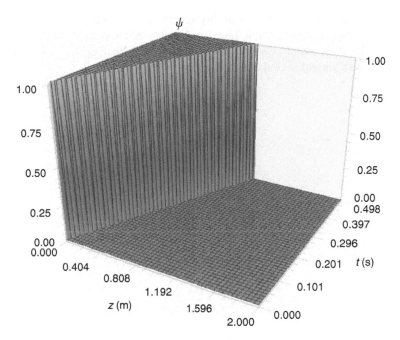

FIGURE 7.6 Temperature profiles against length and time.

7.3.3 COMPLETE DIMENSIONLESS FORM

This problem can be put in complete dimensionless form. The advantages of applying the complete dimensionless form have already been described.

To find a dimensionless length, the variable z should be divided by another length or a parameter with unit of length. Of course, this parameter should involve properties or characteristics of the problem. This is called reference length. As a first option, the reference could be the tube diameter or a given arbitrary length of

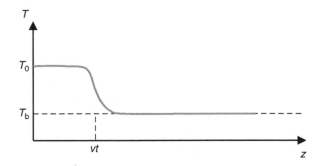

FIGURE 7.7 More realistic temperature profile.

the heating tube. The second possibility for reference length could be the group $\frac{\lambda}{\rho C_p v}$. Such a choice is reasonable here just because all properties and the velocity are considered constant. Let us call this group L. Bearing this in mind, the dimensionless variables could be defined as

$$\zeta = \frac{z}{L} \tag{7.40}$$

$$\tau = \frac{v}{L}t \tag{7.41}$$

where

$$L = \frac{\lambda}{\rho C_p v} \tag{7.41a}$$

Using the above, Equation 7.22 becomes

$$\frac{\partial \psi}{\partial \tau} + \frac{\partial \psi}{\partial \zeta} = 0 \tag{7.42}$$

The boundary conditions given by Equations 7.23 and 7.24 become

$$\psi(0,\zeta) = 0, \quad \zeta \geq 0 \tag{7.43}$$

$$\psi(\tau,0) = 1, \quad \tau > 0 \tag{7.44}$$

Laplace transform applied to any variable leads to the solution given by

$$\psi(\tau,\zeta) = u(\tau - \zeta) \tag{7.45}$$

This general solution is illustrated in Figure 7.8.

The interpretation of Figure 7.8 follows the same lines as for Figure 7.6.

One should notice the simplicity and elegance of Equation 7.42. It also tells something about the nature of the present heat transfer process, which is given by the equivalence of time and space regarding the path to solution. The same is reflected by the similarity of Figures 7.4 and 7.5, and by the symmetry shown in Figure 7.8.

7.3.4 Considerations on Possible Application of Similarity

Bearing in mind the wide range of applications of the method of similarity—described in Appendix F—one might consider applying it to the present problem. On the other hand, this method better fits problems where the dependent variable is a continuous function of the independent ones, which is not the present case. Despite this, the trial will bring interesting discussions regarding applications of this method.

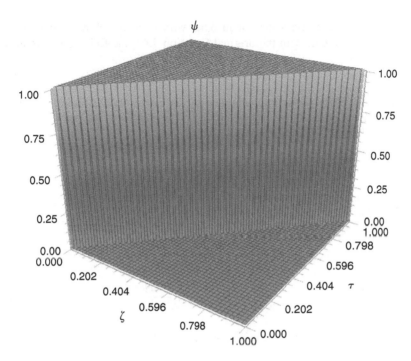

FIGURE 7.8 Dimensionless temperature (ψ) profiles against dimensionless length (ζ) and time (τ).

Let us apply the generalized method of similarity, as described in Section F.2. The first step is to define the following transformations of variables:

$$\bar{\tau} = a^{b_1}\tau \tag{7.46}$$

$$\bar{\zeta} = a^{b_2}\zeta \tag{7.47}$$

$$\bar{\psi} = a^{c}\psi \tag{7.48}$$

Using Equations 7.46 through 7.48, the following changes of variables are found:

$$\frac{\partial\psi}{\partial\tau} = \frac{\partial\bar{\psi}}{\partial\bar{\tau}}\frac{\partial\psi}{\partial\bar{\psi}}\frac{\partial\bar{\tau}}{\partial\tau} = a^{b_1-c}\frac{\partial\bar{\psi}}{\partial\bar{\tau}} \tag{7.49}$$

$$\frac{\partial\psi}{\partial\zeta} = \frac{\partial\bar{\psi}}{\partial\bar{\zeta}}\frac{\partial\psi}{\partial\bar{\psi}}\frac{\partial\bar{\zeta}}{\partial\zeta} = a^{b_2-c}\frac{\partial\bar{\psi}}{\partial\bar{\zeta}} \tag{7.50}$$

Replacing them in Equation 7.42, one gets

$$a^{b_1-c}\frac{\partial\bar{\psi}}{\partial\bar{\tau}} + a^{b_2-c}\frac{\partial\bar{\psi}}{\partial\bar{\zeta}} = 0 \tag{7.51}$$

To preserve the invariance, or in other words, for the above equation to be conformally invariant with the original Equation 7.42, the following is required:

$$b_1 - c = b_2 - c \tag{7.52}$$

which is satisfied as long as b_1 equals b_2 for any c. With this, the similarity variable can be deduced as

$$\omega = \frac{\tau}{\zeta^{b_2/b_1}} = \frac{\tau}{\zeta} \tag{7.53}$$

There is no need for transforming the dependent variable ψ.

The new changes of variables are

$$\frac{\partial \psi}{\partial \tau} = \frac{d\psi}{d\omega} \frac{\partial \omega}{\partial \tau} = \frac{1}{\zeta} \frac{d\psi}{d\omega} \tag{7.54}$$

$$\frac{\partial \psi}{\partial \zeta} = \frac{d\psi}{d\omega} \frac{\partial \omega}{\partial \zeta} = -\frac{\tau}{\zeta^2} \frac{d\psi}{d\omega} \tag{7.55}$$

Using these in Equation 7.42, it is possible to write

$$(1 - \omega) \frac{d\psi}{d\omega} = 0 \tag{7.56}$$

A striking simplicity, which at first glance provides no useful information because either

$$\omega = 1 \tag{7.57}$$

or

$$\frac{d\psi}{d\omega} = 0 \tag{7.58}$$

Therefore,

$$\omega = C \tag{7.59}$$

with constant C to be determined by boundary conditions.

Actually, both Equations 7.57 and 7.59 are solutions. This is because of the following:

1. As observed in Figure 7.8, the derivative of function ψ is always zero, which is attested by Equation 7.57. Therefore, this function can only be a constant.

2. As required by boundary conditions given by Equations 7.43 and 7.44, the value of the constant is either 1 or 0. In fact, $\psi = 0$ for points at $\tau = 0$, and $\psi = 1$ for points at $\zeta = 0$. This can be also seen from Figure 7.8.
3. For the points different from those above, the solution follows Equation 7.57, and according to Equation 7.53

$$\zeta = \tau \tag{7.60}$$

This is represented by the straight line in the ζ–τ plane (see Figure 7.8).
4. Line $\zeta = \tau$ demarks the boundary between the region where $\psi = 1$ or $\psi = 0$.

Regardless of these interesting details, rigorously speaking, the method of similarity could not provide a general solution for this case because

- Equation 7.56 is not a complete or self-contained description of the dependence among the variables involved in the problem, such as that achieved by Equation 7.45.
- Boundary conditions given by Equations 7.43 and 7.44 could not be condensed in a single condition involving just the dependent variable (ψ) and the new independent one (ω).

As seen, if one tries to follow the generalized method presented in Appendix F, no success would be achieved, basically because the boundary conditions are not compatible with any combination in the form $\omega = \tau^a \zeta^b$ (a and b are constants). On the other hand, the following combination might be tried:

$$\omega = \tau - \zeta \tag{7.61}$$

This would lead to

$$\frac{\partial \psi}{\partial \tau} = \frac{d\psi}{d\omega}\frac{\partial \psi}{\partial \tau} = \frac{d\psi}{d\omega} \tag{7.62}$$

$$\frac{\partial \psi}{\partial \zeta} = \frac{d\psi}{d\omega}\frac{\partial \psi}{\partial \zeta} = -\frac{d\psi}{d\omega} \tag{7.63}$$

The above equation satisfies the partial differential equation given by Equation 7.42. It is also satisfied if

$$\frac{d\psi}{d\omega} = 0 \tag{7.64}$$

This would provide a similar solution as given by Equation 7.59, or

$$\psi = C \tag{7.65}$$

Again, looking at Equation 7.45, the above is also a solution because variable ψ acquires only constant values: 0 or 1. Therefore, if one applies the conditions given by Equations 7.43 and 7.44, the solution for $\psi = \psi(\omega) = \psi(\tau - \zeta)$ would be the same as in Equation 7.45.

7.3.5   CONSIDERATIONS ON POSSIBLE APPLICATION OF METHOD OF WEIGHTED RESIDUES

The method of weighted residues (MWR) can be applied to solve a vast range of partial differential equations. Consequently, one may wonder about the possibility of employing it in the present case.

Consider the dimensionless form of the boundary condition problem formed by Equations 7.42 through 7.44. According to Appendix E, the first step is to choose a trial function as an approximate solution. In the case of partial differential equations, the approximate solution would be a composition of trial functions, one for each independent variable. Additionally, one among these would play the part of constant, while the other would be explicit and, if possible, satisfy its respective boundary conditions. For instance, we may write

$$\bar{\psi}_n(x) = \sum_{j=1}^{n} C_j \phi_j + \phi_0 \tag{7.66}$$

Consequently, there are the two following possibilities: $\phi_0(\zeta)$, $\phi_j(\zeta)$, $C_j(\tau)$ or $\phi_0(\tau)$, $\phi_j(\tau)$, $C_j(\zeta)$. Any alternative would lead to equivalent approximate solutions. However, the number of approximation levels (n) to achieve a given small deviation between the exact and approximate solution varies. Usually, it is difficult to visualize the best option. Let us try, for instance

$$\bar{\psi}_n(x) = \phi_0(\zeta) + \sum_{j=1}^{n} C_j(\tau) \phi_j(\zeta) \tag{7.67}$$

The second task is to choose functions ϕ that satisfy the respective boundary conditions.

Remembering Equation 7.44, one might choose $\phi_0(\zeta) = 1$, $\phi_j(\zeta) = \zeta^j$. Thus

$$\bar{\psi}_n(x) = 1 + \sum_{j=1}^{n} C_j(\tau) \zeta^j \tag{7.68}$$

7.3.5.1 First Approximation

With the above, the first approximation is

$$\bar{\psi}_1(x) = 1 + C_1(\tau)\zeta \tag{7.69}$$

According to Equation 7.42, the residue becomes

$$\Lambda_1 = \zeta C_1'(\tau) + C_1(\tau) \tag{7.70}$$

From now, any particular MWR can be applied. For instance, method of moments requires the weighting function (Appendix E) as

$$W_1 = \zeta \tag{7.71}$$

Hence, the following is imposed to minimize the residue:

$$\int_{\zeta=0}^{\zeta=1} \zeta \Lambda_1 \, d\zeta = \int_{\zeta=0}^{\zeta=1} (\zeta^2 C_1' + \zeta C_1) \, d\zeta = 0 \tag{7.72}$$

Of course, the solution should cover the whole range of the integrated variable (ζ). Nonetheless, this would lead to indeterminations. The choice of upper limit as 1 seems adequate because it demarks the region where the variations occur. If one is not sure about this, the limit might be expanded. On the other hand, the smaller the region of integration, the closer the approximated solution would be to the exact one at each level of approximation.

Equation 7.72 leads to

$$2C_1' = -3C_1 \tag{7.73}$$

and

$$C_1(\tau) = K_1 \exp\left(-\frac{3}{2}\tau\right) \tag{7.74}$$

Here K_1 is a constant.

Thus, the first approximation becomes

$$\bar{\psi}_1(x) = 1 + K_1\zeta \exp\left(-\frac{3}{2}\tau\right) \tag{7.75}$$

Despite the simplicity, this solution cannot conform to boundary condition given by Equation 7.43. Therefore, let us try another alternative for the approximate solution in which the variables change roles, or

$$\bar{\psi}_n(x) = \sum_{j=1}^{n} C_j(\zeta)\tau^j \tag{7.76}$$

Notice that the above equation satisfies the condition given by Equation 7.43.

Using the same procedure as before and the method of moments, one would obtain the first approximation as

$$\bar{\psi}_1(x) = K_1 \tau \exp\left(-\frac{3}{2}\zeta\right) \tag{7.77}$$

Unfortunately, the above form cannot satisfy the condition given by Equation 7.44 for every value of variable τ.

This sort of problem is inevitable when one tries to apply MWR to cases of discontinuous functions, as the present one. This is so because MWR always assumes continuous functions to approach the solution.

Regardless of the powerfulness of MWR to obtain excellent approximations, this example was very useful to show its limitations. Another example of such limitations is present in Section 7.4.3.

7.4 PLUG-FLOW REACTOR

A pure substance B is continuously flowing through a tubular plug-flow reactor, as illustrated in Figure 7.9.

Species A is continuously added through an injection grid at cross section $z = 0$. Such a grid is formed by a network of small-diameter tubes with porous

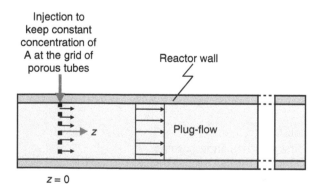

FIGURE 7.9 Scheme of plug-flow reactor where reactant A is injected at the entrance.

walls. The component A is injected into the porous tubes and then into the main reactor stream. It is desired to know the concentrations of reactants and products at any instant and position in the reactor. We are also interested in the rates of reactant consumption.

To ensure clarity, let us list the assumptions made here:

1. Pure component B is flowing through the reactor. At a given instance ($t = 0$), reactant A is injected into the reactor, at position $z = 0$. The injection is made into the network of porous tubes, which constitute the grid. The concentration at the external surfaces of porous tubes instantaneously acquires a given value, which of course remains constant. Additionally, the mechanism for mass transfer into the reactor main stream with reactant B is dictated by diffusion and convection. More realistic models are shown in the Chapters 8 and 9. However, this first attack demonstrates several important aspects of the plug-flow reactor.

2. Reaction between A and B can be considered a first-order irreversible. It is represented by

$$A + B \rightarrow C$$

3. All involved chemical components are in the liquid phase and their mixtures at any proportion occur in a single phase. Therefore, the reactants A and B and the product C do not form another physical phase (gas or solid). In addition, even the liquids are completely mixable, i.e., do not naturally segregate, for instance, in cases of oil and water.

4. No mixing enthalpy change is involved and the occurring reaction is neither exothermic nor endothermic. Therefore, isothermal conditions throughout the reactor can be ensured. Of course, it is very difficult to find such a situation. However, it may also be approached if a slightly endothermic or exothermic reaction takes place, and heating or cooling through the wall is provided and controlled to ensure constant temperature throughout the reactor length.

5. No matter what the composition of the fluid is, it can be considered Newtonian with constant density and viscosity.

6. Diffusivity of component A into B or vice versa is constant. It also remains constant, no matter how far the reaction between them progresses. Product C does not interfere in this diffusivity. This condition is only approximately possible.

7. Plug flow is approached throughout the reactor. The justification for this can be found in the previous section.

8. No swirl or angular velocity component is verified in the reactor.

9. Velocity of fluids in the reactor is high enough to allow neglecting the mass diffusion transport when compared with the convective one.

The consequences of this approximation should become clear during the treatment below. However, such an assumption will lead to sharp variations of concentration, which cannot really occur in the real process. Despite this, the conclusions reached here are useful for the understanding of the similar processes and applicable in control strategies for several industrial systems and equipment. A more realistic approach is given in Chapter 10.

Since plug-flow regime is assumed, the momentum equations do not provide important information. In addition, isothermal condition leads to no usefulness of the energy differential balances. Therefore, only mass transfer or continuity for species would provide useful information.

Owing to assumption 5 above, it is possible to depart from Equation A.41. Assumption 7 implies no variation of concentration in radial (r) or angular (θ) coordinates as well as no velocity components in these directions. After these simplifications and by using the molar basis, Equation A.41 becomes

$$\frac{\partial \tilde{\rho}_A}{\partial t} + v_z \frac{\partial \tilde{\rho}_A}{\partial z} = D_{AB} \frac{\partial^2 \tilde{\rho}_A}{\partial z^2} + \tilde{R}_A \tag{7.78}$$

Owing to assumption 9, the diffusion transport, represented by the first term on the right-hand side, would be assumed much smaller than the convective term (second term on the left-hand side). Thus Equation 7.78 can be rewritten as

$$\frac{\partial \tilde{\rho}_A}{\partial t} + v \frac{\partial \tilde{\rho}_A}{\partial z} + k \tilde{\rho}_A = 0 \tag{7.79}$$

To simplify the notation, the velocity component v_z was replaced by just v, which in turn, is a constant due to simplification of assumption 7 combined with assumption 5. One should also notice that the typical form of first-order reaction rate was already included.

Usually, the reaction coefficient k is a function of temperature and given by the Arrhenius equation:

$$k = k_0 \exp\left(-\frac{\tilde{E}_R}{\tilde{R}T}\right) \tag{7.80}$$

Thus, for the present isothermal process, k is constant.

The boundary conditions for the problem are

$$\tilde{\rho}_A(0,z) = 0, \quad z \geq 0 \tag{7.81}$$

and

$$\tilde{\rho}_A(t,0) = \tilde{\rho}_{A,0}, \quad t \geq 0 \tag{7.82}$$

Here, $\rho_{A,0}$ is the concentration of component A at the grid, or $z = 0$, at any positive time.

In this way, the solution of Equation 7.79 under boundary conditions using Equations 7.81 and 7.82 would lead to the desired information.

The convenience of working with dimensionless variables has been shown. Bearing this in mind let us consider the following:

$$f(t,z) = \frac{\tilde{\rho}_A(t,z)}{\tilde{\rho}_{A,0}} \tag{7.83}$$

Using Equation 7.83 into Equations 7.79, 7.81, and 7.82 leads to

$$\frac{\partial f}{\partial t} + v\frac{\partial f}{\partial z} + kf = 0 \tag{7.84}$$

$$f(0,z) = 0, \quad z \geq 0 \tag{7.85}$$

$$f(t,0) = 1, \quad t \geq 0 \tag{7.86}$$

Despite not being complete dimensionless, the transformation normalizes the problem, i.e., the main variable becomes dimensionless with dominium between 0 and 1.

7.4.1 Solution by Laplace Transform

As seen in Appendix D, Laplace transform can be applied to partial differential equations.

Owing to the condition given by Equation 7.85, it is convenient to apply the transform regarding variable t, or

$$\psi(s,z) = L\{f(t,z)\} \tag{7.87}$$

The derivatives of f would become

$$L\left\{\frac{\partial f}{\partial t}\right\} = s\psi(s,z) - f(0,z) = s\psi \tag{7.88}$$

and

$$L\left\{\frac{\partial f}{\partial z}\right\} = \frac{d\psi}{dz} \tag{7.89}$$

Notice that the condition given by Equation 7.85 was used to write the last identity of Equation 7.88.

Therefore, Equation 7.84 can be written as

$$s\psi + v\frac{d\psi}{dz} + k\psi = 0 \tag{7.90}$$

which is an ordinary differential equation on variable z. In addition, it is a separable equation with the following solution:

$$\psi = a \exp\left(-\frac{s+k}{v}z\right) \qquad (7.91)$$

Here parameter a is constant and can be determined by the application of the boundary condition given by Equation 7.86, which has not been used yet. However, before that, it should be also transformed to give

$$\psi(s,0) = \frac{1}{s} \qquad (7.92)$$

Using this, Equation 7.91 leads to

$$a = \frac{1}{s} \qquad (7.93)$$

Finally, Equation 7.91 is written as

$$\psi = \frac{1}{s}\exp\left(-\frac{s+k}{v}z\right) = \frac{1}{s}\exp\left(-s\frac{z}{v}\right)\exp\left(-\frac{kz}{v}\right) \qquad (7.94)$$

Observing Equation D.17, one may recognize the t-shift given by the two terms on the extreme right-hand side. Therefore

$$f(t,z) = u\left(t - \frac{z}{v}\right)\exp\left(-\frac{kz}{v}\right) \qquad (7.95)$$

The above equation shows that the concentration (or its dimensionless form f) assumes an exponential decaying behavior for increasing values of the space coordinate z, and this is illustrated in Figure 7.10.

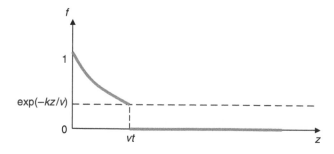

FIGURE 7.10 Dimensionless concentration along a plug-flow reactor.

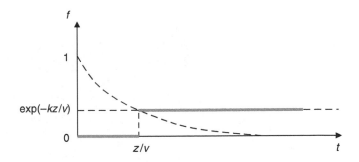

FIGURE 7.11 Dimensionless concentration against time in a plug-flow reactor.

As the component A is injected at position $z = 0$, a wave front with that species advances through the reactor with velocity (v). For positions ahead of the wave front ($z > vt$), the fluid has no concentration of A. Behind this position ($z < vt$), and starting at $z = 0$, the concentration follows a decaying curve.

Conversely, a graph for showing the dependence of function f regarding time is seen at Figure 7.11.

For a better understanding of this behavior, one should imagine an observer measuring the concentration of component A at a given position z of the reactor. It would take time equal to z/v, after the injection of A, for a person to measure an instantaneous jump of its concentration at position z. However, the measured concentrations of A would be smaller than the one at the feeding position ($\tilde{\rho}_{A,0}$) or equal to $\tilde{\rho}_{A,0} \exp\left(-\frac{kz}{v}\right)$. After that instant, the concentration of A at point z would remain constant and equal to the above value.

Figure 7.12 shows the combination of Figures 7.10 and 7.11 for the particular case where $v = 2$ m s^{-1} and $k = 2$ s^{-1}.

The effect of increasing the fluid velocity can be appreciated by Figure 7.13, which was plotted for the particular case where $v = 10$ m s^{-1} and $k = 2$ s^{-1}. It is easy to verify that higher velocities tend to shorten the time at which the wave front with component A reaches a given position z in the reactor.

The effect of increasing the reaction rate can be observed in Figure 7.14, which was drawn for $v = 2$ m s^{-1} and $k = 10$ s^{-1}. A steeper decline of concentration is observed through the length of the reactor when compared with the original Figure 7.12.

7.4.2 COMPLETE DIMENSIONLESS FORM

The convenience of complete dimensionless form can be demonstrated again. For this form, in addition to the concentration of A, dimensionless forms should be found for the independent variables (t and z). This would provide a general representation of the solution, applicable to similar problems using any system of units.

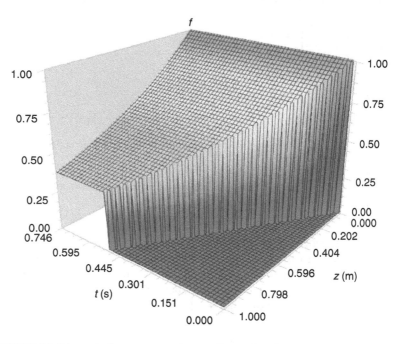

FIGURE 7.12 Dimensionless concentration profiles against time and length in a plug-flow reactor; case when $v = 2$ m s^{-1} and $k = 2$ s^{-1}.

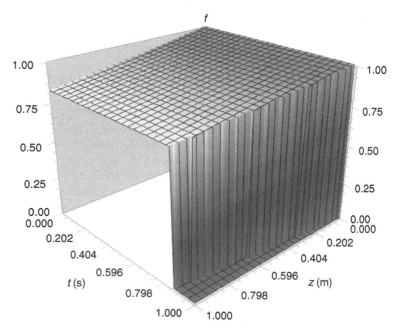

FIGURE 7.13 Dimensionless concentration profiles against time and length in a plug-flow reactor; case when $v = 10$ m s^{-1} and $k = 2$ s^{-1}.

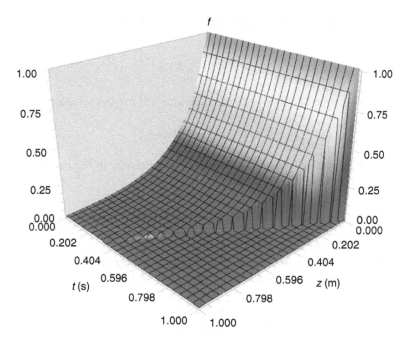

FIGURE 7.14 Dimensionless concentration profiles against time and length in a plug-flow reactor; case when $v = 2$ m s^{-1} and $k = 10$ s^{-1}.

Let the following be the new independent dimensionless variables:

$$\tau = kt \tag{7.96}$$

$$\zeta = \frac{k}{v}z \tag{7.97}$$

Using Equations 7.96 and 7.97 in Equation 7.84 gives the following:

$$\frac{\partial f}{\partial \tau} + \frac{\partial f}{\partial \zeta} + f = 0 \tag{7.98}$$

As seen, the differential equation acquires a simpler form to work with. This even decreases the probability of mistakes during the solution.

The boundary conditions, given by Equations 7.85 and 7.86, become

$$f(0,\zeta) = 0, \quad \zeta \geq 0 \tag{7.99}$$

$$f(\tau,0) = 1, \quad \tau \geq 0 \tag{7.100}$$

Following the similar route of Laplace transform as before, one gets

$$\psi = \frac{1}{s}\exp(-s\zeta)\exp(-\zeta) \qquad (7.101)$$

The solution can be obtained after the inversion and can be written as

$$f(\tau,\zeta) = u(\tau - \zeta)\exp(-\zeta) \qquad (7.102)$$

The graphical representation for this solution is presented in Figure 7.15. One should notice the following:

- The graphic is general for any similar case of reactor. There is no need to specify the fluid velocity (v) or reaction constant (k). The representation is valid for any situation.
- The range of variables τ and ζ have been chosen between 0 and 1. This is not necessary, and the reader might expand or decrease the limits as felt convenient.
- Nonetheless, the unit value for each variable has special significance. For instance, $\tau = 1$ or $t = 1/k$ represents what is called the "reactor time

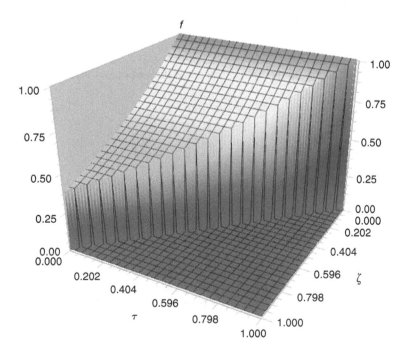

FIGURE 7.15 Dimensionless concentration profiles against dimensionless time and length in a plug-flow reactor.

constant" and is related to the reaction velocity. The higher the velocity, the lower the time constant of a reactor.

- On the spatial point of view, the value $\zeta = 1$ or $z = v/k$ represents the length reached by the reaction wave front after a reactor time constant.

7.4.3 SOLUTION BY METHOD OF WEIGHTED RESIDUES

After the trial in the previous section, the present problem allows a good opportunity to other possibilities of trial functions and would lead to interesting comments regarding the application of MWR.

Using the material introduced in Appendix E, we seek to solve Equation 7.98 under conditions given by Equations 7.99 and 7.100.

There are a great number of possible trial functions. However, one has to bear in mind that integrations will be involved and, therefore, the simplest forms are preferable. Besides, there is no guarantee that complicated trial functions would lead to a quicker solution, or fewer steps to reach an acceptable approximation. In addition, it is very convenient if the trial function implicitly satisfies the boundary conditions. For instance, the form

$$1 - \exp\left(-\frac{\tau}{\zeta}\right)$$

satisfies both conditions given by Equations 7.99 and 7.100. Keeping this in mind, let us try the following trial function:

$$f_n = 1 - \exp\left(-\frac{\tau}{\zeta}\right) + \sum_{j=1}^{n} \tau^j C_j(\zeta) \qquad (7.103)$$

Of course, any approximation f_n satisfies conditions given by Equations 7.99 and 7.100, as long as $C_j(0) = 0$. Functions $C_j(\zeta)$ would be set at each level of approximation, as follows.

Using the above into Equation 7.98, the residue, at level n, is written as

$$\Lambda_n = 1 - \frac{1}{\zeta^2} \exp\left(-\frac{\tau}{\zeta}\right) + \sum_{j=1}^{n} \tau^{j-1} \left[jC_j(\zeta) + \tau C_j(\zeta) + \tau \frac{dC_j}{d\zeta} \right] \qquad (7.104)$$

An approximated solution would be sought within the following dominium:

$$0 \leq t \leq 1 \quad \text{and} \quad 0 \leq z \leq 1 \qquad (7.105)$$

However, approximations could be imagined for other ranges, including wider ones. For a while, the present is convenient because it covers an important region of this process, as shown in Figure 7.15.

7.4.3.1 First Approximation

The first approximation would lead to the following residue:

$$\Lambda_1 = 1 - \frac{1}{\zeta^2}\exp\left(-\frac{\tau}{\zeta}\right) + C_1(\zeta) + \tau C_1(\zeta) + \tau\frac{dC_1}{d\zeta} \qquad (7.106)$$

Due to the complexity of this function, let us apply the simplest method, namely collocation. According to Appendix E, the residue should be equated to zero at a chosen point in the dominium of variable τ. One simple possibility would be to choose the middle value, or $\tau = 1/2$. Hence, the following differential equation would result:

$$2 - \frac{2}{\zeta^2}\exp\left(-\frac{1}{2\zeta}\right) + 3C_1(\zeta) + \frac{dC_1}{d\zeta} = 0 \qquad (7.107)$$

The condition $C_1(0) = 0$, which was commented before, should also be satisfied.

As discussed in Appendix B, Equation 7.107 can be solved by variation of parameters. By this method, function $C_1(\zeta)$ is written as the product of two other functions, or

$$C_1(\zeta) = u(\zeta)v(\zeta) \qquad (7.108)$$

Using Equation 7.108 into Equation 7.107, the following is obtained:

$$u(v' + 3v) + u'v + g(\zeta) = 0 \qquad (7.109)$$

where

$$g(\zeta) = 2 - \frac{2}{\zeta^2}\exp\left(-\frac{1}{2\zeta}\right) \qquad (7.110)$$

Without loss of generality, the following can be established from Equation 7.109:

$$v' + 3v = 0 \qquad (7.111)$$

This results into

$$v = e^{-3\zeta} \qquad (7.112)$$

Therefore, Equation 7.109 would lead to

$$u = -\int_0^\zeta e^{3x} g(x)\,dx \qquad (7.113)$$

The definite integral was applied here because the solution

$$C_1(\zeta) = -e^{3\zeta} \int\limits_0^\zeta e^{3x} g(x)\, dx \tag{7.114}$$

would always satisfy the condition $C_1(0) = 0$.

After this, the first approximation becomes

$$f_1(\tau,\zeta) = 1 - \exp\left(-\frac{\tau}{\zeta}\right) - 2\tau e^{3\zeta} \int\limits_0^\zeta e^{3x}\left[1 - \frac{1}{x^2}e^{-\frac{1}{2x}}\right] dx \tag{7.115}$$

Of course, the complexity would increase for higher approximations. Therefore, it is clear that one should be very careful in applying MWR for cases of discontinuous behavior, such as the present case of function $f(\tau,\zeta)$.

EXERCISES

1. Solve the differential equation given in Equation 7.35 and apply the boundary condition given by Equation 7.36 to arrive at Equation 7.38.

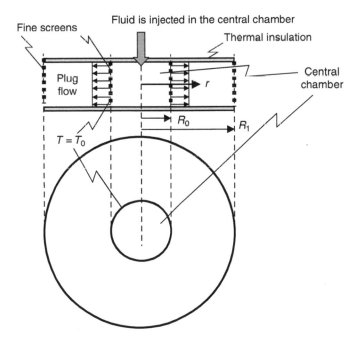

FIGURE 7.16 Heating of a fluid injected between two disks.

2. Solve the following boundary value problems by any method:

(a) $\dfrac{\partial y}{\partial x} + \dfrac{\partial y}{\partial t} = 0;$ $y(0,x) = 0,$ $y(t,0) = 1,$ $x \geq 0,$ and $t \geq 0$

(b) $\dfrac{\partial y}{\partial x} + x\dfrac{\partial y}{\partial t} = 0;$ $y(0,x) = 1,$ $y(t,0) = 2,$ $x \geq 0,$ and $t \geq 0$

(c) $t\dfrac{\partial y}{\partial x} + \dfrac{\partial y}{\partial t} = 0;$ $y(0,x) = 0,$ $y(t,0) = 1,$ $x \geq 0,$ and $t \geq 0$

3. Solve the problem presented in Section 7.4 by Laplace transform in the space (z) variable rather than time.

4. Repeat the problem presented in Section 7.3 for a plug flow inside the space between two circular disks, as shown by Figure 7.16. Therefore, determine the temperature profile as a function of radius (r) and time (t) in the region $R_0 \leq r \leq R_1$. Assume that the temperature in the fluid is T_b and at instant $t = 0$ it becomes T_0 at the injecting screen or at $r = R_0$.

5. Show the impossibility of solving Equation 7.98 using the method of similarity described in Appendix F.

8 Problems 212; Two Variables, 1st Order, 2nd Kind Boundary Condition

8.1 INTRODUCTION

This chapter presents methods to solve problems with two independent variables involving first-order differential equation and second-kind boundary condition. Therefore, the problems fall within the category of partial differential equations. Mathematically, this class of cases can be summarized as $f\left(\phi, \omega_1, \omega_2, \frac{\partial \phi}{\partial \omega_1}, \frac{\partial \phi}{\partial \omega_2}\right)$, second-kind boundary condition.

8.2 HEATING OF FLOWING LIQUID

Consider a similar heating method for a flowing liquid as explained in Section 7.3. However, instead of heating the fluid instantaneously at temperature T_0, the electrical resistance at the grid delivers a constant heat flux (Figure 8.1). This is a much more realistic and feasible condition. Actually, mathematically speaking, this is equivalent to imposing a derivative of temperature in the fluid at the grid or position $z = 0$.

With exception of assumption 8, the simplifying assumptions are the same as adopted in Section 7.3. To avoid asking the reader to leave the present page as well as for the sake of clarity, let us list all the assumptions here as follows:

1. Plug-flow regime.
2. Grid does not interfere in the flow. In fact, grids are often employed to provide a flat velocity profile for the flow.
3. Tube is very long and all possible border (inlet or outlet) effects are negligible, at least in the region of interest.
4. Tube is perfectly insulated.
5. Liquid is Newtonian and its properties are approximately constant.
6. No energy dissipation due to friction is observed.

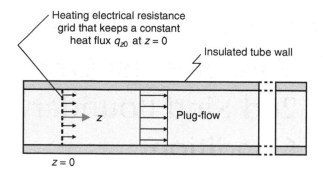

FIGURE 8.1 Scheme of a system delivering a constant heat flux to a flowing liquid.

7. Before the heating system is turned on, the fluid is at uniform temperature T_b.
8. Once the heating starts working, the grid delivers a constant flux of
   energy (q_{z0}) at $z = 0$. Again, this is a more realistic situation than the
   imposed situation at Section 7.3 because the power delivered to an
   electrical resistance can be controlled.
9. Rate of heat transfer by convection is much higher than the rate by
   conduction. This is reasonable for cases where the velocity of the fluid is
   relatively high.

Using Equation A.35 and the above assumptions, it is possible to arrive at

$$\frac{\partial T}{\partial t} + v \frac{\partial T}{\partial z} = 0 \tag{8.1}$$

Again, the axial velocity component (v_z) is written just as v.

It should be noticed that no source term (R_Q) is added here. The electrical
heated grid inputs a localized source of energy and this would be possible to
consider as a boundary condition, as shown below.

The boundary conditions for the problem are

$$T(0,z) = T_b, \quad z \geq 0 \tag{8.2}$$

and

$$\left. \frac{\partial T}{\partial z} \right|_{z=0} = -\frac{q_{z0}}{\lambda}, \quad t > 0 \tag{8.3}$$

Notice that the flux q_{z0} is a given constant here, which can be measured by the
power consumed by the electrical resistance.

As seen, Equation 8.3 characterizes a second-kind boundary condition.

According to the explanation given in Section 7.3, it is convenient to work with dimensionless variables, which are given by

$$\psi = \frac{T - T_b}{T_b} \tag{8.4}$$

With this, Equation 8.2 becomes

$$\frac{\partial \psi}{\partial t} + v \frac{\partial \psi}{\partial z} = 0 \tag{8.5}$$

The boundary conditions are written as

$$\psi(0,z) = 0, \quad z \geq 0 \tag{8.6}$$

$$\left. \frac{\partial \psi}{\partial z} \right|_{z=0} = -\frac{q_{z0}}{\lambda T_b} = -a, \quad t > 0 \tag{8.7}$$

Here, parameter a is assumed as constant.

8.2.1 SOLUTION BY LAPLACE TRANSFORM

The same reasoning as in Section 7.3 leads to the following transform:

$$\Psi(s,z) = L\{\psi(t,z)\} \tag{8.8}$$

After the application of Equation 8.8, the transform of Equation 8.5 is

$$s\Psi - \psi(0,z) + v \frac{d\Psi}{dz} = 0 \tag{8.9}$$

Using the condition given by Equation 8.6, Equation 8.9 can be written as

$$\frac{d\Psi}{dz} = -\frac{s}{v} \Psi \tag{8.10}$$

As seen, this is a separable ordinary differential equation, and the integration leads to

$$\ln \Psi = -\frac{s}{v} z + \ln C_1 \tag{8.11}$$

or

$$\Psi = C_1 \exp\left(-\frac{s}{v} z\right) \tag{8.12}$$

The Laplace transform of the boundary condition given by Equation 8.7 is

$$\frac{d\Psi}{dz}\bigg|_{z=0} = -\frac{a}{s} \tag{8.13}$$

From Equation 8.12, one gets

$$\Psi = \frac{av}{s^2} \exp\left(-\frac{s}{v}z\right) \tag{8.14}$$

Again, a t-shift (see Appendix D) is involved and the inversion of the above equation gives

$$\psi(t,z) = av\left(t - \frac{z}{v}\right)u\left(t - \frac{z}{v}\right) \tag{8.15}$$

The dimensionless temperature profile against time and position is presented at Figure 8.2 for the case where $v = 2$ m s^{-1} and $a = 1$ m^{-1}.

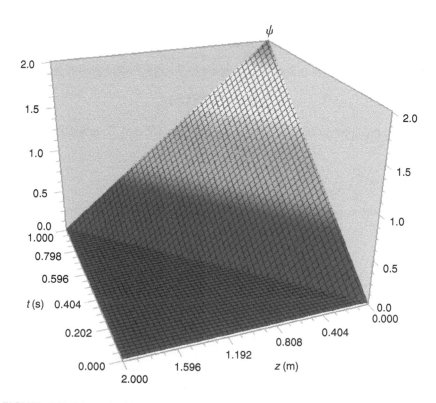

FIGURE 8.2 Dimensionless temperature (ψ) profile against time and position ($v = 2$ m s^{-1}, $a = 1$ m^{-1}).

Notice that the wave front reaches a position z at instant z/v. Take, for example, the position 1.0 m. As $v = 2$ m s^{-1}, the wave would reach it at 0.5 s. Before this, the temperature is unaffected, and after 0.5 s it increases steadily. Of course, there will be a limit regarding the temperature when either the fluid reaches boiling conditions or the resistance material melts.

Reversibly, for a given instant, say 0.5 s, the dimensionless temperature at the grid (or $z = 0$) is 1.0 and decreases for positions above this. Of course, the temperature does not vary for position where the wave front has not yet reached, or $z = 1.0$.

The heating rate imposed by the electrical resistances at the grid is governed by parameter a given by Equation 8.7. For instance, if the heating rate doubles, the slope of the plane shown in Figure 8.2 would be twice as steep. This is shown by Figure 8.3.

8.2.2 COMPLETE DIMENSIONLESS FORM

The advantages of working with dimensionless variables have already been described. For this, one might observe that the dimension of parameter a is the

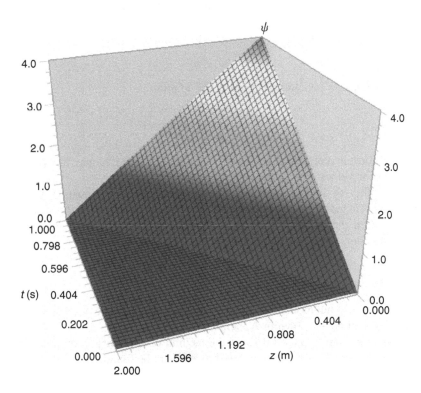

FIGURE 8.3 Dimensionless temperature (ψ) profile against time and position ($v = 2$ m s^{-1}, $a = 2$ m^{-1}).

inverse of length or m^{-1} in SI system. Therefore, a possible set of dimensionless variables might be

$$\tau = avt \tag{8.16}$$

and

$$\zeta = az \tag{8.17}$$

Using Equations 8.16 and 8.17, Equation 8.5 can be written as

$$\frac{\partial \psi}{\partial \tau} + \frac{\partial \psi}{\partial \zeta} = 0 \tag{8.18}$$

The boundary conditions given by Equations 8.6 and 8.7 would be transformed into

$$\psi(0,\zeta) = 0, \quad \zeta \geq 0 \tag{8.19}$$

$$\left. \frac{\partial \psi}{\partial \zeta} \right|_{\zeta=0} = -1, \quad \tau > 0 \tag{8.20}$$

Using Laplace transform as before, the solution is

$$\psi(\tau,\zeta) = (\tau - \zeta)\, u(\tau - \zeta) \tag{8.21}$$

This new form is represented by Figure 8.4, which does not require any particular definition for velocity (v) or the heating parameter (a).

The effect of higher velocities can be observed from Equation 8.16 and Figure 8.4. If, for instance, at given instant (t) and parameter a the velocity is doubled, then the value of τ doubles. Therefore, the temperature wave would reach a position twice as far than before.

8.2.3 COMMENTS ON SIMILARITY

This is an opportune moment to make some comments on the application of the method of similarity, shown in Appendix F.

Deviating from Equation 8.18, if the method of similarity is applied (details are left as exercise) the following new variable can be found:

$$\omega = \frac{\tau}{\zeta} \tag{8.22}$$

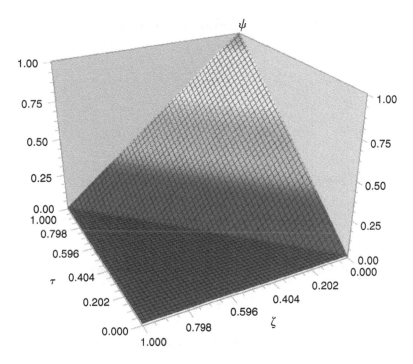

FIGURE 8.4 Dimensionless temperature (ψ) profile against dimensionless time (τ) and position (ζ).

Therefore,

$$\frac{\partial \psi}{\partial \tau} = \frac{\mathrm{d}\psi}{\mathrm{d}\omega} \frac{\partial \omega}{\partial \tau} = \frac{1}{\zeta} \frac{\mathrm{d}\psi}{\mathrm{d}\omega} \tag{8.23}$$

$$\frac{\partial \psi}{\partial \zeta} = \frac{\mathrm{d}\psi}{\mathrm{d}\omega} \frac{\partial \omega}{\partial \zeta} = -\frac{\tau}{\zeta^2} \frac{\mathrm{d}\psi}{\mathrm{d}\omega} = -\frac{\omega}{\zeta} \frac{\mathrm{d}\psi}{\mathrm{d}\omega} \tag{8.24}$$

With this, Equation 8.18 can be written as

$$(1 - \omega)\frac{\mathrm{d}\psi}{\mathrm{d}\omega} = 0 \tag{8.25}$$

Boundary conditions given by Equations 8.19 and 8.20 become

$$\psi(\omega = 0) = 0 \tag{8.26}$$

$$\frac{\mathrm{d}\psi}{\mathrm{d}\omega}\bigg|_{\omega \to \infty} = 0 \tag{8.27}$$

On the other hand, Equation 8.25 provides the following:

$$\omega = 1 \tag{8.28}$$

or

$$\frac{d\psi}{d\omega} = 0 \tag{8.29}$$

Equation 8.29 leads to

$$\psi = C \tag{8.30}$$

where C is a constant.

One should notice that Equation 8.28 defines a line where

$$\tau = \zeta \tag{8.31}$$

Equation 8.30 satisfies Equation 8.27. It would also satisfy Equation 8.26 if $C=0$.

Thus, the two possibilities given by Equations 8.30 and 8.31 are coherent with the solution found before. This can be seen by inspecting Figure 8.4 where the line given by Equation 8.31 defines the boundary between two distinct behaviors of function ψ: a constant value and an inclined plane.

Despite the above interesting details, the method of similarity could not provide a true general solution for this case because of the following causes:

1. Equation 8.25 is not a complete or self-contained description of the dependence among the variables involved in the problem, such as achieved by Equation 8.21.
2. Boundary conditions given by Equations 8.19 and 8.20 could not be condensed into a single condition involving just the dependent variable (ψ) and the new independent one (ω).

8.3 PLUG-FLOW REACTOR

The problem set at Section 7.4 is presented again, however, with a different condition at the grid, as shown in Figure 8.5.

Such a grid is formed by a network of small-diameter tubes with porous walls. The component A is injected into the porous tubes and then into the main reactor stream. Similar to the situation in Section 7.4, the concentration of A is zero for the entire reactor before the injection starts. The only difference between both situations is that now the injection is such that the diffusion transfer dictates the

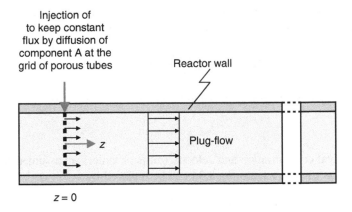

Injection of
to keep constant
flux by diffusion of
component A at the
grid of porous tubes

Reactor wall

z

Plug-flow

$z = 0$

FIGURE 8.5 Injection of component A that keeps its flux constant by diffusion at the grid of porous tubes.

mass transfer of component A into the main stream. Details and consequences of this are described and discussed ahead.

The assumptions made in Section 7.4 continue to be valid, except the first one, which is now written as follows:

Pure component B is flowing through the reactor. At a given instance ($t = 0$), reactant A is injected into the reactor at position $z = 0$. The injection is made through a network of porous tubes, which constitute the grid. The mass transfer of component A from the porous tube surfaces into the reactor main stream is dictated by diffusion. Therefore, the amount of species A injected into the grid tubes is just to make up for the amount diffused through the pores. Hence, the concentration of A at $z = 0$ is not constant, but just the diffusion rate of A transfer at that position. This is a more realistic situation than the assumed in Section 7.4 and no instantaneous jump of concentration is imposed.

Since the same considerations regarding the flow and kinetics of Section 7.4 are maintained, the problem is governed by a similar equation as Equation 7.79, or

$$\frac{\partial \tilde{\rho}_A}{\partial t} + v \frac{\partial \tilde{\rho}_A}{\partial z} + k \tilde{\rho}_A = 0 \tag{8.32}$$

As before, the velocity component v_z was replaced just by v. Notice that the typical form of first-order reaction rate is already included.

The following are the boundary conditions for the problem:

$$\tilde{\rho}_A(0,z) = 0, \quad z \geq 0 \tag{8.33}$$

and

$$\tilde{N}_A(t,0) = \tilde{N}_{A,0}, \quad t \geq 0 \tag{8.34}$$

Here $\tilde{N}_{A,0}$ is a constant. The above can be written in terms of concentration. For that, Equation A.54 provides the flux for a single direction (z) as

$$\tilde{N}_A = -D_{AB}\tilde{\rho}\frac{\partial x_A}{\partial z} + x_A(\tilde{N}_A + \tilde{N}_B) \qquad (8.35)$$

From Equation A.56 it is possible to obtain

$$\tilde{N}_A + \tilde{N}_B = \tilde{\rho}\tilde{v} \qquad (8.36)$$

Since the total concentration and velocity (mass or molar) are assumed constants, the condition given by Equation 8.34 can be replaced by

$$\left(-D_{AB}\frac{\partial \tilde{\rho}_A}{\partial z} + \tilde{\rho}_A\tilde{v}\right)\Bigg|_{z=0} = \tilde{N}_{A0}, \quad t > 0 \qquad (8.37)$$

However, according to the first assumption mentioned above, the convective part of the flux is assumed much smaller than the diffusion contribution and the total flux is given just by

$$\left(-D_{AB}\frac{\partial \tilde{\rho}_A}{\partial z}\right)\Bigg|_{z=0} = \tilde{N}_{A0}, \quad t > 0 \qquad (8.38)$$

Following the variables given in Section 7.4, convenient changes of variables are applied to simplify the handling of equations as well to generalize solutions. For that, let the new variables be

$$\psi(t,z) = \frac{\tilde{\rho}_A(t,z)}{\tilde{N}_{A0}}v \qquad (8.39)$$

$$\tau = kt \qquad (8.40)$$

$$\zeta = \frac{k}{v}z \qquad (8.41)$$

Using Equations 8.39 through 8.41, the original equation, Equation 8.32, becomes

$$\frac{\partial \psi}{\partial \tau} + \frac{\partial \psi}{\partial \zeta} + \psi = 0 \qquad (8.42)$$

Boundary conditions given by Equations 8.33 and 8.38 become

$$\psi(0,\zeta) = 0, \quad \zeta \geq 0 \qquad (8.43)$$

$$\left(-a\frac{\partial \psi}{\partial \zeta}\right)\Bigg|_{\zeta=0} = 1, \quad \tau > 0 \qquad (8.44)$$

Here

$$a = D_{AB}\frac{k}{v^2} \tag{8.45}$$

8.3.1 SOLUTION BY LAPLACE TRANSFORM

The method of Laplace transform might be used here as well. For that, the transform regarding the dimensionless time is applied, or

$$\Psi(s,\zeta) = L\{\psi(\tau,\zeta)\} \tag{8.46}$$

From Equation 8.46 and the condition given by Equation 8.43, the transform of Equation 8.42 becomes

$$s\Psi + \frac{d\Psi}{d\zeta} + \Psi = 0 \tag{8.47}$$

This is a separable equation with the following solution:

$$\Psi = C\exp[-(s+1)\zeta] \tag{8.48}$$

The transform of the boundary condition given by Equation 8.44 is

$$\left(-a\frac{d\Psi}{d\zeta}\right)\Big|_{\zeta-0} = \frac{1}{s} \tag{8.49}$$

Using Equation 8.49 in Equation 8.48, one obtains

$$\Psi = \frac{\exp[-(s+1)\zeta]}{as(s+1)} \tag{8.50}$$

It should be noticed that (see tables at Appendix D)

$$L^{-1}\left\{\frac{1}{s(s+1)}\right\} = 1 - e^{-\tau} \tag{8.51}$$

Using the time-shift property (Appendix D), the final solution becomes

$$\psi(\tau,\zeta) = \frac{1}{a}e^{-\zeta}(1 - e^{-\tau})\,u(\tau - \zeta) \tag{8.52}$$

The graph representing this relation for the case where $a = 1$ is shown in Figure 8.6, and for the case where $a = 2$ in Figure 8.7.

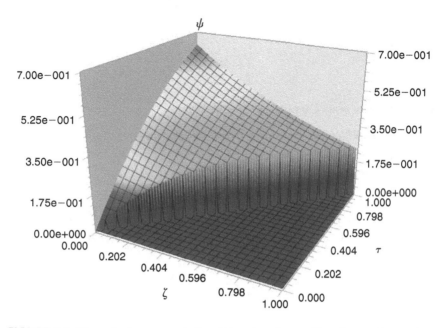

FIGURE 8.6 Dimensionless concentration (ψ) against time and length; case for $a = 1$.

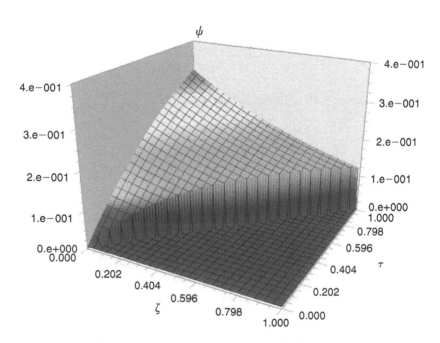

FIGURE 8.7 Dimensionless concentration (ψ) against time and length; case for $a = 2$.

The following are worth noticing from those graphs:

1. At a given instant (or its dimensionless form τ), the concentration profile (ψ) is described by a decreasing exponential regarding the position (ζ). Therefore, at the reactor entrance ($\zeta = 0$) the concentration is maximum. Of course, the decrease is because of the reaction with component B. However, owing to plug-flow regime, there will be positions without reactant A. This explains the abrupt decrease to zero concentration. Again, a wave front is established.

2. For a given position (ζ) in the reactor, the concentration would be zero until reached by the wave front. After a sudden increase, the concentration keeps growing due to continuous injection of component A at reactor entrance ($\zeta = 0$).

3. Graphs show that the derivative of component A concentration is constant at the reactor entrance ($\zeta = 0$ or $z = 0$). This was imposed by the condition given by Equation 8.44. Because of the reaction, the concentration of A at $z = 0$ should be constantly increased to maintain the positive derivative at the grid.

4. There is an asymptotic behavior of concentration against time. In other words, the process tends to steady-state regime.

5. Notice that the edge of advancing wave front occurs at the line given by $\zeta = \tau$. Now, consider a point on the edge, for instance $\zeta = \tau = 0.5$. In addition, let us assume a velocity equal to 1 m s and k equal to 1 s^{-1}. Therefore, the point would be at 0.5 m from the entrance and 0.5 s after the injection of reacting component A. Given their definitions by Equations 8.40 and 8.41, if the reaction parameter (k) is doubled ($k = 2$ s^{-1}), and we keep the same fluid velocity and the same values of ζ and τ, the real position and time of the wave-front edge would be halved, or $z = 0.25$ m and $t = 0.25$ s. In other words, if the reaction rate is increased, the wave front retracts. This is the expected behavior as faster reaction takes shorter distance or less time to completely consume reactant A.

EXERCISES

1. Solve the following boundary value problems by any method:

   (a) $\dfrac{\partial y}{\partial x} + \dfrac{\partial y}{\partial t} = 0$; $y'(0,x) = 0$, $y(t,0) = 1$, $x \geq 0$, and $t \geq 0$

   (b) $\dfrac{\partial y}{\partial x} + x\dfrac{\partial y}{\partial t} = 0$; $y'(0,x) = 1$, $y(t,0) = 2$, $x \geq 0$, and $t \geq 0$

   (c) $t\dfrac{\partial y}{\partial x} + \dfrac{\partial y}{\partial t} = 0$; $y'(0,x) = 0$, $y(t,0) = 1$, $x \geq 0$, and $t \geq 0$

2. From the solution presented in Section 8.2, deduce the heat flux delivered by the electrical resistance to the fluid stream.

3. Solve the problem presented in Section 8.3 in the case of a second-order irreversible reaction in which the reaction rate can be written as $k\tilde{\rho}_A^2$. Use any method, including approximate ones if necessary.

9 Problems 213; Two Variables, 1st Order, 3rd Kind Boundary Condition

9.1 INTRODUCTION

This chapter presents methods to solve problems with two independent variables involving first-order differential equations and third-kind boundary conditions. Therefore, the problems fall in the category of partial differential equations. Mathematically, this class of problems can be summarized as $f\left(\phi, \omega_1, \omega_2, \frac{\partial \phi}{\partial \omega_1}, \frac{\partial \phi}{\partial \omega_2}\right)$, third-kind boundary condition.

9.2 HEATING OF FLOWING LIQUID

Consider a heating system similar to the system studied in Sections 7.3 and 8.2. However, now the situation is even more realistic because the heat flux delivered by the electrical resistance is given by the convective heat transfer between the grid and the passing fluid. This is illustrated in Figure 9.1.

With the exception of assumption 8, the simplifying assumptions are the same as those adopted in Section 7.3. Again, to avoid asking the reader to leave the present page as well as for the sake of clarity, let us list all the assumptions here as follows:

1. Velocity profile is approximately flat. In other words, a plug-flow regime is assumed. Obviously, this is an approximation, however, useful for a first attack to the problem.
2. Grid does not interfere in the flow. Actually, grids are often employed to provide a flat velocity profile for the flow.
3. Tube is very long so that no border (inlet or outlet) effects are to be considered, at least in the region of interest.
4. Tube is perfectly insulated.

235

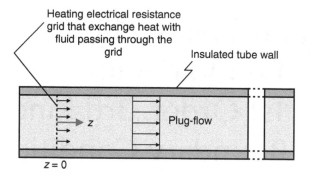

FIGURE 9.1 Scheme of heating a flowing liquid; combination heat transfer by conduction and convection at the grid.

5. Liquid is a Newtonian fluid and its properties (for instance, density) can be taken as constants. This is assumed despite the heating, or temperature variations. Of course, it might be a strong approximation if large variations of temperature are to be expected. This assumption would be invalid in case of gases.
6. No energy dissipation due to friction is observed.
7. Before the heating system is turned on, the fluid is at uniform temperature T_b.
8. Once the heating starts working, the grid delivers a flux of energy at $z = 0$, which depends on the convective heat transfer between the grid and the fluid. To simplify the problem, it is also assumed that the grid is kept at a constant temperature T_g. The solution would provide the heat flux to keep this temperature at the surface of the grid wiring. This is a more realistic and feasible situation than that imposed in Section 7.3 or Section 8.2 because the heat flux is not constant but depends on the conditions to allow the energy transfer between wire and fluid.
9. Rate of heat transfer by convection is much higher than the rate of heat transfer by conduction. This is reasonable for cases where the velocity of the fluid is relatively high.

Again, applying Equation A.35 and using the above assumptions, it is possible to write

$$\frac{\partial T}{\partial t} + v\frac{\partial T}{\partial z} = 0 \tag{9.1}$$

Here the axial velocity component (v_z) is written as just v.
   The boundary conditions for the problem are

$$T(0,z) = T_b, \quad z \geq 0 \tag{9.2}$$

and

$$q_z|_{z=0} = -\lambda \left.\frac{\partial T}{\partial z}\right|_{z=0} = \alpha[T_g - T(t,0)], \quad t > 0 \qquad (9.3)$$

Thus, this is a third-kind boundary condition, and one should be careful to ensure its coherence. As seen, the fluid is heated when in contact with the grid; however, once passing through the grid (or $z = 0$) its temperature should decrease. The heat flux occurs in the positive z direction, therefore the derivative of fluid temperature is negative leading to a positive term related to conduction. Consequently, the term related to convection should be positive here as well since $T(t,0)$ is smaller than T_g.

Let us now work with dimensionless variables. Bearing in mind that the ratio α/λ has dimensions of inverse length, the following changes would be possible choices:

$$\psi = \frac{T - T_b}{T_g - T_b}, \qquad (9.4)$$

$$\tau = avt, \qquad (9.5)$$

and

$$\zeta = az \qquad (9.6)$$

Here

$$a = \frac{\alpha}{\lambda} \qquad (9.7)$$

Using Equations 9.4 through 9.6, Equation 9.1 becomes

$$\frac{\partial \psi}{\partial \tau} + \frac{\partial \psi}{\partial \zeta} = 0 \qquad (9.8)$$

Notice that the dimensionless temperature adopted here differs from those used in previous cases. This alternative provides simple forms for boundary conditions given by Equations 9.2 and 9.3, as shown below:

$$\psi(0,\zeta) = 0, \quad \zeta \geq 0 \qquad (9.9)$$

$$\left.\frac{\partial \psi}{\partial \zeta}\right|_{\zeta=0} = \psi(\tau,0) - 1, \quad \tau > 0 \qquad (9.10)$$

Obviously, several other choices for the dimensionless variable are possible. The guideline is always to arrive at the simplest forms of boundary conditions.

Whenever possible, one should try to arrive at an initial condition that sets value zero for the independent variable. This usually leads to simpler differential equations, particularly after the application of Laplace transform, as demonstrated ahead.

9.2.1 Solution by Laplace Transform

Consider the transform

$$\varphi(s,\zeta) = L\{\psi(t,\zeta)\} \tag{9.11}$$

According to Appendix D, Equation 9.8 leads to

$$s\varphi - \psi(0,\zeta) + \frac{d\varphi}{d\zeta} = 0 \tag{9.12}$$

Using the condition given by Equation 9.9, Equation 9.12 can be written as

$$\frac{d\varphi}{d\zeta} = -s\varphi \tag{9.13}$$

As seen, the zero initial condition simplifies the differential equation. Actually, this is a separable equation, the solution of which is

$$\ln \varphi = -s\zeta + \ln C \tag{9.14}$$

or

$$\varphi = C \exp(-s\zeta) \tag{9.15}$$

The Laplace transform of the boundary condition given by Equation 9.10 is

$$\frac{d\varphi}{d\zeta}\bigg|_{\zeta=0} = \varphi(s,0) - \frac{1}{s} \tag{9.16}$$

Using this in Equation 9.15, one gets

$$\varphi = \frac{1}{s(1+s)} \exp(-s\zeta) \tag{9.17}$$

The inversion starts with the term multiplying the exponential (in Equation 9.17), or

$$L^{-1}\left\{\frac{1}{s(1+s)}\right\} = 1 - e^{-\tau} \tag{9.18}$$

Using the t-shift property (see Appendix D) the complete inversion is

$$\psi(\tau,\zeta) = [1 - e^{-(\tau - \zeta)}]\, u(\tau - \zeta) \qquad (9.19)$$

Figure 9.2 illustrates the above solution.

If this graph is compared with Figures 7.8 and 8.4, it becomes clear how more realistic this model is in relation to the former approaches. It shows that fluid temperature tends to the grid temperature, or $\psi = 1$. In addition, the derivative of temperature at the grid ($\zeta = 0$) is not constant, but decreases when the fluid approaches this limiting temperature.

Despite this, neglecting the conduction term when writing Equation 9.1 leads to the possibility of sudden change of temperature at the points reached by the heating front-wave. This is not possible because no natural process allows instantaneous change of property values or even their derivatives. If the conduction term, or second derivative in the right-hand side of Equation 7.15, were included, a smoother increase of temperature would be obtained. This is much more in accordance with reality.

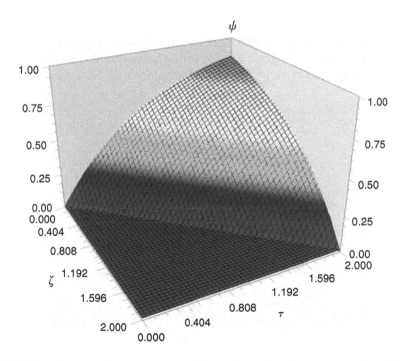

FIGURE 9.2 Dimensionless temperature (ψ) profile against dimensionless time (τ) and position (ζ).

9.3 DYNAMIC PLUG-FLOW REACTOR

The problem set at Sections 7.4 and 8.3 is posed again, however, with a different condition at the grid, as shown in Figure 9.3.

Such a grid is formed by a network of small-diameter tubes with porous walls. The component A is injected into the porous tubes and then into the main reactor stream. Similar to the situation in Sections 7.4 and 8.3, the concentration of A is zero for the entire reactor before the injection starts. However now, the injection is such that the proper combination of diffusion and convection dictates the mass transfer of component A into the main stream.

Again, it is desired to determine the component A concentration profile and consumption rate.

Despite the similarity of most assumptions as before, for the sake of clarity it is interesting to repeat them as follows:

1. Pure component B is flowing through the reactor. At a given instance ($t=0$), reactant A is injected into the reactor at position $z=0$. The injection is made through a network of porous tubes, which constitutes the grid. The mass transfer of component A from the porous tube surfaces into the reactor main stream is dictated by the correct combination of diffusion and convection. This is a more realistic situation than that assumed in Section 7.4, because no instantaneous jump of concentration is imposed. It is also more truthful than the picture at Section 8.3 since the rate is not given only by the diffusion contribution.

2. Reaction between A and B is represented by a first-order irreversible reaction, or

$$A + B \rightarrow C$$

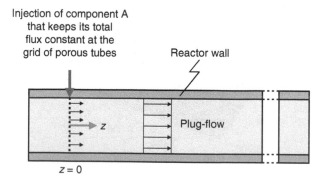

FIGURE 9.3 Plug-flow reactor where the rate of component A injection is dictated by its diffusion-convective combination mass transfer at $z = 0$.

3. All involved chemical components are in the liquid phase and their mixtures at any proportion occur in a single phase. In addition, the liquids are completely miscible, i.e., do not naturally separate from each other such as oil and water.
4. No mixing enthalpy change is involved and the occurring reaction is neither exothermic nor endothermic and the conditions are such that the isothermal reactor can be ensured. Of course, it is very difficult to find such a situation. However, this approximation is valid if the reaction is slightly endothermic or exothermic and the heat transfer between the reactor and environment is such that constant temperature throughout the reactor length can be achieved.
5. No matter what the composition of the fluid, it can be considered Newtonian with constant density and viscosity.
6. Diffusivity of component A into B or vice versa is constant. It also remains constant, no matter how far the reaction between them progress. Product C does not interfere in this diffusivity.
7. Plug-flow regime. Of course, this is a strong approximation and can only be assumed as a first approach to the problem of tubular reactor.
8. No swirl or angular velocity component is verified in the reactor.
9. Velocity of fluids in the reactor is high enough to allow neglecting the mass diffusion transport when compared with the convective one. This assumption will become clear below.

Since the same considerations regarding the flow and kinetics of Section 7.4 are maintained, particularly that the total convection is much higher than the transport by diffusion, the differential equation governing the problem is given by a similar equation as Equation 7.79:

$$\frac{\partial \tilde{\rho}_A}{\partial t} + v\frac{\partial \tilde{\rho}_A}{\partial z} + k\tilde{\rho}_A = 0 \qquad (9.20)$$

As before, the velocity component v_z was replaced by just v.

The boundary conditions for this case are

$$\tilde{\rho}_A(0,z) = 0, \quad z \geq 0 \qquad (9.21)$$

and

$$\tilde{N}_A(t,0) = \tilde{N}_{A,0}, \quad t \geq 0 \qquad (9.22)$$

The last condition can be written in terms of concentration. Using Equation A.54 in a single direction (z)

$$\tilde{N}_A = -D_{AB}\tilde{\rho}\frac{\partial x_A}{\partial z} + x_A(\tilde{N}_A + \tilde{N}_B) \qquad (9.23)$$

From Equation A.56

$$\tilde{N}_A + \tilde{N}_B = \bar{\rho}\tilde{v} \tag{9.24}$$

As the total concentration and velocity (mass or molar) are assumed constant, the condition given by Equation 9.22 can be replaced by

$$\left(-D_{AB}\frac{\partial \tilde{\rho}_A}{\partial z} + \tilde{\rho}_A\tilde{v}\right)\Bigg|_{z=0} = \tilde{N}_{A0}, \quad t > 0 \tag{9.25}$$

The molar flux at the grid, on the right-hand side of the equation, is constant. Notice that the diffusivity coefficient is also assumed constant. The same treatment can be made using mass-based variables.

Remembering the convenience of dimensional variables, the new variables are defined below:

$$\psi(t,z) = \frac{\tilde{\rho}_A(t,z)}{\tilde{N}_{A0}}v \tag{9.26}$$

$$\tau = kt \tag{9.27}$$

$$\zeta = \frac{k}{v}z \tag{9.28}$$

Using Equations 9.26 through 9.28, the original equation (Equation 9.20) can be written as

$$\frac{\partial \psi}{\partial \tau} + \frac{\partial \psi}{\partial \zeta} + \psi = 0 \tag{9.29}$$

The boundary condition given by Equation 9.21 becomes

$$\psi(0,\zeta) = 0, \quad \zeta \geq 0 \tag{9.30}$$

The boundary condition given by Equation 9.22 or Equation 9.25 can be represented by

$$\left(-a\frac{\partial \psi}{\partial \zeta} + \psi\right)\Bigg|_{\zeta=0} = 1, \quad \tau > 0 \tag{9.31}$$

Here

$$a = D_{AB}\frac{k}{v^2} \tag{9.32}$$

In addition, in writing Equation 9.31, the following was used:

$$\frac{\tilde{v}}{v} = 1 \tag{9.33}$$

This is valid for a plug-flow regime where the diffusivity of one species in relation to another is neglected. Therefore, if v is a constant it is also equal to the individual velocity of each component in the process (v_j). Using Equation A.52, one arrives at Equation 9.33.

9.3.1 SOLUTION BY LAPLACE TRANSFORM

Let us transform with respect to the dimensionless time variable, or

$$\varphi(s,\zeta) = L\{\psi(\tau,\zeta)\} \tag{9.34}$$

Applying Equation 9.34 in Equation 9.29 and using the condition given by Equation 9.30, it is possible to write

$$s\varphi + \frac{d\varphi}{d\zeta} + \varphi = 0 \tag{9.35}$$

This is a separable equation with the following solution:

$$\varphi = C\exp\left[-(s+1)\zeta\right] \tag{9.36}$$

The constant can be obtained using the boundary condition given by Equation 9.31, which after transformation becomes

$$\left(-a\frac{d\varphi}{d\zeta} + \varphi\right)\Bigg|_{\zeta=0} = \frac{1}{s} \tag{9.37}$$

After obtaining C, the final form is

$$\varphi = \frac{\exp\left[-(s+1)\zeta\right]}{s[a(s+1)+1]} \tag{9.38}$$

Separation of partial fractions can be used by setting

$$\frac{1}{s[a(s+1)+1]} = \frac{A}{s} + \frac{B}{a(s+1)+1} \tag{9.39}$$

After equating the numerator to 1, constants A and B are found:

$$A = \frac{1}{a+1} \text{ and } B = -\frac{a}{a+1}$$

Therefore, Equation 9.38 becomes

$$\varphi = \frac{1}{a+1}\frac{\exp[-(s+1)\zeta]}{s} - \frac{a}{a+1}\frac{\exp[-(s+1)\zeta]}{a(s+1)+1} \qquad (9.40)$$

According to Appendix D, the inversions of those two terms are

$$L^{-1}\left\{\frac{1}{a+1}\frac{\exp[-(s+1)\zeta]}{s}\right\} = L^{-1}\left\{\frac{e^{-\zeta}}{a+1}\frac{e^{-s\zeta}}{s}\right\} = \frac{e^{-\zeta}}{a+1}u(\tau-\zeta) \quad (9.41)$$

$$L^{-1}\left\{\frac{1}{a+1}\frac{\exp[-(s+1)\zeta]}{a(s+1)+1}\right\} = L^{-1}\left\{\frac{e^{-\zeta}}{a(a+1)}\frac{e^{-s\zeta}}{s+\left(\frac{a+1}{a}\right)}\right\}$$

$$= \frac{e^{-\zeta}}{a(a+1)}\exp\left(-\frac{a+1}{a}\tau\right)u(\tau-\zeta) \qquad (9.42)$$

Therefore the inverse of Equation 9.40 is

$$\psi(\tau,\zeta) = \frac{e^{-\zeta}}{a+1}\left[1 - \frac{\exp\left(-\frac{a+1}{a}\tau\right)}{a}\right]u(\tau-\zeta) \qquad (9.43)$$

The graph representing the relation for the case where $a=1$ is presented in Figure 9.4, and for the case where $a=2$ in Figure 9.5.

It is interesting to verify the similarity between Figures 9.4 and 9.5 with Figures 8.6 and 8.7. Similar to Section 8.3, the following is worth noticing in Figures 9.4 and 9.5:

1. For a given instant (or τ), a profile of concentration (ψ) can be described as decreasing exponential with respect to the position (ζ). Therefore, at the reactor entrance ($\zeta=0$) the concentration is the maximum for this instant. Of course, the decrease is due to the reaction with component B. However, because of the plug-flow regime, there will be positions not affected by the concentration of reactant A, or reached by the concentration wave-front.
2. For a given position (ζ) in the reactor, the concentration would be zero until reached by the wave front after a period (τ). Then, the concentration increases abruptly and continues exponentially to reach the value at the reactor entrance ($\zeta=0$).
3. Different from the case presented in Section 8.3, the condition given by Equation 9.25 (or Equation 9.31) does not impose constant derivative of component A at the reactor entrance ($\zeta=0$ or $z=0$). Despite this, the

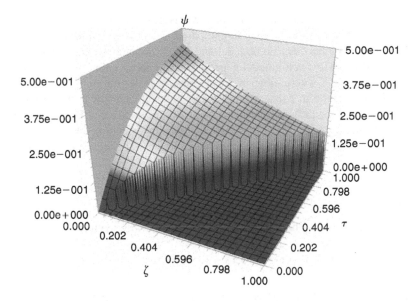

FIGURE 9.4 Dimensionless concentration (ψ) against dimensionless time (τ) and space (ζ) in a plug-flow reactor; case for $a = 1$.

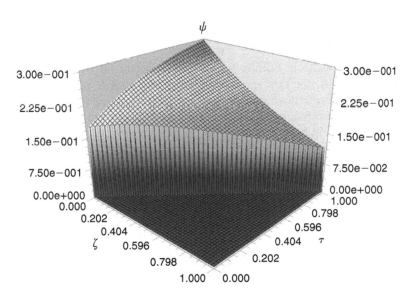

FIGURE 9.5 Dimensionless concentration (ψ) against dimensionless time (τ) and space (ζ) in a plug-flow reactor; case for $a = 2$.

derivative does not vary too much at this position. This shows that the diffusion term is sizable when compared to a convective term.

4. Comparison between Figures 9.4 and 9.5 demonstrates that reaction rate and diffusivity, combined by parameter a, play similar roles because component A would either react or disperse faster through the reactor.

5. Abrupt change in concentration is not possible and this is the result of neglecting the diffusion term at Equation 7.78 to arrive at Equation 7.79 or Equation 9.20. This term would provide a smooth change in the concentration for points around the wave front. This more realistic model is discussed in chapters ahead.

EXERCISES

1. Solve the following boundary value problems by any method:

(a) $\dfrac{\partial y}{\partial x} + \dfrac{\partial y}{\partial t} = 0$; $y'(0,x) = y(0,x) - 1$, $y(t,0) = 1$, $x \geq 0$, and $t \geq 0$

(b) $\dfrac{\partial y}{\partial x} + x\dfrac{\partial y}{\partial t} = 0$; $y'(0,x) = y(1,x)$, $y(t,0) = 2$, $x \geq 0$, and $t \geq 0$

(c) $t\dfrac{\partial y}{\partial x} + \dfrac{\partial y}{\partial t} = 0$; $y'(0,x) = y(0,x)$, $y(t,0) = 1$, $x \geq 0$, and $t \geq 0$

2. From the solution presented in Section 9.2, deduce the heat flux delivered by the electrical resistance to the fluid stream.

3. Rework the solution presented in Section 9.2 using the following new dimensionless temperature: $\psi = \dfrac{T - T_g}{T_b - T_g}$

4. Solve the problem in Section 9.2 using the method of weighted residuals. Obtain, at least, the first approximation by submethods of collocation and Galerkin. Compare these with the exact solution.

5. What sort of information would the method of similarity provide when applied to the problem in Section 9.2.

6. Try to solve the problem in Section 9.3 by any method of weighted residuals. Arrive, at least, to a first approximation.

10 Problems 221; Two Variables, 2nd Order, 1st Kind Boundary Condition

10.1 INTRODUCTION

This chapter presents methods to solve problems with two independent variables involving second-order differential equation and first-kind boundary condition. Therefore, the problems fall in the category of partial differential ones. Mathematically, this class of problems can be summarized as $f\left(\phi,\omega_1,\omega_2,\frac{\partial^2\phi}{\partial\omega_i^2},\frac{\partial^2\phi}{\partial\omega_i\partial\omega_j},\frac{\partial\phi}{\partial\omega_i}\right)$, first-boundary condition.

These are a very important class of transport phenomena problems, and it will be very useful to demonstrate the applications of several analytical and approximate mathematical methods. When compared to previous chapters, the present one deals with more realistic situations. This is so because the second derivatives would be included, which are related to the following:

- Diffusion in mass transfer
- Thermal conduction in heat transfer
- Viscosity resistance in momentum transfer

The above terms were neglected in Chapters 7 through 9, which led to strong approximations with abrupt jumps in concentration and temperature profiles. It will also be an opportunity to demonstrate the similarity of various transport phenomena. For instance, after proper considerations, many solutions obtained for heat transfer can be applied at mass transfer problems and vice versa. For this, concentration is equivalent to temperature, diffusivity to thermal conductivity, and mass transfer coefficient to heat transfer coefficient. With this, Sherwood and Biot numbers would be interchangeable. Of course, one should be careful about these similarities because proper conditions must be observed before applications.

10.2   HEATING AN INSULATED ROD OR A SEMI-INFINITE BODY

Figure 10.1 illustrates the heating process of a semi-infinite cylindrical solid.

Before the heating, the rod is at a constant and known temperature T_b. Then, temperature T_0 is imposed at surface $z = 0$. It is desired to determine the temperature profile in the rod as a function of position and time, as well as the rate of heat transfer to the rod. In addition, at the end of this section, it is demonstrated how the solution achieved here can be applied to a similar situation involving any semi-infinite body. This is shown at the end of the present section.

To simplify and allow an analytical solution, the following is assumed:

1. Rod material is uniform.
2. Thermal conductivity and density of the rod remain constant. Of course, this is an approximation valid for relative small temperature gradients in the rod.
3. No phase change of the solid material of the rod is observed.
4. Insulation around the rod is perfect, i.e., no heat transfer occurs at surfaces coated with insulation material.
5. Rod is long enough to consider its extreme right position unaffected by the imposed changes of temperature at its left extreme (at $z = 0$). As the left end is heated to keep the temperature at T_0, a heating wave front travels from the left to the right in the bar. Therefore, the solution presented here is valid for the period preceding the instant when the temperature of the right end starts being modified.

As the rod is insulated on its sides, no heat transfer is possible at the radial direction, or

$$q_r = -\lambda \frac{\partial T}{\partial r} = 0 \tag{10.1}$$

Therefore, no temperature variation in the radial direction will be observed as well.

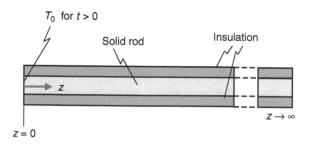

FIGURE 10.1 Insulated semi-infinite cylindrical solid rod.

Because of assumption 1, the angular derivative of temperature is zero. Using Equation A.35 and other simplifications described above, it is possible to write

$$\rho C_p \frac{\partial T}{\partial t} = \lambda \frac{\partial^2 T}{\partial z^2} \tag{10.2}$$

The boundary conditions are as follows:

$$T(0,z) = T_b, \quad z > 0 \tag{10.3}$$

$$T(t,0) = T_0, \quad t > 0 \tag{10.4}$$

$$T(t,\infty) = T_b, \quad t > 0 \tag{10.5}$$

Assumption 5 justifies the last condition and the symbol ∞ refers to positions far from $z = 0$.

As seen, Equation 10.2 combined with conditions given by Equations 10.3 through 10.5 forms a second-order partial differential equation with first-kind boundary condition.

Before applying any method to solve this boundary value problem, let us change variables to work with dimensionless ones. The conveniences of such methods have already been commented.

The chosen dimensionless temperature is

$$\psi = \frac{T - T_b}{T_0 - T_b} \tag{10.6}$$

Regardless of the possibility of finding dimensionless variables for time and space, for now the solution using just the above change is shown below.

Applying Equation 10.6 the following is obtained:

$$\frac{\partial T}{\partial t} = (T_0 - T_b) \frac{\partial \psi}{\partial t} \tag{10.7}$$

$$\frac{\partial^2 T}{\partial z^2} = (T_0 - T_b) \frac{\partial^2 \psi}{\partial z^2} \tag{10.8}$$

Therefore, Equation 10.2 becomes

$$\frac{\partial \psi}{\partial t} = a^2 \frac{\partial^2 \psi}{\partial z^2} \tag{10.9}$$

where

$$a = \sqrt{\frac{\lambda}{\rho C_p}} \tag{10.10}$$

The boundary conditions are written now as

$$\psi(0,z) = 0, \quad z > 0 \tag{10.11}$$

$$\psi(t,0) = 1, \quad t > 0 \tag{10.12}$$

$$\psi(t,\infty) = 0, \quad t > 0 \tag{10.13}$$

As shown, a convenient change of variables brings about easier to handle and simpler forms.

10.2.1 SOLUTION BY LAPLACE TRANSFORM

As seen in Appendix D, the partial differential equation can be transformed into an ordinary equation by the correct application of Laplace transform.

Because of the condition given by Equation 10.11, a convenient transform for this case is

$$\Psi(s,z) = L\{\psi(t,z)\} \tag{10.14}$$

Applying this at Equation 10.9, one arrives at the following:

$$s\Psi = a^2 \frac{d^2\Psi}{dz^2} \tag{10.15}$$

The transforms of other two conditions are

$$\Psi(s,0) = \frac{1}{s} \tag{10.16}$$

$$\Psi(s,\infty) = 0 \tag{10.17}$$

The general solution of Equation 10.15 is

$$\Psi = C_1 e^{\frac{z}{a}\sqrt{s}} + C_2 e^{-\frac{z}{a}\sqrt{s}} \tag{10.18}$$

Since the temperature is finite, applying the condition given by Equation 10.17 to Equation 10.18 requires C_1 to be zero. Using the other remaining condition, the following is achieved:

$$\Psi = \frac{1}{s} e^{-\frac{z}{a}\sqrt{s}} \tag{10.19}$$

Using the tables at Appendix D, the inverse is

$$\psi = \frac{T - T_b}{T_0 - T_b} = \mathrm{erfc}\left(\frac{z}{2a\sqrt{t}}\right) = \mathrm{erfc}\left(\frac{z}{2\sqrt{t\frac{\lambda}{\rho C_p}}}\right) \tag{10.20}$$

The error function is defined as

$$\text{erf}(z) = \frac{2}{\sqrt{\pi}} \int_0^z e^{-\eta^2} d\eta \qquad (10.21)$$

Its complementary form is

$$\text{erfc}(z) = 1 - \text{erf}(z) = \frac{2}{\sqrt{\pi}} \int_z^\infty e^{-\eta^2} d\eta \qquad (10.22)$$

A series representation of this function is given by [1]

$$\text{erfc}(x) = 1 - \frac{2}{\sqrt{\pi}} \left(x - \frac{x^3}{3 \cdot 1!} + \frac{x^5}{5 \cdot 2!} - \frac{x^7}{7 \cdot 3!} + \cdots \right) \qquad (10.23)$$

Therefore, it is possible to recognize a few basic properties of this function, such as

$$\text{erfc}(0) = 1 \qquad (10.24)$$

and

$$\text{erfc}(\infty) = 0 \qquad (10.25)$$

From this, it is easy to see that

1. Because of the property given by Equation 10.25, Equation 10.20 satisfies boundary conditions given by Equations 10.11 and 10.13.
2. Because of the property given by Equation 10.24, the condition in Equation 10.12 is also satisfied by Equation 10.20.

10.2.2 HEAT FLUX TO THE ROD

Using the solution for the temperature profile, it is possible to calculate the heat flux necessary to keep the left surface ($z = 0$) of the rod at the desired constant temperature. This is given by

$$q_z|_{z=0} = -\lambda \frac{\partial T}{\partial z}\bigg|_{z=0} \qquad (10.26)$$

Using Equation 10.20, one would obtain

$$q_z|_{z=0} = -\lambda(T_0 - T_b)\frac{\partial\left[\mathrm{erfc}\left(z/2\sqrt{t(\lambda/\rho C_p)}\right)\right]}{\partial z}\Bigg|_{z=0}$$

$$= -\lambda(T_0 - T_b)\left[\frac{\partial\left(\dfrac{z}{2\sqrt{t(\lambda/\rho C_p)}}\right)}{\partial z}\frac{\partial}{\partial u}\left(\frac{2}{\sqrt{\pi}}\int\limits_{u=\frac{z}{2\sqrt{t(\lambda/\rho C_p)}}}^{u=\infty}e^{-u^2}du\right)\right]_{z=0}$$

$$= -\lambda(T_0 - T_b)\frac{1}{\sqrt{\pi t(\lambda/\rho C_p)}}\left[\left(e^{-u^2}\right)_{u=\frac{z}{2\sqrt{t(\lambda/\rho C_p)}}}^{u=\infty}\right]_{z=0} = (T_0 - T_b)\sqrt{\frac{\lambda\rho C_p}{\pi t}}$$

$$(10.27)$$

Here, u is just the integration variable.

Again, it is very important to exercise critical analysis of every solution, and as advised in Section 1.1.2, it has been shown that Equation 10.20 satisfies the boundary conditions. In addition, the solution is physically consistent. For instance, in the above example, heat flux decreased with time, increases for bodies with higher thermal conductivity, and increases with the difference of temperatures.

10.2.3 COMPLETE DIMENSIONLESS FORM

To facilitate the representation of temperature as a function of space and time, the following dimensionless forms are proposed:

$$\zeta = \frac{z}{L} \tag{10.28}$$

and

$$\tau = t\frac{\lambda}{\rho C_p L^2} \tag{10.29}$$

Here, the length L is a chosen value, for instance the unit of length in the measurement system. In the case of SI units, one might choose $L = 1$ m. After applying these new variables, Equation 10.20 is written as

$$\psi = \frac{T - T_b}{T_0 - T_b} = \mathrm{erfc}\left(\frac{\zeta}{2\sqrt{\tau}}\right) \tag{10.30}$$

Figure 10.2 represents the dimensionless temperature (ψ) distribution in the rod against dimensionless time (τ). The range covered here for dimensionless length and time is enough to provide a good idea of the temperature profiles in the rod.

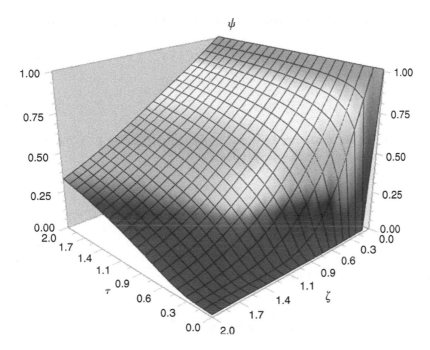

FIGURE 10.2 Representation of dimensionless temperature (ψ) against the rod of dimensionless length (ζ) and time (τ).

From the figure, the following is worth mentioning:

1. For positions at the left extreme of the rod ($\zeta = 0$), the temperature remains as T_0 (or $\psi = 1$). As expected, the temperature decreases for positions away from this.
2. At any given position in the bar, the temperature increases with time, but the rate of variation decreases with time. Of course, for infinite time, the temperature would reach T_0 (or $\psi = 1$) at all points in the rod.

10.2.4 SOLUTION BY SIMILARITY

As commented in Appendix F, there is a reasonable range of partial differential equations solvable by the method of similarity. The present case falls in that range. This is so because, after choosing the appropriate new, combined variables, it is possible to coalesce the boundary conditions given by Equations 10.11 and 10.13.

It is worthwhile to use the complete dimensionless forms, as given by Equations 10.6, 10.28, and 10.29. After applying these to Equation 10.9, one gets

$$\frac{\partial \psi}{\partial \tau} = \frac{\partial^2 \psi}{\partial \zeta^2} \tag{10.31}$$

The boundary conditions given by Equations 10.11 through 10.13 become

$$\psi(0,\zeta) = 0, \quad \zeta > 0 \tag{10.32}$$

$$\psi(\tau,0) = 1, \quad \tau > 0 \tag{10.33}$$

$$\psi(\tau,\infty) = 0, \quad \tau > 0 \tag{10.34}$$

According to the generalized method of similarity (Section F.2), the following transformations of variables are proposed:

$$\bar{\tau} = a^{b_1}\tau \tag{10.35}$$

$$\bar{\zeta} = a^{b_2}\zeta \tag{10.36}$$

$$\bar{\psi} = a^c\psi \tag{10.37}$$

Therefore, it is possible to write

$$\frac{\partial \psi}{\partial \tau} = \frac{\partial \bar{\tau}}{\partial \tau}\frac{\partial \psi}{\partial \bar{\psi}}\frac{\partial \bar{\psi}}{\partial \bar{\tau}} = a^{b_1}a^{-c}\frac{\partial \bar{\psi}}{\partial \bar{\tau}} \tag{10.38}$$

$$\frac{\partial \psi}{\partial \zeta} = \frac{\partial \bar{\zeta}}{\partial \zeta}\frac{\partial \psi}{\partial \bar{\psi}}\frac{\partial \bar{\psi}}{\partial \bar{\zeta}} = a^{b_2}a^{-c}\frac{\partial \bar{\psi}}{\partial \bar{\zeta}} \tag{10.39}$$

$$\frac{\partial^2 \psi}{\partial \zeta^2} = \frac{\partial \bar{\zeta}}{\partial \zeta}\frac{\partial}{\partial \bar{\zeta}}\frac{\partial \bar{\psi}}{\partial \bar{\zeta}} = a^{2b_2}a^{-c}\frac{\partial^2 \bar{\psi}}{\partial \bar{\zeta}^2} \tag{10.40}$$

Using these into Equation 10.31, one gets

$$a^{b_1}a^{-c}\frac{\partial \bar{\psi}}{\partial \bar{\tau}} = a^{2b_2}a^{-c}\frac{\partial^2 \bar{\psi}}{\partial \bar{\zeta}^2} \tag{10.41}$$

The above would be an invariant of Equation 10.31 if

$$b_1 = 2b_2 \tag{10.42}$$

Parameter c can assume any finite real value.

Therefore, the new combined variables would be

$$\omega = \frac{\zeta}{\tau^{b_2/b_1}} = \frac{\zeta}{\sqrt{\tau}} \tag{10.43}$$

To simplify, let us arbitrate $c=0$, which leads to

$$f(\omega) = \psi \tag{10.44}$$

Applying these to Equation 10.31, the following is obtained:

$$-\frac{\omega}{2}\frac{df}{d\omega} = \frac{d^2f}{d\omega^2} \tag{10.45}$$

This is an ordinary differential equation, which can be transformed into a sequence of two first-order separable equations, and therefore is given by a straightforward solution:

$$f(\omega) = C_1 \int \exp\left(-\frac{\omega^2}{4}\right) d\omega + C_2 \tag{10.46}$$

Using Equations 10.43 and 10.44, the boundary conditions given by Equations 10.32 and 10.34 coalesce into

$$f(\omega \to \infty) = 0 \tag{10.47}$$

The condition given by Equation 10.33 becomes

$$f(0) = 1 \tag{10.48}$$

Applying Equations 10.47 and 10.48 to Equation 10.46, the following is obtained:

$$\psi(\tau,\zeta) = f(\omega) = \frac{\displaystyle\int_\omega^\infty \exp\left(-\frac{\omega^2}{4}\right) d\omega}{\displaystyle\int_0^\infty \exp\left(-\frac{\omega^2}{4}\right) d\omega} \tag{10.49}$$

Keeping in mind the definition of complementary error function and its properties (as given above), the final solution is

$$\psi(\tau,\zeta) = \operatorname{erfc}\left(\frac{\omega}{2}\right) = \operatorname{erfc}\left(\frac{\zeta}{2\sqrt{\tau}}\right) \tag{10.50}$$

This is also given by Equation 10.30, therefore following the same consequences discussed before.

Finally, it is important to notice that the above solution is applicable to the case of a semi-infinite body, as illustrated in Figure 10.3. The whole body is initially at temperature T_b and suddenly its flat face is set at temperature T_0.

Owing to indefinite dimensions of the body in all directions but one (z), no variations of temperature in direction x or y would occur. Therefore, the only two independent variables that remain are t and z.

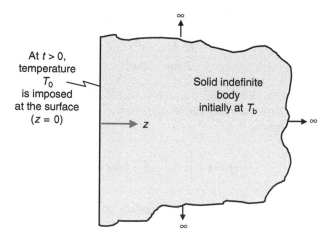

FIGURE 10.3 Semi-infinite body with a face kept at constant temperature.

10.3 SUDDEN MOTION OF A PLATE

The horizontal solid plate immersed in a fluid, as shown in Figure 10.4, is at rest and suddenly starts moving with a given velocity u.

One is interested to know the drag force applied by the fluid on the plate and vice versa.

Of course, this is the most simplified situation of unsteady state regime of the common problem of resistance to the movement of bodies in fluids. However, its fundamentals are useful and are applied to complex problems in the naval, automobile, and airplane industries, not to mention pipes, mixers, and all sorts of equipment or systems where relative movement of fluids and solid surfaces occur.

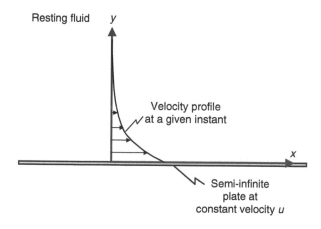

FIGURE 10.4 Illustration for the problem of moving plate immersed in a fluid.

For this simple situation, the following are assumed:

1. Plate is very wide and long or indefinite in the x and z direction. Because of this, the velocity component in the x direction varies only with time and direction y.
2. Plate is thin enough to allow negligible edge effects. Therefore, with the exception of the velocity component in the x direction, all others are either zero or negligible.
3. Plate and the fluid are in thermal equilibrium and any dissipation of energy by viscous flow is negligible.
4. Fluid is Newtonian with constant density and viscosity.
5. Process is considered isobaric, or at least the effects of pressure variation can be neglected.

As always, in problems of momentum transfer, the first thing to apply is the conservation of mass, represented by Equation A.1. Because of assumptions 1, 2, and 4 this equation provides

$$\frac{\partial v_x}{\partial x} = 0 \qquad (10.51)$$

The next step is to apply the momentum conservations at each direction. However, as the movement involves just velocity in the x direction, the only equation that might lead to some information is that which deals with the conservation in this direction. Owing to assumption 4, Equation A.7 can be applied. After the first assumptions 1, 2, 4, and 5, it is possible to simplify the equation to

$$\frac{\partial v_x}{\partial t} = v \frac{\partial^2 v_x}{\partial y^2} \qquad (10.52)$$

10.3.1 BOUNDARY CONDITIONS

From the above equation, just one condition regarding time and two involving the vertical direction (y) are required to completely define the boundary value problem. These are

$$v_x(0,y) = 0, \quad -\infty \leq y \leq \infty \qquad (10.53)$$

$$v_x(t,0) = u, \quad t > 0 \qquad (10.54)$$

$$v_x(t,\infty) = 0, \quad t > 0 \qquad (10.55)$$

Of course, the notation in Equation 10.55 should be understood as a limit for positions far from the plate surface.

10.3.2 Solution by Laplace Transform

Let us apply the transform to variable t, or

$$\varphi(s,y) = L\{v_x(t,y)\} \tag{10.56}$$

According to Section D.4, Laplace transform of Equation 10.52 provides

$$s\varphi(s,y) - v_x(0,y) = \nu \frac{d^2\varphi}{dy^2} \tag{10.57}$$

Note that the last term of the equation is an ordinary derivative. This can be written because s is not a variable of the physical problem but only a parameter of the transform. Therefore, the partial differential problem becomes an ordinary one.

The condition given by Equation 10.53 applied to the previous equation gives

$$\frac{d^2\varphi}{dy^2} = \frac{s}{\nu}\varphi(s,y) \tag{10.58}$$

According to Appendix B, its solution is

$$\varphi(s,y) = C_1 e^{-y\sqrt{\frac{s}{\nu}}} + C_2 e^{y\sqrt{\frac{s}{\nu}}} \tag{10.59}$$

In order to determine the integration constants, the boundary conditions given by Equations 10.54 and 10.55 should also be transformed leading to

$$\varphi(s,0) = \frac{u}{s} \tag{10.60}$$

and

$$\varphi(s,\infty) = 0 \tag{10.61}$$

Applying the condition given by Equation 10.61 to Equation 10.59 leads to the conclusion that C_2 is equal to zero. Then, applying the condition given by Equation 10.60 to Equation 10.59 provides

$$C_1 = \frac{u}{s} \tag{10.62}$$

Hence

$$\varphi(s,y) = \frac{u}{s} e^{-y\sqrt{\frac{s}{\nu}}} \tag{10.63}$$

The inverse can be found in the tables presented at Appendix D, and the solution is

$$v_x(t,y) = u \text{ erfc}\left(\frac{y}{2\sqrt{\nu t}}\right) \tag{10.64}$$

Equation 10.25 ensures that the above solution satisfies conditions given by Equations 10.53 and 10.55, while Equation 10.24 shows that the condition given by Equation 10.54 is also satisfied.

Owing to the similarity of Equations 10.64 and 10.50, Figure 10.2 can be used to represent the velocity profile if

$$\psi = \frac{v_x}{u}, \quad \zeta = y, \quad \tau = \nu t$$

Of course here, neither τ nor ζ are dimensionless variables. However, the graph has the same form as in the previous case and the following should be noticed:

1. At the beginning of the process ($\tau = 0$), the fluid was at rest ($\psi = 0$).
2. As imposed by the boundary conditions, at the plate surface ($\zeta = 0$) the velocity remains equal to u, and therefore $\psi = 1$.
3. For positions far from the plate surface (large ζ), the fluid remains static ($\psi = 0$).
4. As time increases (large τ), all positions will tend to velocity equal to the plate ($\psi = 1$).
5. For a given position ζ above the plate surface, the velocity of fluids with higher viscosity (ν) would be represented by lines at larger values of τ, and therefore by greater values of velocities (ψ). In other words, as viscosity increases, the momentum transfer between plate and fluid, as well as throughout the fluid, becomes more effective. This can also be seen by Newton law ($\tau_{yx} = -\mu \partial v_x / \partial y$). Greater viscosities allow the same shear stress with smaller gradients of velocity.

10.3.3 Solution by Similarity

According to the method shown in Appendix F, the application of the generalized method of similarity starts by assuming the transformations of variables as

$$\bar{t} = a^{b_1} t \tag{10.65}$$

$$\bar{y} = a^{b_2} y \tag{10.66}$$

$$\bar{v}_x = a^c v_x \tag{10.67}$$

Therefore

$$\frac{\partial v_x}{\partial t} = \frac{\partial \bar{t}}{\partial t} \frac{\partial v_x}{\partial \bar{v}_x} \frac{\partial \bar{v}_x}{\partial \bar{t}} = a^{b_1 - c} \frac{\partial \bar{v}_x}{\partial \bar{t}} \tag{10.68}$$

$$\frac{\partial v_x}{\partial y} = \frac{\partial \bar{y}}{\partial y} \frac{\partial v_x}{\partial \bar{v}_x} \frac{\partial \bar{v}_x}{\partial \bar{y}} = a^{b_2 - c} \frac{\partial \bar{v}_x}{\partial \bar{y}} \tag{10.69}$$

$$\frac{\partial^2 v_x}{\partial y^2} = \frac{\partial \bar{y}}{\partial y} \left(a^{b_2 - c} \frac{\partial \bar{v}_x}{\partial \bar{y}} \right) = a^{2b_2 - c} \frac{\partial^2 \bar{v}_x}{\partial \bar{y}^2} \tag{10.70}$$

Using these, Equation 10.52 becomes

$$a^{b_1 - c} \frac{\partial \bar{v}_x}{\partial \bar{t}} = a^{2b_2 - c} \nu \frac{\partial^2 \bar{v}_x}{\partial \bar{y}^2} \tag{10.71}$$

The condition for invariance is obtained by comparing Equations 10.71 and 10.52, or

$$b_1 - c = 2b_2 - c \tag{10.72}$$

This leads to

$$b_1 = 2b_2, \quad \text{for any } c \tag{10.73}$$

Of course, the simplest alternative is to set $c = 0$. According to Appendix F, the new variables are

$$\omega = \frac{y}{t^{b_1/b_2}} = \frac{y}{t^{1/2}} \tag{10.74}$$

and

$$f(\omega) = \frac{v_x(t,y)}{t^0} = v_x(t,y) \tag{10.75}$$

Therefore

$$\frac{\partial v_x}{\partial t} = \frac{df}{d\omega} \frac{d\omega}{\partial t} = -\frac{1}{2} y t^{-\frac{3}{2}} \frac{df}{d\omega} \tag{10.76}$$

$$\frac{\partial v_x}{\partial y} = t^{-\frac{1}{2}} \frac{df}{d\omega} \tag{10.77}$$

$$\frac{\partial^2 v_x}{\partial y^2} = t^{-1} \frac{d^2 f}{d\omega^2} \tag{10.78}$$

Applying these into Equation 10.52 provide

$$f'' = -\frac{\omega}{2\nu}f' \tag{10.79}$$

This is a separable equation on the derivative of f; therefore, it is possible to write

$$\ln(f') = -\frac{\omega^2}{4\nu} + \ln C_1 \tag{10.80}$$

$$f' = C_1 \exp\left(-\frac{\omega^2}{4\nu}\right) \tag{10.81}$$

$$f = C_1 \int_0^\omega \exp\left(-\frac{u^2}{4\nu}\right) du + C_2 \tag{10.82}$$

It should be noticed that defining limits for the above integral would not impose any loss of generality.

The boundary conditions for f can be deduced from the conditions given by Equations 10.53 through 10.55. For this the definition of combined variables (Equation 10.74) is used, which provides the following:

- Condition given by Equation 10.53, or at $t=0$ leads to

$$f(\infty) = 0 \tag{10.83}$$

- Condition given by Equation 10.54, or at $y=0$ leads to

$$f(0) = u \tag{10.84}$$

- Condition given by Equation 10.55, or for y tending to infinity, just repeats Equation 10.83.

The possibility of condensing conditions, such as Equations 10.53 and 10.55, demonstrates one of the requirements that allow the application of the method of similarity. The conditions given by Equations 10.83 and 10.84 permit determining single values for each constant in Equation 10.82. After this, the following can be written:

$$\frac{f}{u} = 1 - \frac{\displaystyle\int_0^\omega \exp\left(-\frac{u^2}{4\nu}\right) du}{\displaystyle\int_0^\infty \exp\left(-\frac{u^2}{4\nu}\right) du} \tag{10.85}$$

Using the definition for the complementary error function, the solution given by Equation 10.64 is reproduced. The details are left as an exercise for the reader.

10.3.4 SHEAR STRESS

Using the previous solution for the velocity profile, it is easy to calculate the stress between plate surface and fluid, which is given by

$$\tau_{yx}|_{y=0} = -\mu \frac{\partial v_x}{\partial y}\bigg|_{y=0} \tag{10.86}$$

Using Equation 10.64, one would obtain

$$\tau_{yx}|_{y=0} = -\mu u \frac{\partial \left[\text{erfc}\left(\frac{y}{2\sqrt{vt}}\right)\right]}{\partial y}\Bigg|_{y=0}$$

$$= -\mu u \left[\frac{\partial (y/2\sqrt{vt})}{\partial y} \frac{\partial}{\partial u}\left(\frac{2}{\sqrt{\pi}} \int\limits_{z=(y/2\sqrt{vt})}^{z=\infty} e^{-z^2} dz\right)\right]_{y=0}$$

$$= -\mu u \frac{1}{\sqrt{\pi vt}} \left[\left(e^{-z^2}\right)_{z=(y/2\sqrt{vt})}^{z=\infty}\right]_{y=0} = u\sqrt{\frac{\rho\mu}{\pi t}} \tag{10.87}$$

Here, z is just the integration variable.

The above shows that the force (stress times plate area) applied to move the plate is proportional to the plate velocity, the square root of viscosity times density, and inversely proportional to the square root of time. Since, at infinite time, the whole fluid would be at velocity u, no force would be required to maintain the plate at that velocity.

10.4 HEATING A FLOWING LIQUID

This is a more complete attack to the problem presented in Section 7.3. Therefore, it will lead to a solution much more in accordance with reality.

As seen, the governing equation would be

$$\rho C_p \left(\frac{\partial T}{\partial t} + v_z \frac{\partial T}{\partial z}\right) = \lambda \frac{\partial^2 T}{\partial z^2} \tag{10.88}$$

All discussions made in Section 7.3 are valid, except the assumption that total convection is much larger than energy diffusion. Therefore, Equation 10.88 should be solved without further simplification. In view of this, one condition related to time and two related to space (axial direction z) are needed. These are

$$T(0,z) = T_b, \quad z > 0 \tag{10.89}$$

$$T(t,0) = T_0, \quad t > 0 \tag{10.90}$$

$$T(t,\infty) = T_b, \quad t > 0 \tag{10.91}$$

Here the symbol ∞ indicates positions far from the region affected by the heating.

The change to dimensionless variables is always interesting, and the same as proposed in Section 7.3 is adopted here, or

$$\psi = \frac{T - T_b}{T_0 - T_b} \tag{10.92}$$

Additionally, the following changes are useful to transform all variables into dimensionless ones

$$\zeta = \frac{zv}{D_T} \tag{10.93}$$

$$\tau = \frac{tv^2}{D_T} \tag{10.94}$$

The thermal diffusivity is defined by

$$D_T = \frac{\lambda}{\rho C_p} \tag{10.95}$$

The changes in differentials of temperature against time and space are shown by Equations 7.20 and 7.21. The second differential is given by

$$\frac{\partial^2 T}{\partial z^2} = \frac{\partial}{\partial z}\left[(T_0 - T_b)\frac{\partial \psi}{\partial z}\right] = (T_0 - T_b)\frac{\partial^2 \psi}{\partial z^2} \tag{10.96}$$

Using all the above, Equation 10.88 is written as

$$\frac{\partial \psi}{\partial \tau} + \frac{\partial \psi}{\partial \zeta} = \frac{\partial^2 \psi}{\partial \zeta^2} \tag{10.97}$$

The transformation leads to an elegant and simple form.

The boundary conditions given by Equations 10.89 through 10.91 become

$$\psi(0,\zeta) = 0, \quad \zeta > 0 \tag{10.98}$$

$$\psi(\tau,0) = 1, \quad \tau > 0 \tag{10.99}$$

$$\psi(\tau,\infty) = 0, \quad \tau > 0 \tag{10.100}$$

10.4.1 APPLICATION OF LAPLACE TRANSFORM

Looking at the boundary conditions, it seems convenient to transform with regard to variable time, or

$$\Psi(s,\zeta) = L\{\psi(\tau,\zeta)\} \tag{10.101}$$

This is so because, if applied to Equation 10.97 together with the condition given by Equation 10.98, one arrives at

$$\frac{d^2\Psi}{d\zeta^2} - \frac{d\Psi}{d\zeta} - s\Psi = 0 \tag{10.102}$$

The above equation is a linear ordinary differential second-order one. The solution is indicated in Appendix B and is given by

$$\Psi = C_1 \exp(a_1\zeta) + C_2 \exp(a_2\zeta) \tag{10.103}$$

Here

$$a_1 = \frac{1 + \sqrt{1 + 4s}}{2} \tag{10.104}$$

and

$$a_2 = \frac{1 - \sqrt{1 + 4s}}{2} \tag{10.105}$$

Transforms of boundary conditions given by Equations 10.99 and 10.100 are

$$\Psi(s,0) = \frac{1}{s} \tag{10.106}$$

and

$$\Psi(s,\infty) = 0 \tag{10.107}$$

To apply the above, it is interesting to remember that solutions are real numbers and depend on ζ. Additionally, since s is a real and positive parameter (see Appendix D regarding parameter s), a_1 would be positive and a_2 negative. Having these in mind, the condition given by Equation 10.107 applied to Equation 10.103 forces $C_1 = 0$, and the condition given by Equation 10.106 leads to $C_2 = 1/s$. Therefore

$$\Psi = \frac{1}{s}\exp{(a_2\zeta)} = \frac{1}{s}\exp\left(\frac{\zeta}{2}\right)\exp\left(-\frac{\zeta}{2}\sqrt{1+4s}\right)$$

$$= \exp\left(\frac{\zeta}{2}\right)\frac{1}{s}\exp\left(-\zeta\sqrt{s+\frac{1}{4}}\right) \qquad (10.108)$$

From any table of transforms of Appendix D, one would find the following inversion:

$$L^{-1}\left[\exp{(-\zeta\sqrt{s})}\right] = \frac{\zeta}{2\sqrt{\pi\tau^3}}\exp\left(-\frac{\zeta^2}{4\tau}\right) \qquad (10.109)$$

Applying the "s-shift" theorem, it is possible to write

$$L^{-1}\left[\exp\left(-\zeta\sqrt{s+\frac{1}{4}}\right)\right] = e^{-\frac{\tau}{4}}\frac{\zeta}{2\sqrt{\pi\tau^3}}\exp\left(-\frac{\zeta^2}{4\tau}\right) \qquad (10.110)$$

Finally

$$L^{-1}\left[\exp\left(-\zeta\sqrt{s+\frac{1}{4}}\right)\Big/ s\right] = \frac{\zeta}{2\sqrt{\pi}}\int_0^\tau \frac{\exp\left(-\frac{\zeta^2}{4x}-\frac{1}{4}x\right)}{x^{3/2}}dx \qquad (10.111)$$

Here x is the dummy variable for the integration. Using Appendix D, the former inverse can be also written as

$$L^{-1}\left[\exp\left(-\zeta\sqrt{s+\frac{1}{4}}\right)\Big/ s\right] = \frac{1}{2}\left[e^{\frac{\zeta}{2}}\operatorname{erfc}\left(\frac{\zeta}{2\sqrt{\tau}}+\frac{\sqrt{\tau}}{2}\right)+e^{-\frac{\zeta}{2}}\operatorname{erfc}\left(\frac{\zeta}{2\sqrt{\tau}}-\frac{\sqrt{\tau}}{2}\right)\right]$$

$$(10.112)$$

Therefore, the final solution is

$$\psi(\tau,\zeta) = \frac{1}{2}\left[e^{\zeta}\operatorname{erfc}\left(\frac{\zeta}{2\sqrt{\tau}}+\frac{\sqrt{\tau}}{2}\right)+\operatorname{erfc}\left(\frac{\zeta}{2\sqrt{\tau}}-\frac{\sqrt{\tau}}{2}\right)\right] \qquad (10.113)$$

Notice that the boundary conditions are met, or

1. At the beginning of the process, or when τ is equal to 0, the parameters inside the first complementary error functions lead to infinite value and the complementary error functions tend to zero. Thus, ψ tends to zero or T tends to T_b, which satisfies the boundary condition given by Equation 10.89 or its dimensionless form in Equation 10.98.

2. Same as above occurs for remote positions (very large ζ), therefore satisfying the condition given by Equation 10.91 or 10.100.
3. At position z equal to zero (or $\zeta = 0$), the error functions and all exponential ones would become equal to 1. Consequently, ψ would be equal to 1, and T equals T_0, which agree with the condition given by Equation 10.90 or 10.99.
4. Solution also shows that the same also occurs for time (or τ) tending to infinity. In this case, the parameter of first complementary error function tends to infinite values and the function to zero. The parameter in the second one tends to $-\infty$. However, the error function is an odd one, or

$$\mathrm{erf}(-x) = -\mathrm{erf}(x) \tag{10.114}$$

Hence

$$\mathrm{erfc}(-x) = 1 + \mathrm{erf}(x) \tag{10.115}$$

and

$$\lim_{x \to -\infty} \mathrm{erfc}(x) = 2 \tag{10.116}$$

Thus, in Equation 10.113

$$\lim_{\tau \to \infty} \psi(\tau,\zeta) = 1 \tag{10.117}$$

Figure 10.5 illustrates the behavior of temperature against time and space using the dimensionless forms.

Additionally, it is interesting to notice the following:

1. Definitions of dimensionless variables by Equations 10.93 and 10.94 show that, for any given positive values of position (z) and time (t), increases in the fluid velocity lead to larger values for ζ and τ. On the other hand, Figure 10.5 shows that the temperature decreases for larger ζ and increases for greater τ. Therefore, the final effect of velocity on temperature at a given position is not easy to see. To circumvent this difficulty, one can write Equation 10.113 using Equations 10.93 and 10.94 to explicitly show variables and parameters of the process. The result is

$$\psi(\tau,\zeta) = \frac{1}{2}\left[e^{\frac{zv}{D_T}} \mathrm{erfc}\left(\frac{z}{2\sqrt{D_T t}} + v\frac{\sqrt{t/D_T}}{2} \right) + \mathrm{erfc}\left(\frac{z}{2\sqrt{D_T t}} - v\frac{\sqrt{t/D_T}}{2} \right) \right]$$

$$\tag{10.118}$$

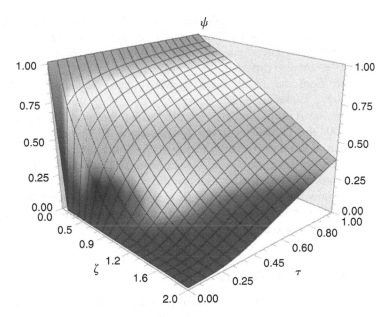

FIGURE 10.5 Dimensionless temperature (ψ) as function of dimensionless time (τ) and space (ζ).

Here it is possible to verify the limit of dimensionless temperature when the velocity tends to infinity. From the discussion above, (item 4) for very large periods, it is easy to see that ψ would tend to 1, or the temperature at all points would tend to T_0. As long as it is ensured that all fluid passing through the grid is heated to the temperature T_0, increases in the flow rate (or fluid velocity) tend to increase the convective term and, therefore, any point in the tube tends to approach temperature T_0. Of course, such a heating system is an abstraction, because a limitless power input would be necessary.

2. From Equations 10.24 and 10.118, it is possible to see that increases in the thermal diffusivity (D_T) lead to the complementary error functions to approach 1. In addition, the multiplying exponential would also tend to 1. Therefore, the dimensionless temperature would again tend to 1. This is also simple to understand because high thermal diffusion contributes toward the equalization of temperature at every point inside the tube. Of course, the final value would be the same as the imposed temperature (T_0) at $z = 0$.

3. It is interesting to compare the present situation and the one set in Section 7.3, where no diffusion term was considered. The following should be noticed:

    a. Dimensionless form (ψ) of the dependent variable (T) is the same and given by Equations 7.19 and 10.92, respectively. However, the dimensionless forms of independent variable related to space ζ are different and given by Equations 7.40 and 10.93, respectively. According to Equations 7.41 and 10.94, the variables related to time (τ) are also different.

b. Apart from the second-order derivative term, Equation 10.97 is similar to Equation 7.42.

c. Again, apart from the extra boundary condition given by Equation 10.100, the conditions given by Equations 10.98 and 10.99 are similar to those provided by Equations 7.43 and 7.44.

d. Remarkable improvement in the mathematical modeling of the present process is verified by comparing Figures 7.8 and 10.5. They show how the inclusion of the diffusion term leads to a much more realistic picture of temporal and spatial temperature variation throughout the tube.

10.5　PLUG-FLOW REACTOR

This is an improvement over the treatment presented in Section 7.4 for a plug-flow reactor and illustrated in Figure 7.9. For this, the previous assumption (9) of negligible effect of diffusion term in relation to the convection is dropped. In other words, the second derivative of concentration in the space is included. Hence, after the application of other assumptions listed in Section 7.4, Equation A.41 leads to

$$\frac{\partial \rho_A}{\partial t} + v \frac{\partial \rho_A}{\partial z} = D_{AB} \frac{\partial^2 \rho_A}{\partial z^2} + R_A \tag{10.119}$$

The component of velocity in axial (z) direction is just written as v. Likewise in Section 7.4, a first-order and irreversible reaction is assumed. Hence, its rate is given by

$$R_A = k\rho_A \tag{10.120}$$

Therefore, the unit for reaction coefficient (k) is s^{-1}.

The boundary conditions are

$$\rho_A(0,z) = 0, \quad z \geq 0 \tag{10.121}$$

and

$$\rho_A(t,0) = \rho_{A,0}, \quad t > 0 \tag{10.122}$$

Here, $\rho_{A,0}$ is the concentration at the grid at any instant after the start of the injection process ($t > 0$).

Even though these two conditions were necessary and sufficient for the previous solution shown in Section 7.4, the present treatment introduces the diffusion term, or a second derivative. Thus, another condition on variable z should be found, and this is derived from the fact that, during its travel through the reactor, component A is continuously consumed. Consequently, its concentration should tend to zero and the following must be valid:

$$\rho_A(t,\infty) = \lim_{z \to \infty} \rho_A = 0, \quad t > 0 \tag{10.123}$$

Here, the symbol ∞ means a limit for z tending to infinity, or at positions very far from the injection point. Therefore, the solution presented here would be valid for regions unaffected by the change in concentration.

Again, it is advisable to work with dimensionless variables. One possibility is to set

$$F(t,z) = \frac{\rho_A(t,z)}{\rho_{A,0}} \tag{10.124}$$

$$\tau = kt \tag{10.125}$$

$$\zeta = \frac{kz}{v} \tag{10.126}$$

After applying the above, Equation 10.119 becomes

$$\frac{\partial F}{\partial \tau} + \frac{\partial F}{\partial \zeta} - \beta \frac{\partial^2 F}{\partial \zeta^2} + F = 0 \tag{10.127}$$

where

$$\beta = \frac{D_{AB}k}{v^2} \tag{10.128}$$

The boundary conditions are rewritten as

$$F(0,\zeta) = 0, \quad \zeta > 0 \tag{10.129}$$

$$F(\tau,0) = 1, \quad \tau > 0 \tag{10.130}$$

$$F(\tau,\infty) = 0, \quad \tau > 0 \tag{10.131}$$

Usually, there is a series of possible changes in variables to achieve a dimensionless set. For instance, in the present case the following could be applied: dimensionless time $= t\, v/L$ and dimensionless space $= z/L$. The length L could be an arbitrarily set value. On the other hand, the forms in Equations 10.124 through 10.126 do not need such choices and include parameters better related to the process. For instance, the time required to consume 1 kmol of reactant A is $1/k$.

10.5.1 SOLUTION BY LAPLACE TRANSFORM

The following transform is chosen:

$$\psi(s,z) = L\{F(\tau,\zeta)\} \tag{10.132}$$

The transform of time derivative combined with the condition given by Equation 10.129 is

$$L\left\{\frac{\partial F}{\partial \tau}\right\} = s\psi(s,\zeta) - F(0,\zeta) = s\psi \tag{10.133}$$

The other derivatives are transformed as follows:

$$L\left\{\frac{\partial F}{\partial \zeta}\right\} = \frac{d\psi}{d\zeta} \tag{10.134}$$

and

$$L\left\{\frac{\partial^2 F}{\partial \zeta^2}\right\} = \frac{d^2\psi}{d\zeta^2} \tag{10.135}$$

Using the above in Equation 10.127, one gets

$$\frac{d^2\psi}{d\zeta^2} - \frac{1}{\beta}\frac{d\psi}{dz} - \frac{s+1}{\beta}\psi = 0 \tag{10.136}$$

This is a second-order ordinary linear differential equation with constant coefficients, and according to Appendix B, its solution can be found by setting

$$\psi = e^{a\zeta} \tag{10.137}$$

If the above is used at Equation 10.136, the following is achieved:

$$a^2 - \frac{1}{\beta}a - \frac{s+1}{\beta} = 0 \tag{10.138}$$

Here, parameter a is constant.

The solution of this second-order polynomial leads to

$$a_1 = \frac{\frac{1}{\beta} + \left(\frac{1}{\beta^2} + 4\frac{s+1}{\beta}\right)^{1/2}}{2} \tag{10.139}$$

and

$$a_2 = \frac{\frac{1}{\beta} - \left(\frac{1}{\beta^2} + 4\frac{s+1}{\beta}\right)^{1/2}}{2} \qquad (10.140)$$

Therefore

$$\psi = b_1 \exp(a_1\zeta) + b_2 \exp(a_2\zeta) \qquad (10.141)$$

The condition related to time has already been used to write Equation 10.133. The transformed boundary conditions in ζ are

$$\psi(s,0) = \frac{1}{s} \qquad (10.142)$$

and

$$\psi(s,\infty) = 0 \qquad (10.143)$$

Applying Equations 10.142 and 10.143 into Equation 10.141, one obtains

$$b_1 + b_2 = \frac{1}{s} \qquad (10.144)$$

and

$$b_1 = 0 \qquad (10.145)$$

This last conclusion comes from the fact that a_1, as given by Equation 10.139, is always a positive number. Consequently, when ζ tends to very large values, the first exponential of Equation 10.141 would tend to ∞ as well. The only way to avoid this is to apply Equation 10.145. Therefore, from Equation 10.144

$$b_2 = \frac{1}{s} \qquad (10.146)$$

Hence, Equation 10.141 becomes

$$\psi = \frac{1}{s} \exp(a_2\zeta) = \exp\left(\frac{\zeta}{2\beta}\right) \frac{\exp\left[-2\zeta\left(\frac{s+b_3}{\beta}\right)^{1/2}\right]}{s} \qquad (10.147)$$

where

$$b_3 = 1 + \frac{1}{4\beta} \qquad (10.148)$$

The details of the above maneuver are left as an exercise to the reader.

From Appendix D, one would find the following inversion:

$$L^{-1}\left\{\exp\left(-2\zeta\sqrt{\frac{s}{\beta}}\right)\right\} = \frac{\zeta}{\sqrt{\pi\beta\tau^3}}\exp\left(-\frac{\zeta^2}{\beta\tau}\right) \tag{10.149}$$

Applying the s-shift theorem, it is possible to write

$$L^{-1}\left\{\exp\left(-2\zeta\sqrt{\frac{s+b_3}{\beta}}\right)\right\} = e^{-b_3\tau}\frac{\zeta}{\sqrt{\pi\beta\tau^3}}\exp\left(-\frac{\zeta^2}{\beta\tau}\right) \tag{10.150}$$

Finally

$$L^{-1}\left\{\frac{\exp\left(-2\zeta\sqrt{\frac{s+b_3}{\beta}}\right)}{s}\right\} = \frac{\zeta}{\sqrt{\pi\beta}}\int_0^\tau \frac{\exp\left(-\frac{\zeta^2}{\beta x} - b_3 x\right)}{x^{3/2}}\,dx \tag{10.151}$$

Here x is a dummy variable for the integration.

Similarly as in Section 10.4, the above can be set as

$$L^{-1}\left\{\frac{\exp\left(-2\zeta\sqrt{\frac{s+b_3}{\beta}}\right)}{s}\right\} = \frac{1}{2}\left[e^{2\zeta\sqrt{\frac{b_3}{\beta}}}\operatorname{erfc}\left(\frac{\zeta}{\sqrt{\beta\tau}} + \sqrt{b_3\tau}\right)\right.$$
$$\left. + e^{-2\zeta\sqrt{\frac{b_3}{\beta}}}\operatorname{erfc}\left(\frac{\zeta}{\sqrt{\beta\tau}} - \sqrt{b_3\tau}\right)\right] \tag{10.152}$$

Returning to Equation 10.147, one gets

$$F(t,z) = \frac{e^{\frac{\zeta}{2\beta}}}{2}\left[e^{2\zeta\sqrt{\frac{b_3}{\beta}}}\operatorname{erfc}\left(\frac{\zeta}{\sqrt{\beta\tau}} + \sqrt{b_3\tau}\right) + e^{-2\zeta\sqrt{\frac{b_3}{\beta}}}\operatorname{erfc}\left(\frac{\zeta}{\sqrt{\beta\tau}} - \sqrt{b_3\tau}\right)\right] \tag{10.153}$$

It is important to check the solution by, for instance, verifying if the above result satisfies conditions given by Equations 10.129 through 10.131. According to Equation 10.25, for $\tau = 0$ the complementary error functions would be zero, which shows that the condition given by Equation 10.129 is satisfied. For ζ tending to infinity, the same is achieved and the condition given by Equation 10.131 is satisfied. Finally, remembering that the error is an odd function [or erf $(-x) = -\operatorname{erf}(x)$], for $\zeta = 0$, Equation 10.153 becomes

$$F(\tau,0) = \frac{1}{2}\left[1 - \operatorname{erf}\left(\sqrt{b_3\tau}\right) + 1 - \operatorname{erf}\left(-\sqrt{b_3\tau}\right)\right]$$
$$= \frac{1}{2}\left[1 - \operatorname{erf}\left(\sqrt{b_3\tau}\right) + 1 + \operatorname{erf}\left(\sqrt{b_3\tau}\right)\right] = 1 \tag{10.154}$$

This agrees with Equation 10.130.

Equation 10.153 is represented in Figure 10.6 for the specific case of $\beta = 1$ and Figure 10.7 illustrates the dimensionless concentration profile for $\beta = 10$. From Figure 10.6, it is possible to observe that

1. At any instant, the dimensionless concentration (F) of component A at the injection point $(z = 0$ or $\zeta = 0)$ would be equal 1.
2. For a given position ahead of the entrance $(\zeta = 0)$, the concentration of A increases with time (τ).
3. After a very long period, the steady-state regime would be achieved. The concentration at any point of the reactor can be deduced by obtaining the limit for expression in Equation 10.153, or

$$
F(\infty, z) = \lim_{\tau \to \infty} \left\{ \frac{e^{\frac{\zeta}{2\beta}}}{2} \left[e^{2\zeta \sqrt{\frac{b_3}{\beta}}} \operatorname{erfc}\left(\frac{\zeta}{\sqrt{\beta \tau}} + \sqrt{b_3 \tau} \right) + e^{-2\zeta \sqrt{\frac{b_3}{\beta}}} \operatorname{erfc}\left(\frac{\zeta}{\sqrt{\beta \tau}} - \sqrt{b_3 \tau} \right) \right] \right\}
$$

$$
= \frac{e^{\frac{\zeta}{2\beta}}}{2} \left[0 + 2 e^{-2\zeta \sqrt{\frac{b_3}{\beta}}} \right] = \exp\left[-\frac{\zeta}{2\beta} \left(2\sqrt{1 + 4\beta} - 1 \right) \right]
$$

$$
= \exp\left[-\frac{zv}{2 D_{AB}} \left(2\sqrt{1 + 4\frac{k D_{AB}}{v^2}} - 1 \right) \right] \tag{10.155}
$$

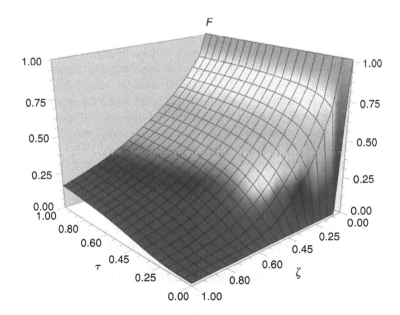

FIGURE 10.6 Concentration profiles in the tubular reactor for the case of $\beta = 1$.

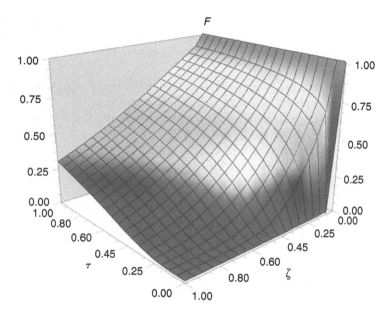

FIGURE 10.7 Concentration profiles in the tubular reactor for the case of $\beta = 10$.

From all this the following can be observed:

- Concentration of component A decays exponentially through the reactor ($z > 0$).
- Decreases in the reaction rate—which can be traced to the coefficient k—lead to slower decay of species A concentration throughout the reactor. Figures 10.6 and 10.7 can be used to verify the effect of parameter k. As seen by Equations 10.125 and 10.126, k equally affects the dimensionless variables of time (τ) and space (ζ). Therefore, for a given real time (t) and position (z), both terms (τ and ζ) would increase at the same ratio. For instance, using Figure 10.6, let us take $\tau = \zeta = 0.20$. From the figure, the value of F would be around 0.57. If k is multiplied by 4, and real time and position remains constant, we would have $\tau = \zeta = 0.80$. In this case, it is possible to observe that F would be around 0.20. Hence, and as expected, the concentration of component A would decrease at the same instant and position in the reactor.
- Diffusivity plays an inverse role of k. Therefore, higher diffusivity tends to slow the decrease of concentration for positions ahead in the reactor. This is easy to imagine because higher diffusivities tend to spread the concentrations more evenly throughout the reactor. The effect of diffusivity is easier to verify than k from the above figures because it just affects parameter β, as defined by Equation 10.128. To exemplify, take for instance dimensionless position $\zeta = 1.0$ and time $\tau = 1.0$. Comparing

Figures 10.6 and 10.7 it is possible to see that increases in diffusivity (or β from 1 to 10) lead to higher dimensionless concentrations (or F approximately from 0.15 to 0.30). Therefore, higher diffusivities lead to larger concentrations at the same values of ζ and τ.

10.6 TEMPERATURE PROFILE IN A RECTANGULAR PLATE

This is the classical problem of steady-state heat transfer in a plate, as illustrated in Figure 10.8.

Three side faces of the plate are kept at known temperature T_0 while the superior one is maintained at constant value T_1. It is desired to obtain the temperature profile throughout the plate.

For this example, the following is assumed:

1. Plate material is solid and, within the range of temperatures observed here, there is no phase change. Therefore, no velocity field is present.
2. No variation of temperature on the orthogonal coordinate (z) to the plane x–y. This condition is possible if the body is indefinite at this direction or if the plate is perfectly isolated at the two main surfaces (x–y planes).
3. Despite variations of temperature, the plate density and thermal conductivity remain constant, or at least approximately constant.
4. No chemical reactions or any other energy source term is involved.

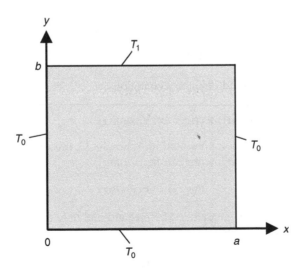

FIGURE 10.8 Rectangular plate with constant temperatures at the edges.

From the above, the energy balance shown by Equation A.31 (or Equation A.34) can be simplified to

$$\frac{\partial^2 T}{\partial x^2} + \frac{\partial^2 T}{\partial y^2} = 0 \tag{10.156}$$

The following new dimensionless variables are proposed:

$$\theta = \frac{T - T_0}{T_1 - T_0} \tag{10.157}$$

$$\chi = \frac{x}{a} \tag{10.158}$$

$$\sigma = \frac{y}{b} \tag{10.159}$$

For the simpler case where $a = b$, Equation 10.156 can be written as

$$\frac{\partial^2 \theta}{\partial \chi^2} + \frac{\partial^2 \theta}{\partial \sigma^2} = 0 \tag{10.160}$$

Hence, for the complete definition of the boundary value problem, two conditions for each independent variable are required. From Figure 10.8, it is possible to set the following:

$$\theta(0,\sigma) = 0, \quad 0 < \sigma < 1 \tag{10.161}$$

$$\theta(\chi,1) = 1, \quad 0 < \chi < 1 \tag{10.162}$$

$$\theta(1,\sigma) = 0, \quad 0 < \sigma < 1 \tag{10.163}$$

$$\theta(\chi,0) = 0, \quad 0 < \chi < 1 \tag{10.164}$$

Therefore, all are first-kind boundary conditions.

10.6.1 SOLUTION BY SEPARATION OF VARIABLES

According to the discussion presented in Appendix G, this technique is based on the assumption that the following can be written:

$$\theta(\chi,\sigma) = F(\chi)G(\sigma) \tag{10.165}$$

Of course, this cannot be guaranteed beforehand and only trial would tell. Given this, Equation 10.160 leads to

$$\frac{F''(\chi)}{F(\chi)} = -\frac{G''(\sigma)}{G(\sigma)} = C \tag{10.166}$$

The equality between groups involving functions of independent variables forces each side of the equation to be equal to a constant C. Consequently, two ordinary differential equations are generated, and one of them is

$$F''(\chi) - CF(\chi) = 0 \tag{10.167}$$

The general solution for such an equation presents the following possibilities:

1. According to Appendix B, if C is positive, or equal to γ^2 (where γ is a real parameter), the solution would be

$$F(\chi) = C_1 e^{-\gamma\chi} + C_2 e^{\gamma\chi} \tag{10.168}$$

   However, from Equations 10.161, 10.163, and 10.165, one would conclude that

$$F(0) = 0 \tag{10.169}$$

$$F(1) = 0 \tag{10.170}$$

   The above conditions would force $C_1 = C_2 = 0$, i.e., the trivial solution for $F(\chi)$ and, consequently, to θ as well. Therefore, the above is not an acceptable choice.

2. If C is equal to zero, the solution would be

$$F(\chi) = C_1 \chi + C_2 \tag{10.171}$$

   After the application of boundary conditions, the trivial solution is forced again.

3. If C is negative or equal to $-\gamma^2$ (γ real), the solution would be

$$F(\chi) = C_1 \sin(\gamma\chi) + C_2 \cos(\lambda\chi) \tag{10.172}$$

Application of the condition given by Equation 10.169 leads to $C_2 = 0$ and the condition given by Equation 10.170 to

$$0 = C_1 \sin(\gamma) \tag{10.173}$$

If $\gamma = n\pi$ ($n = 1, 2, \ldots$), C_1 might be different from zero, therefore avoiding the trivial solution. Thus

$$F_n(\chi) = C_n \sin(n\pi\chi) \tag{10.174}$$

The index n is used here to emphasize the multiple determinations of the solution.

Returning to Equation 10.166, the solution for function G would be

$$G_n(\sigma) = A_n e^{-n\pi\sigma} + B_n e^{n\pi\sigma} \qquad (10.175)$$

The condition given by Equation 10.164 leads to $G(0) = 0$, and the above equation would require that

$$A_n = -B_n \qquad (10.176)$$

Therefore, Equation 10.175 can be written as

$$G_n(\sigma) = A_n^* \sinh(n\pi\sigma) \qquad (10.177)$$

At the present, the condition given by Equation 10.162 does not allow any conclusion.

Combining the solutions for functions F and G given by Equations 10.174 and 10.177, a particular form for the dimensionless temperature can be arrived at

$$\theta_n(\chi,\sigma) = K_n \sin(n\pi\chi) \sinh(n\pi\sigma) \qquad (10.178)$$

The general solution would be a linear combination of all those particular ones, or

$$\theta(\chi,\sigma) = \sum_{n=1}^{\infty} K_n \sin(n\pi\chi) \sinh(n\pi\sigma) \qquad (10.179)$$

Applying the condition given by Equation 10.162 it is possible to write

$$1 = \sum_{n=1}^{\infty} K_n \sinh(n\pi) \sin(n\pi\chi) \qquad (10.180)$$

As seen in Appendix G, this is a Fourier half-range expansion of function 1. Therefore

$$K_n = \frac{2}{\sinh(n\pi)} \int_0^1 \sin(n\pi\chi)d\chi = \frac{2}{n\pi \sinh(n\pi)}[1 - \cos(n\pi)]$$

$$= \frac{2}{n\pi \sinh(n\pi)}[1 - (-1)^n], \quad n = 1, 2, 3, \dots \qquad (10.181)$$

Finally, the temperature profile in the plate is given by

$$\theta(\chi,\sigma) = \frac{2}{\pi} \sum_{n=1}^{\infty} \frac{1 - (-1)^n}{n} \frac{\sin(n\pi\chi) \sinh(n\pi\sigma)}{\sinh(n\pi)} \qquad (10.182)$$

The above solution is illustrated in Figure 10.9. The figure is self-explanatory.

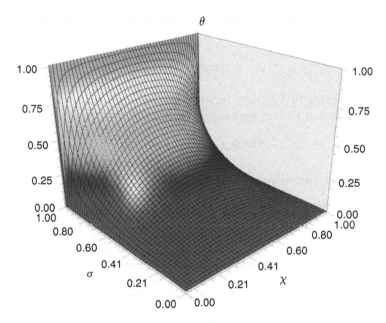

FIGURE 10.9 Dimensionless temperature (θ) profile in the plate as a function of dimensionless coordinates χ and σ.

10.6.2 Solution by the Method of Weighted Residuals

As commented before, MWR can also be applied to partial differential equations.

The first step is to select trial functions for the solution of boundary problems formed by Equation 10.160 and conditions given by Equations 10.161 through 10.164.

For cases of partial differential equations, a good choice for the trial function should

1. Be simple, such as polynomials. In the present case, the polynomials could be set using either χ or σ as an independent variable. The amount of work involved might depend on a particular choice.
2. Satisfy the boundary conditions.

From the above, a sensible alternative for the trial functions could be

$$\theta_n = \sum_{j=1}^{n} \sin(j\pi\chi)\psi_j(\sigma) \qquad (10.183)$$

As seen, the above function satisfies the conditions given by Equations 10.161 and 10.163. In addition, if

$$\psi_j(0) = 0 \tag{10.184}$$

the condition given by Equation 10.164 would be satisfied.

Using Equation 10.160, the residual is given by

$$\Lambda(\sigma,\psi_n) = (\theta_n)_{\chi\chi} + (\theta_n)_{\sigma\sigma} \tag{10.185}$$

10.6.2.1 First Approximation

From Equation 10.183, the first approximation is

$$\theta_1 = \sin(\pi\chi)\psi_1(\sigma) \tag{10.186}$$

Using the two previous equations, the residual becomes

$$\Lambda_1 = \sin(\pi\chi)\psi_1'' - \pi^2 \sin(\pi\chi)\psi_1 \tag{10.187}$$

Thus, the condition to annul the residual can be achieved here without the need for applying any specific weighted residual submethod.

The solution for

$$\psi_1'' = \pi^2\psi_1 \tag{10.188}$$

is

$$\psi_1(\sigma) = C_{11} \sinh(\pi\sigma) + C_{12} \cosh(\pi\sigma) \tag{10.189}$$

Employing the condition given by Equation 10.184, $C_{12} = 0$. Consequently, the first approximation can be written as

$$\theta_1 = C_{11} \sin(\pi\chi) \sinh(\pi\sigma) \tag{10.190}$$

Nonetheless, this brings a problem because the condition given by Equation 10.162 cannot be satisfied. Let us go ahead with further approximations.

10.6.2.2 Further Approximations

Further approximations would require the application of a branch method of MWR. Let us apply the collocation method.

A more general choice for equidistant collocation points would be $\dfrac{m}{n+1}$, $m = 1, 2, \ldots n$, where n is the approximation order. Let us exemplify with the third approximation, which from Equation 10.183 is written as

$$\theta_3 = \sin(\pi\chi)\psi_1(\sigma) + \sin(2\pi\chi)\psi_2(\sigma) + \sin(3\pi\chi)\psi_3(\sigma) \tag{10.191}$$

Using Equation 10.185, the residual becomes

$$\Lambda_3 = \sin(\pi\chi)\psi_1'' - \pi^2 \sin(\pi\chi)\psi_1 + \sin(2\pi\chi)\psi_2'' - 4\pi^2 \sin(2\pi\chi)\psi_2$$
$$+ \sin(3\pi\chi)\psi_3'' - 9\pi^2 \sin(3\pi\chi)\psi_3 \tag{10.192}$$

The collocation points would be 1/4, 1/2, and 3/4. Applying this to the above equation and setting it to zero, one would arrive at the following differential equations:

$$\frac{\sqrt{2}}{2}\left[\psi_1'' - \pi^2\psi_1\right] + \psi_2'' - 4\pi^2\psi_2 + \frac{\sqrt{2}}{2}\left[\psi_3'' - 9\pi^2\psi_3\right] = 0 \tag{10.193}$$

$$\left[\psi_1'' - \pi^2\psi_1\right] - \left[\psi_3'' - 9\pi^2\psi_3\right] = 0 \tag{10.194}$$

$$\frac{\sqrt{2}}{2}\left[\psi_1'' - \pi^2\psi_1\right] - \left[\psi_2'' - 4\pi^2\psi_2\right] + \frac{\sqrt{2}}{2}\left[\psi_3'' - 9\pi^2\psi_3\right] = 0 \tag{10.195}$$

Such a system can be reduced to

$$\psi_1'' - \pi^2\psi_1 = 0 \tag{10.196}$$

$$\psi_2'' - 4\pi^2\psi_2 = 0 \tag{10.197}$$

$$\psi_3'' - 9\pi^2\psi_3 = 0 \tag{10.198}$$

After the application of the condition given by Equation 10.184, the solution for Equation 10.196 is given by Equation 10.190. The solutions for the others are similar.

Of course, it is easy to see that the whole process would lead to a general solution, which can be written as

$$\theta_n = \sum_{j=1}^{n} C_j \sin(j\pi\chi) \sinh(j\pi\sigma) \tag{10.199}$$

In order to satisfy the condition given by Equation 10.162, the following should happen:

$$\sum_{j=1}^{n} C_j \sin(j\pi\chi) \sinh(j\pi) = 1 \tag{10.200}$$

When n tends to infinite values, the above reproduces the problem arrived at by Equations 10.179 and 10.180, or an odd half-range expansion of function 1. Therefore, the result would be the same as that given by Equation 10.182.

This exemplifies how, in few occasions, the method of weighted residual might even lead to the exact solution.

10.7 HEATING A LIQUID FILM

A liquid film runs on an inclined plate, as shown in Figure 10.10.

The fluid, at temperature T_0, enters a section of the plate—beginning at $z=0$—which is kept at constant temperature T_s. It is desired to determine the temperature profile in the liquid film as well as the heat transfer rate between it and the plate.

For this example, the following is assumed:

1. Steady-state regime.
2. Liquid fluid is Newtonian with constant physical properties, among them density and viscosity. Of course, the assumption of constant properties becomes increasingly critical with the magnitude of temperature differences in the liquid film.
3. Laminar flow. As discussed in Section 5.2, this is also an approximation and valid for very smooth solid surfaces and within a relatively short length of plate.
4. Plate is very wide and long and the studied region is far from the borders.
5. Flow is already developed at position $z=0$.
6. Since the fluid density is assumed constant—which is the same to say that the fluid is incompressible—its thickness (δ) is constant, at least within the plate length where the present solution is approximately valid.
7. Shear stress at the interface between the fluid and air is negligible. Owing to the very low viscosity of air when compared with usual fluids, this is a very reasonable approximation.

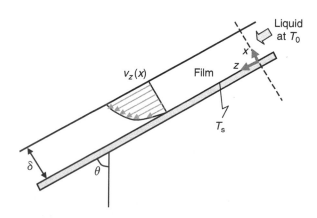

FIGURE 10.10 Flowing film over a plate kept at constant temperature.

8. Negligible dissipation of energy due to viscous flow. This is reasonable, given the relatively low velocity field found in the present situation.

The mass and momentum conservation equations would provide the velocity profile. Since the density and viscosity are considered constant, the solution achieved in Section 5.2 is valid and after the change in coordinates' position, it is possible to write

$$v_z = \frac{\rho g}{2\mu} \cos \theta x^2 \qquad (10.201)$$

Assumption 2 allows the use of Equation A.34 to describe energy conservation. After the assumptions listed above, it is possible to write

$$\rho C_p \left(v_z \frac{\partial T}{\partial z} \right) = \lambda \left(\frac{\partial^2 T}{\partial x^2} + \frac{\partial^2 T}{\partial z^2} \right) \qquad (10.202)$$

Equation 10.202 can be further simplified by noticing that the temperature variation in the x direction should be much higher than the variation in the z direction. Therefore, it becomes

$$v_z \frac{\partial T}{\partial z} = D_T \frac{\partial^2 T}{\partial x^2} \qquad (10.203)$$

Here D_T is the thermal diffusivity term and is given by Equation 10.95. Using Equation 10.201, the equation to solve becomes

$$x^2 \frac{\partial T}{\partial z} = b \frac{\partial^2 T}{\partial x^2} \qquad (10.204)$$

where

$$b = \frac{2D_T \mu}{\rho g \cos \theta} = \frac{2\mu\lambda}{\rho^2 g C_p \cos \theta} \qquad (10.205)$$

Three boundary conditions are needed to achieve the solution. From Figure 10.10 it is possible to verify that two of them are

$$T(x,0) = T_0, \quad 0 < x \le \delta \qquad (10.206)$$

$$T(0,z) = T_s, \quad z > 0 \qquad (10.207)$$

The third condition would be related to coordinate x. For instance, one possibility would be to base it on correlations for the heat transfer at the film surface or at $x = \delta$. However, this would lead to a second- or even third-kind boundary condition. To keep the problem within the scope of the present chapter, let us assume that the solution is sought for small contact time between the liquid and the plate. This would allow a solution for layers of liquid near the plate for which the position at the film surface would be considered very far, or at $x = \infty$. These far positions would remain at T_0 during the period of heating or liquid-plate contact. Therefore, it would be possible to write the third condition as

$$T(\infty, z) = T_0, \quad z > 0 \tag{10.208}$$

As always, it is worthwhile to change variables to facilitate the solution. In the present case, the following have been selected:

$$\psi = \frac{T - T_0}{T_s - T_0} \tag{10.209}$$

$$\chi = \frac{x}{\delta} \tag{10.210}$$

$$\zeta = \frac{z}{\delta} \tag{10.211}$$

Using these, Equation 10.204 becomes

$$\chi^2 \frac{\partial \psi}{\partial \zeta} = \beta \frac{\partial^2 \psi}{\partial \chi^2} \tag{10.212}$$

where

$$\beta = \frac{b}{\delta^3} = \frac{2 D_T \mu}{\rho g \delta^3 \cos \theta} = \frac{2 \mu \lambda}{\rho^2 g C_p \delta^3 \cos \theta} \tag{10.213}$$

The boundary conditions become

$$\psi(\chi, 0) = 0, \quad 0 < \chi \le \delta \tag{10.214}$$

$$\psi(0, \zeta) = 1, \quad \zeta > 0 \tag{10.215}$$

$$\psi(\infty, \zeta) = 0, \quad \zeta > 0 \tag{10.216}$$

One should notice that two among these conditions might coalesce into a single condition. This suggests the application of similarity, also known as combination of variables.

10.7.1 SOLUTION BY SIMILARITY

From Appendix F, the application of the method of similarity starts by proposing new variables, such as

$$\bar{\chi} = a^{b_1}\chi \tag{10.217}$$

$$\bar{\zeta} = a^{b_2}\zeta \tag{10.218}$$

$$\bar{\psi} = a^c\psi \tag{10.219}$$

Therefore

$$\frac{\partial\psi}{\partial\chi} = \frac{\partial\bar{\psi}}{\partial\bar{\chi}}\frac{\partial\bar{\chi}}{\partial\chi}\frac{\partial\psi}{\partial\bar{\psi}} = a^{b_1+c}\frac{\partial\bar{\psi}}{\partial\bar{\chi}} \tag{10.220}$$

$$\frac{\partial\psi}{\partial\zeta} = \frac{\partial\bar{\psi}}{\partial\bar{\zeta}}\frac{\partial\bar{\zeta}}{\partial\zeta}\frac{\partial\psi}{\partial\bar{\psi}} = a^{b_2+c}\frac{\partial\bar{\psi}}{\partial\bar{\zeta}} \tag{10.221}$$

$$\frac{\partial^2\psi}{\partial\chi^2} = a^{b_1+c}\frac{\partial\bar{\chi}}{\partial\chi}\frac{\partial}{\partial\bar{\chi}}\left(\frac{\partial\bar{\psi}}{\partial\bar{\chi}}\right) = a^{2b_1+c}\frac{\partial^2\bar{\psi}}{\partial\bar{\chi}^2} \tag{10.222}$$

Following the same procedure for other involved derivatives and substituting them into Equation 10.211, one would arrive at

$$a^{b_2-2b_1+c}\bar{\chi}^2\frac{\partial\bar{\psi}}{\partial\bar{\zeta}} = a^{2b_1+c}\beta\frac{\partial^2\bar{\psi}}{\partial\bar{\chi}^2} \tag{10.223}$$

To assure the invariance condition it is necessary that

$$2b_1 + c = b_2 - 2b_1 + c \tag{10.224}$$

or

$$b_1 = \frac{b_2}{4}, \quad \text{for any } c \tag{10.225}$$

Thus, the new combined variables would be

$$\eta = \frac{\chi}{\zeta^{1/4}} \tag{10.226}$$

and

$$f(\eta) = \frac{\psi}{\zeta^c} \tag{10.227}$$

Without any loss of generality, the choice of $c = 0$ is convenient. Consequently, one would arrive at $f(\eta) = \psi$. After this, the following is obtainable:

$$\frac{\partial \psi}{\partial \chi} = \frac{df}{d\eta}\frac{\partial \eta}{\partial \chi} = \zeta^{-1/4}\frac{df}{d\eta} \tag{10.228}$$

$$\frac{\partial^2 \psi}{\partial \chi^2} = \frac{\partial \eta}{\partial \chi}\frac{d}{d\eta}\left(\zeta^{-1/4}\frac{df}{d\eta}\right) = \zeta^{-1/2}\frac{d^2 f}{d\eta^2} \tag{10.229}$$

$$\frac{\partial \psi}{\partial \zeta} = \frac{df}{d\eta}\frac{\partial \eta}{\partial \zeta} = -\frac{1}{4}\chi\zeta^{-5/4}\frac{df}{d\eta} \tag{10.230}$$

Applying these at Equation 10.212, the following is obtained:

$$\frac{d^2 f}{d\eta^2} = -\frac{1}{4\beta}\eta^3\frac{df}{d\eta} \tag{10.231}$$

Consider a new variable defined as

$$\varphi = \frac{df}{d\eta} \tag{10.232}$$

Hence, it is possible to write Equation 10.231 as

$$\varphi = C_1 \exp\left(-\frac{1}{16\beta}\eta^4\right) + C_2 \tag{10.233}$$

and

$$f = C_1 \int \exp\left(-\frac{1}{16\beta}\eta^4\right)d\eta + C_2 \tag{10.234}$$

Using Equations 10.226 and 10.227, the boundary conditions given by Equations 10.214 through 10.216 can be written as

$$f(\eta \to \infty) = 0 \tag{10.235}$$

$$f(\eta = 0) = 1 \tag{10.236}$$

Equation 10.235 can be applied at Equation 10.234. Setting 1 limit for the integral, and eliminating the constant C_2, one may write

$$f = C_1 \int_{\eta}^{\infty} \exp\left(-\frac{1}{16\beta}u^4\right)du \tag{10.237}$$

Here, u is just an integration variable.

At this point, the condition given by Equation 10.236 can be applied to Equation 10.237 to provide the remaining integration constant, or

$$C_1 = \frac{1}{\int_0^\infty \exp\left(-\frac{1}{16\beta}u^4\right)du}$$
(10.238)

Finally

$$f(\eta) = \frac{\int_\eta^\infty \exp\left(-\frac{1}{16\beta}u^4\right)du}{\int_0^\infty \exp\left(-\frac{1}{16\beta}u^4\right)du}$$

$$= \psi(\chi,\zeta) = \frac{T-T_0}{T_s-T_0} = \frac{\int_{\chi/\zeta^{1/4}}^\infty \exp\left(-\frac{1}{16\beta}u^4\right)du}{\int_0^\infty \exp\left(-\frac{1}{16\beta}u^4\right)du}$$
(10.239)

The dimensionless temperature profile (ψ) is illustrated in Figures 10.11 and 10.12 for cases of $\beta = 1$ and $\beta = 0.1$.

From these figures, it is possible to verify that temperature equalizes faster in the film layer (x or χ direction) for larger values of parameter β. This is easy to

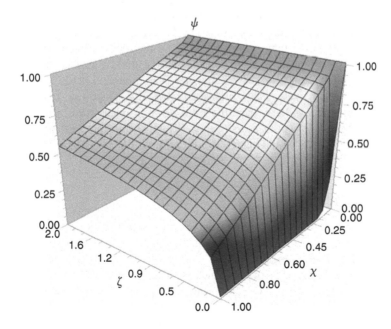

FIGURE 10.11 Dimensionless temperature profiles within liquid film for the case of $\beta = 1$.

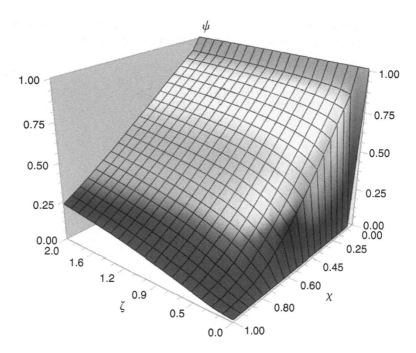

FIGURE 10.12 Dimensionless temperature profiles within liquid film for the case of $\beta = 0.1$.

understand because, as shown by Equation 10.213, such a parameter is proportional to the thermal diffusivity of the flowing liquid. Notice, as well, that even small increases on the film thickness (δ) leads to sharp decreases of parameter β, hence providing larger temperature gradients in the film (x direction).

For the case of water at around 290 K and a film 1 mm thick flowing on a plate with inclination of 60°, the parameter β is around 6×10^{-5}. Such a low value indicates that water has a high thermal inertia. Therefore, large gradients of the temperature would occur in the x direction. In other words, for most of the film the temperature would remain very different from temperature T_s. This is the same to say that ψ would approach zero for positions not too far from the plate surface ($\chi = 0$). Such a case is illustrated in Figure 10.13.

10.8 ABSORBING FLOWING FILM

Several industrial separation processes work by selective absorption of a component from a gas mixture by a liquid. The present discussion exemplifies a simple model for this process.

Consider a liquid film of substance B flowing on a solid vertical wall, as illustrated in Figure 10.14. The film, with constant thickness δ, enters a section ($x \geq 0$) where it comes in contact with a gas mixture. The component A of that mixture is absorbed by component B in the liquid. It is desired to determine the concentration of component A throughout the film as well as its rate of absorption.

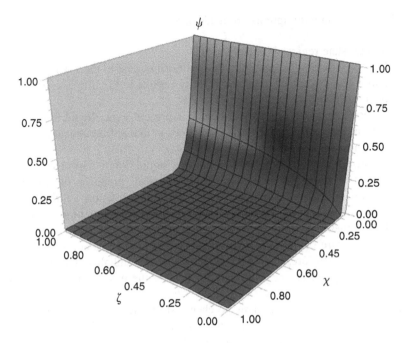

FIGURE 10.13 Dimensionless temperature profiles within liquid water film.

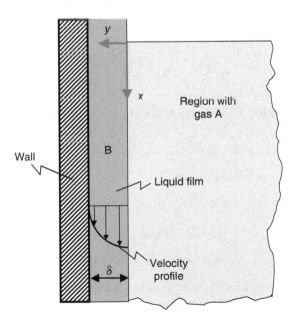

FIGURE 10.14 Liquid film of component B absorbing component A from surrounding gas.

The following assumptions are made here:

1. Steady-state regime is established.
2. Wall is indefinite in the z direction and orthogonal to the x–y plane. The plane is also very tall, and the velocity regime in the film is completely developed at the entering section $x = 0$.
3. At position $x = 0$, only component B is present in the liquid film.
4. Within the studied region, the fluid flows in laminar regime and the film thickness (δ) is constant.
5. Despite any variation in composition, the liquid film remains Newtonian.
6. Density, viscosity, and any other physical property of the film remain constant. Of course, this is an approximation valid for low levels of absorption, or regions with small concentration of species A in liquid phase.
7. Temperature of the film is the same as that of the gas and they do not change. Therefore, it is assumed that either the enthalpy changes due to the absorption are negligible or, as before, the absorption rate is relatively small.
8. Absorption process involves no chemical reaction.
9. Short contact times between the gas and film. Therefore, the solution would be valid only for the region where the absorbing front does not reach too deep into the film.

Considering the above, there are only velocity components in the vertical (x) direction. In addition, all variations of concentration would occur in the x–y plane. Therefore, Equation A.40 becomes

$$v_x \frac{\partial \rho_A}{\partial x} = D_{AB} \left(\frac{\partial^2 \rho_A}{\partial x^2} + \frac{\partial^2 \rho_A}{\partial y^2} \right) \tag{10.240}$$

Owing to assumption 3 above, the following conditions can be set:

$$\rho_A(0,y) = 0, \quad 0 < y \leq \delta \tag{10.241}$$

Additionally, it is possible to write

$$\rho_A(\infty,y) = \rho_{A,sat}, \quad 0 \leq y \leq \delta \tag{10.242}$$

$$\rho_A(x,0) = \rho_{A,sat}, \quad 0 < x \leq L \tag{10.243}$$

$$\rho_A(x,\infty) = 0, \quad 0 < x \leq L \tag{10.244}$$

Here $\rho_{A,sat}$ is the saturation or equilibrium concentration of component A in component B. Notice that the condition given by Equation 10.244 is only possible because of assumption 9. Consequently, the regions near the wall are viewed as too far from the film surface.

For a film region near the surface, it is possible to approximate the velocity profile to

$$v_x = \frac{g\delta}{\nu} y \tag{10.245}$$

This relation can be obtained from the results obtained in Section 5.2 and is proposed as an exercise.

Since the variation of concentration in the horizontal direction is much higher than in the vertical one, it also possible to simplify Equation 10.240 by assuming the following:

$$\frac{\partial^2 \rho_A}{\partial y^2} \gg \frac{\partial^2 \rho_A}{\partial x^2} \tag{10.246}$$

Hence, the governing equation becomes

$$\frac{\partial \rho_A}{\partial x} = \zeta \frac{1}{y} \frac{\partial^2 \rho_A}{\partial y^2} \tag{10.247}$$

where

$$\zeta = \frac{D_{AB}\nu}{g\delta} \tag{10.248}$$

10.8.1 SOLUTION BY SIMILARITY

According to the technique shown in Appendix F, let there be the following new set of variables:

$$\bar{x} = a^{b_1} x \tag{10.249}$$

$$\bar{y} = a^{b_2} y \tag{10.250}$$

$$\bar{\rho}_A = a^c \rho_A \tag{10.251}$$

After these changes of variables, Equation 10.247 becomes

$$a^{c-b_1} \frac{\partial \bar{\rho}_A}{\partial \bar{x}} = a^{c-3b_2} \zeta \frac{1}{\bar{y}} \frac{\partial^2 \bar{\rho}_A}{\partial \bar{y}^2} \tag{10.252}$$

The invariance regarding Equation 10.247 would be observed if $b_1 = 3b_2$ and for any value of c. Consequently, it is possible to set a new variable given by

$$\eta = \frac{y}{x^{1/3}} \tag{10.253}$$

Using the above, Equation 10.247 can be written as

$$-\eta^2 \frac{d\rho_A}{d\eta} = \zeta \frac{d^2\rho_A}{d\eta^2} \qquad (10.254)$$

This is a simple second-order differential equation, according to Appendix B, with a general solution given by

$$\rho_A = C_1 \int_0^\eta \exp\left(-\frac{z^3}{3\zeta}\right) dz + C_2 \qquad (10.255)$$

Here z is just a dummy integration variable. The choice of integration limits does not lead to any loss of generality because the conformity to boundary conditions would be obtained by proper values of the constants.

Employing Equation 10.253, the boundary conditions given by Equations 10.241 and 10.244 coalesce to give

$$\rho_A(\eta \to \infty) = 0 \qquad (10.256)$$

The conditions given by Equations 10.242 and 10.243 can also be condensed into the following:

$$\rho_A(\eta = 0) = \rho_{A,\text{sat}} \qquad (10.257)$$

The final solution is obtained after application of these conditions to Equation 10.255, or

$$\frac{\rho_A}{\rho_{A,\text{sat}}} = \frac{\int_\eta^\infty \exp\left(-\frac{z^3}{3\zeta}\right) dz}{\int_0^\infty \exp\left(-\frac{z^3}{3\zeta}\right) dz} \qquad (10.258)$$

10.8.2 COMMENTS

To facilitate the discussion and handling of equations, the above problem is rewritten using dimensionless variables defined as follows:

$$\chi = \frac{x}{L} \qquad (10.259)$$

$$\omega = \frac{y}{\delta} \qquad (10.260)$$

and

$$\psi = \frac{\rho_A}{\rho_{A,sat}} \tag{10.261}$$

Here L is the length of the wall on which the absorption process takes place and within the region where the assumptions made at the beginning of the present section are valid.

Applying the above, Equation 10.258 can be written as

$$\psi(\chi,\omega) = \frac{\int_{\sigma}^{\infty} \exp\left(-\frac{z^3}{3\xi}\right) dz}{\int_{0}^{\infty} \exp\left(-\frac{z^3}{3\xi}\right) dz} \tag{10.262}$$

Here

$$\xi = \frac{D_{AB}\nu L}{g\delta^4} \tag{10.263}$$

$$\sigma = \frac{\omega}{\chi^{1/3}} \tag{10.264}$$

To illustrate, let $D_{AB} = 1 \times 10^{-5}$ m^2 s^{-1}, $\nu = 1 \times 10^{-6}$, and $\delta = 1$ mm. The value of kinematic viscosity is relative to water at around 293 K. Therefore, $\xi = 1$. Equation 10.262 is valid only for regions near the film surface—or small values of ω—and is represented in Figure 10.15.

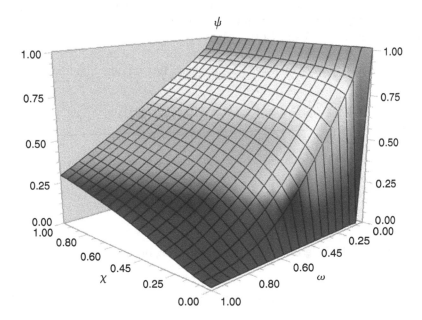

FIGURE 10.15 Concentration profile in the film for $\xi = 1$.

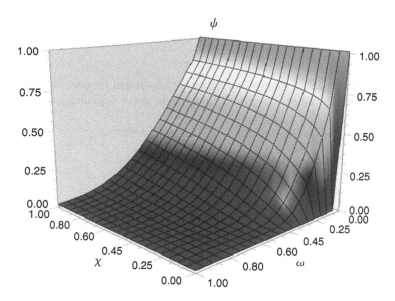

FIGURE 10.16 Concentration profile in the film for $\xi = 0.1$.

It is worthwhile to compare Figure 10.15 with Figure 10.16, plotted for the case when $\xi = 0.1$.

According to Equation 10.263, ξ is proportional to the diffusivity of A into B. Therefore, decreases in diffusivity should lead to lower concentrations of absorbed component A into the film. This is easy to verify from the differences between Figures 10.15 and 10.16.

The rate at which component A is absorbed is given by its mass flux at the film surface. From Equation A.53, the mass flux into the film (direction y) is

$$\mathbf{N}_{A_y} = -D_{AB}\rho \frac{\partial w_A}{\partial y} + w_A(\mathbf{N}_{A_y} + \mathbf{N}_{B_y}) \qquad (10.265)$$

Since component B is stationary in this direction and there is no overall velocity in the y direction, Equation A.55 yields

$$\mathbf{N}_{A_y} + \mathbf{N}_{B_y} = 0 \qquad (10.266)$$

Using these and taking the flux at the film surface, one can write

$$\mathbf{N}_{A_y}(y = 0) = -D_{AB}\rho \frac{\partial w_A}{\partial y}\bigg|_{y=0} \qquad (10.267)$$

Because of assumption 6, the density remains constant and using Equations 10.260 and 10.261, the above becomes

$$N_{A_y}(y=0) = -\frac{D_{AB}\rho_{A,sat}}{\delta}\frac{\partial \psi}{\partial \omega}\bigg|_{\omega=0} \tag{10.268}$$

Using the change of variable defined before it is possible to write

$$N_{A_y}(y=0) = -\frac{D_{AB}\rho_{A,sat}}{\delta}\chi^{-1/3}\frac{d\psi}{d\sigma}\bigg|_{\sigma=0} \tag{10.269}$$

As expected, the rate of absorption is a function of the vertical position (x or χ).

Now, the derivative at Equation 10.269 can be obtained from Equation 10.262, leading to

$$N_{A_y}(y=0) = -\frac{D_{AB}\rho_{A,sat}}{\delta\chi^{1/3}}\frac{1}{\int_0^\infty \exp\left(-\frac{z^3}{3\xi}\right)dz} \tag{10.270}$$

Therefore, the total rate of absorption per unit of film surface area (kg m^{-2} s^{-1}) would be obtained after the integration of flux N_A within the vertical length L or the film, or from $\chi=0$ to $\chi=1$. The result is

$$\overline{N}_{A_y} = -\frac{3D_{AB}\rho_{A,sat}}{2\delta}\frac{1}{\int_0^\infty \exp\left(-\frac{z^3}{3\xi}\right)dz} \tag{10.271}$$

Again, one should be aware that the present results are only valid for a relatively short contact time between the gas and absorbing liquid, as imposed by assumption 9 and the approximation represented by Equation 10.245.

EXERCISES

1. Solve the differential Equation 10.45 to arrive at Equation 10.46.
2. Solve the problem presented in Section 10.2 in the case of a finite bar. To save time, do not use Equation 10.31 with boundary conditions given by Equations 10.32 and 10.33, but replace the boundary condition given by Equation 10.34 by

$$\psi(\tau,1) = 0, \quad \tau > 0$$

   Apply any method.
3. From the results obtained in Section 10.4, deduce a formula to compute the rate of heat transfer between the grid and the fluid.

4. Try to apply the method of similarity to the problem presented in Section 10.4.
5. From the assumptions made at the beginning of Section 10.6 and departing from Equation A.31 or A.34, arrive at Equation 10.156.
6. Using the proposed changes of variables given by Equations 10.157 through 10.159, depart from Equation 10.156 and arrive at Equation 10.160.
7. Using the result for temperature profile in a plate, as given in Section 10.6, obtain the expressions for the heat fluxes at faces $x=0$ and $y=b$.
8. Instead of collocation, try to apply other branches of the method of weighted residuals to solve the problem presented in Section 10.6.
9. From Equation 10.204 and using the proposed change of variables given by Equations 10.209 through 10.211, deduce Equation 10.212.
10. From Section 5.2, show that the approximation indicated by Equation 10.245 is reasonable at regions near the film surface.
11. Deduce Equation 10.252.
12. Work the details to arrive at Equation 10.262.
13. A constant film thickness was assumed to solve the problem presented in Section 10.8. Verify the validity of such an assumption based on the momentum equations applied to that situation.
14. Let there be a semi-infinite slab (Figure 10.17) initially at temperature T_0. The slab, with constant thermal conductivity, is insulated in both larger faces. At a given instant, the surface temperature at $x=0$ is immediately raised to temperature T_1. Determine the temperature profile throughout the slab as a function of space (x) and time.
Solve the problem by Laplace transform and by similarity.
15. Solve the last problem for a situation where the thermal conductivity of the slab material is not constant but is a linear function of the temperature given by

$$\lambda(T) = aT + b$$

with a and b constants. Use any method, including the MWR.

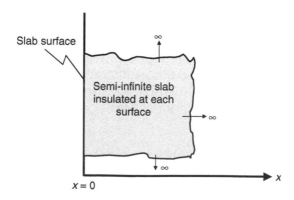

FIGURE 10.17 Heating a semi-infinite slab.

16. The equation that governs the vertical position y of each point in the horizontal direction x of a string against time t under tension is given by

$$\frac{\partial^2 y}{\partial t^2} = c^2 \frac{\partial^2 y}{\partial x^2}$$

where c is a parameter related to the tension and density of the string. The string, with length L, is fixed at each end. The initial form of the string $[y(x,0)]$ and velocity of each point are given by known functions $f(x)$ and $g(x)$, respectively. Solve the above equation by separation of variables to obtain the form of the string at any time, or $y(x,t)$.

17. In the previous problem, imagine a string initially at rest and with the deflection given by $f(x) = 0.01 \sin x$. Draw the form of the string for the following instants: $L/6c$, $L/5c$, $L/3c$, and L/c.

18. Determine the thickness of the moisture concentration profile in a direction against time for a wall with indefinite surface area—or area not limited at directions perpendicular to the thickness. The initial moisture throughout the wall is u_0 and the moisture at the surface is constant and equal to u_e. This value is given by the equilibrium with the ambiance air. Use the similarity method.

19. Use any variation of the method of weighted residual (MWR) to obtain a reasonable approximate solution of the previous problem.

20. Try to solve the problem presented in Section 10.8 using the Laplace transform, method of separation of variables, and method of weighted residual (any variation).

21. Researchers in the area of soil contamination apply various models to verify the extent or depth of a pollutant concentration against time. Of course, nowadays such models are at a very sophisticated level. However, for the sake of illustrating the basic principles of such models, let us consider Figure 10.18, where the surface of a region has been contaminated by a pollutant (A). Aqueous solutions of the contaminant tend to sip and diffuse into the soil. It is

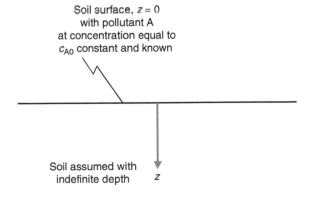

Soil surface, $z = 0$
with pollutant A
at concentration equal to
c_{A0} constant and known

Soil assumed with
indefinite depth z

FIGURE 10.18 Illustration for the problem on soil contamination.

desired to find the concentration of A at any given time and any given depth (z) in the soil. For this, assume the following:

a. The concentration of contaminant at the surface of the soil is known and constant.
b. Initially, no contaminant A is found in the soil.
c. The soil (land and water) can be modeled by a single homogenous substance (call it B) into which component A is soluble.
d. The diffusivity of contaminant into the soil is known. Of course, one of the problems of researchers in the area is to develop equations for the proper calculation of such diffusion parameters.
e. The soil has an indefinite depth.
f. As a first approximation, negligible overall velocity is observed in the soil.

22. Repeat the previous problem assuming now an overall constant velocity in the direction z. Its value is known and given. Actually, this is a more realist picture for situations where, for instance, the contaminant is also carried into the soil due to rain.

REFERENCE

1. Abramovitz, M. and Stegun, I.A., *Handbook of Mathematical Functions*, Dover, New York, 1972.

11 Problems 222; Two Variables, 2nd Order, 2nd Kind Boundary Condition

11.1 INTRODUCTION

This chapter presents methods to solve problems with two independent variables involving second-order differential equations and second-kind boundary conditions. Therefore, the problems fall in the category of partial differential equations. Mathematically, this class of cases can be summarized as $f\left(\phi, \omega_1, \omega_2, \frac{\partial^2 \phi}{\partial \omega_i^2}, \frac{\partial^2 \phi}{\partial \omega_i \partial \omega_j}, \frac{\partial \phi}{\partial \omega_i}\right)$, second-kind boundary condition.

Several examples involving mass, momentum, and heat transfer are described. Unlike in Chapter 10, the examples shown in this chapter involve second-type boundary conditions, which usually lead to more feasible situations than before.

As in the previous chapters, it is important to notice that the solutions presented here for mass transfer can be used for cases of heat transfer, and vice versa. For this, the concentration must be equivalent to temperature, diffusivity to thermal conductivity, and mass transfer coefficient to heat transfer coefficient. With the above conditions satisfied, Sherwood and Biot numbers would be interchangeable.

11.2 HEATING AN INSULATED ROD OR SEMI-INFINITE BODY

Let us consider a finite solid rod as shown in Figure 11.1. Initially, it is at uniform temperature T_b, and at a given instant, the temperature T_0 is imposed at surface $z = 0$, and one end is insulated. It is desired to determine the temperature profile in the rod as a function of position and time as well as the rate of heat transfer to the rod.

It is important to notice that the present problem is more feasible than the one proposed in Section 10.2, where a semi-infinite cylindrical rod was considered. As commented at the end of this section, the solutions achieved here can equally be applied to the problem of a wall with one end at T_0 and the other insulated.

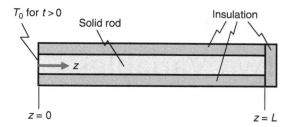

FIGURE 11.1 Finite and insulated cylindrical rod.

The following are assumed:

1. Rod material is uniform.
2. Thermal conductivity and density of the rod remain constant. Of course, and again, this approximation is acceptable for relatively low gradients of temperature in the body.
3. No phase change of the rod is observed.
4. Insulations around the rod are perfect, i.e., no heat transfer is observed at surfaces covered with insulation material.

Following the same considerations as in Section 10.2, it is possible to write

$$q_r = -\lambda \frac{\partial T}{\partial r} = 0 \tag{11.1}$$

Similarly, by combining Equation A.35 with the assumptions described above, it is possible to arrive at

$$\rho C_p \frac{\partial T}{\partial t} = \lambda \frac{\partial^2 T}{\partial z^2} \tag{11.2}$$

However, in the present case, the boundary conditions are

$$T(0,z) = T_b, \quad 0 < z \le L \tag{11.3}$$

$$T(t,0) = T_0, \quad t > 0 \tag{11.4}$$

$$\left. \frac{\partial T}{\partial z} \right|_{z=L} = 0, \quad t \ge 0 \tag{11.5}$$

Of course, the third condition given by Equation 11.5 is imposed due to the insulation at $z = L$.

As can be seen, Equation 11.2 combined with the conditions given by Equations 11.3 through 11.5 forms a second-order differential equation with at least one second-kind boundary condition.

Before applying any method, let us change variables to work with dimensionless ones. The advantages of the method have already been demonstrated.

The chosen dimensionless temperature is defined by

$$\psi = \frac{T - T_b}{T_0 - T_b} \qquad (11.6)$$

The dimensionless length is defined by

$$\zeta = \frac{z}{L} \qquad (11.7)$$

The dimensionless time is defined by

$$\tau = t\frac{\lambda}{\rho C_p L^2} = t\frac{D_T}{L^2} \qquad (11.8)$$

Therefore

$$\frac{\partial T}{\partial t} = (T_0 - T_b)\frac{\lambda}{\rho C_p L^2}\frac{\partial \psi}{\partial \tau} \qquad (11.9)$$

$$\frac{\partial^2 T}{\partial z^2} = \frac{T_0 - T_b}{L^2}\frac{\partial^2 \psi}{\partial \zeta^2} \qquad (11.10)$$

Thus Equation 11.2 becomes

$$\frac{\partial \psi}{\partial \tau} = \frac{\partial^2 \psi}{\partial \zeta^2} \qquad (11.11)$$

Now, the boundary conditions are written as

$$\psi(0,\zeta) = 0, \quad 0 < \zeta \leq 1 \qquad (11.12)$$

$$\psi(\tau,0) = 1, \quad \tau > 0 \qquad (11.13)$$

$$\left.\frac{\partial \psi}{\partial \zeta}\right|_{\zeta=1} = 0, \quad \tau > 0 \qquad (11.14)$$

As shown, a convenient change of variables brings simpler forms to the problem.

11.2.1   SOLUTION BY LAPLACE TRANSFORM

Recalling the condition given by Equation 11.12, a good suggestion would be to apply the transform for variable τ, or

$$\Psi(s,\zeta) = L\{\psi(\tau,\zeta)\} \qquad (11.15)$$

According to the equations presented in Appendix D, it is easy to verify that the transform of Equation 11.11 is

$$s\Psi = \frac{d^2\Psi}{d\zeta^2} \qquad (11.16)$$

From Appendix B, the general solution is

$$\Psi = C_1 \sinh(\zeta\sqrt{s}) + C_2 \cosh(\zeta\sqrt{s}) \qquad (11.17)$$

Transforms of conditions given by Equations 11.13 and 11.14 are as follows:

$$\Psi(s,0) = \frac{1}{s} \qquad (11.18)$$

and

$$\left.\frac{d\Psi}{d\zeta}\right|_{\zeta=1} = 0 \qquad (11.19)$$

Applying Equations 11.18 and 11.19 in Equation 11.17, one would obtain

$$\Psi = -\frac{1}{s}\tanh(\sqrt{s})\sinh(\zeta\sqrt{s}) + \frac{1}{s}\cosh(\zeta\sqrt{s}) \qquad (11.20)$$

The above relation can be written as

$$\Psi = \frac{1}{s}\frac{\cosh(\zeta\sqrt{s})\cosh(\sqrt{s}) - \sinh(\zeta\sqrt{s})\sinh(\sqrt{s})}{\cosh(\sqrt{s})} = \frac{1}{s}\frac{\cosh[(1-\zeta)\sqrt{s}]}{\cosh(\sqrt{s})} \qquad (11.21)$$

Its inversion can be found in the tables in Appendix D, leading to

$$\psi(\tau,\zeta) = 1 + \frac{4}{\pi}\sum_{n=1}^{\infty}\frac{(-1)^n}{(2n-1)}\exp\left[-\frac{(2n-1)^2\pi^2\tau}{4}\right]\cos\left[\frac{(2n-1)\pi(1-\zeta)}{2}\right] \qquad (11.22)$$

The result is illustrated in Figure 11.2.

From the figure, it is clear that the conditions given by Equations 11.12 and 11.13 are satisfied. The condition given by Equation 11.14 is also satisfied due to the zero derivative at position $\zeta = 1$.

Let us now imagine that a thermocouple is installed at any position of the bar, say $\zeta = 0.6$. The time for the temperature to start increasing would be smaller for bars with higher values of thermal conductivity or thermal diffusivity. This can be

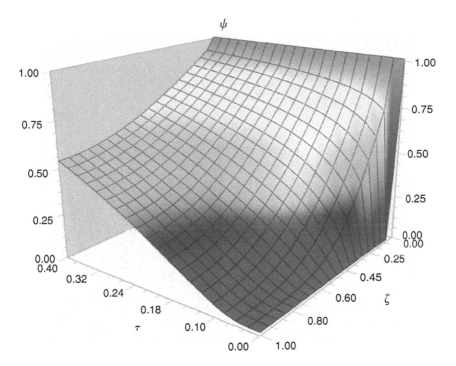

FIGURE 11.2 Dimensionless temperature (ψ) against dimensionless time (τ) and position (ζ) in the heating bar.

seen by examining Equation 11.8 because, for a given dimensionless time (τ), higher conductivity would lead to a lower value of real time (t). The effects of other physical properties can be examined in the same way.

Finally, Figure 11.3 presents the temperature distributions for a larger time span than chosen for the previous figure. It is easy to verify that a temperature $\tau = 2$ would be enough to equalize the temperature in the bar.

It is interesting to compare the result of Section 10.2 with the present one. Despite the fact that in Section 10.2 the rod had an indefinite length, Figure 10.2 resembles Figure 11.2. The major difference appears in the derivative of temperature at a given position far from $z = 0$ (or $\zeta = 0$). However, even in the case of an indefinite (Figure 10.2) bar, it is easy to see the asymptotic tendency toward zero derivatives for large values of z. Of course, the solution presented here is more realistic and applicable to the real world, where bodies are finite.

11.2.2 Solution by Separation of Variables

According to Appendix G, a solution for Equation 11.11 is assumed to be in the following form:

$$\psi(\tau,\zeta) = F(\tau)G(\zeta) \tag{11.23}$$

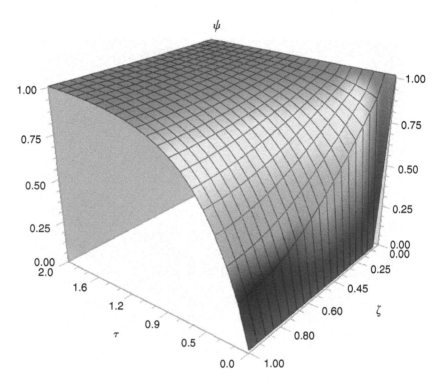

FIGURE 11.3 Dimensionless temperature (ψ) against dimensionless time (τ) and position (ζ) in the heating bar (larger time span than at Figure 11.2).

It would lead to

$$\frac{\mathrm{d}F/\mathrm{d}\tau}{F(\tau)} = \frac{\mathrm{d}^2 G/\mathrm{d}\zeta^2}{G} = C \qquad (11.24)$$

Here C is a constant.

The solution for the first-order differential equation

$$\frac{\mathrm{d}F/\mathrm{d}\tau}{F(\tau)} = C \qquad (11.25)$$

is

$$F(\tau) = C_1 \exp\left(C\tau\right) \qquad (11.26)$$

However, the solution above is unable to satisfy the boundary condition given by Equation 11.12. The way to circumvent this is to redefine the dimensionless temperature variable. Instead of using the form given by Equation 11.6, the following is proposed:

$$\varphi = \frac{T - T_0}{T_b - T_0} \tag{11.27}$$

The differential equation (Equation 11.11) can be solved by just replacing ψ with φ, however, with the following boundary conditions:

$$\varphi(0,\zeta) = 1, \quad 0 < \zeta \leq 1 \tag{11.28}$$

$$\varphi(\tau,0) = 0, \quad \tau > 0 \tag{11.29}$$

$$\left.\frac{\partial \varphi}{\partial \zeta}\right|_{\zeta=1} = 0, \quad \tau > 0 \tag{11.30}$$

Again, assuming Equation 11.23 (which continues to be valid by using φ instead of ψ), one would arrive at Equation 11.24, and Equation 11.25 would result into Equation 11.26. Nonetheless, the condition given by Equation 11.28 cannot be used to determine constants C and C_1. On the other hand, the general solution of ordinary differential equation of variable G depends on whether C is a positive or negative constant. If one tries several possibilities, the only one not leading to incoherence is when C is a real negative constant, or

$$C = -a^2 \tag{11.31}$$

Here parameter a is a real number. Another possibility of arriving at this conclusion is to understand that the temperature in the rod might increase or decrease at given points in the bar but cannot grow or decrease indefinitely. In other words, no matter what the situation, the temperature profile in the bar should converge to steady state. Hence, the term or terms related to time should have a decreasing influence of the process, or an asymptotic approach to zero. Thus

$$F(\tau) = C_1 \exp\left(-a^2\tau\right) \tag{11.32}$$

The solution of Equation 11.24 regarding function G is

$$G(\zeta) = C_2 \sin\left(a\zeta\right) + C_3 \cos\left(a\zeta\right) \tag{11.33}$$

The condition given by Equation 11.29 leads to

$$G(\zeta) = C_2 \sin\left(a\zeta\right) \tag{11.34}$$

Again, Equation 11.34 can be written in a more generalized way as

$$G(\zeta) = \sum_{j=1}^{\infty} C_j \sin\left(a_j\zeta\right) \tag{11.35}$$

Using Equations 11.32 and 11.35, it is possible to write Equation 11.23 as

$$\varphi(\tau,\zeta) = \sum_{j=1}^{\infty} C_j \exp\left(-a_j^2 \tau\right) \sin\left(a_j \zeta\right) \tag{11.36}$$

There is one degree of freedom, which allows to impose

$$a_j = \frac{2\pi}{p} j \tag{11.37}$$

Here p is the period to be determined. Consequently, Equation 11.36 becomes

$$\varphi(\tau,\zeta) = \sum_{j=1}^{\infty} C_j \exp\left(-a_j^2 \tau\right) \sin\left(\frac{2\pi}{p} j \zeta\right) \tag{11.38}$$

The condition given by Equation 11.28 can be used to provide the following:

$$1 = \sum_{j=1}^{\infty} C_j \sin\left(\frac{2\pi}{p} j \zeta\right) \tag{11.39}$$

According to Appendix G, Equation 11.39 can be seen as an odd half-range expansion of function "1," which allows writing the following:

$$C_j = \frac{4}{p} \int_0^{p/2} \sin\left(\frac{2\pi}{p} j x\right) dx = \frac{2}{\pi j}[1 - \cos(\pi j)] = \frac{2}{\pi j}\left[1 - (-1)^j\right] \tag{11.40}$$

With this, Equation 11.38 becomes

$$\varphi(\tau,\zeta) = \frac{2}{\pi} \sum_{j=1}^{\infty} \frac{1 - (-1)^j}{j} \exp\left(-\frac{4\pi^2}{p^2} j^2 \tau\right) \sin\left(\frac{2\pi}{p} j \zeta\right) \tag{11.41}$$

The only condition left to satisfy is Equation 11.30. Employing this, one gets

$$\sum_{j=1}^{\infty} \frac{1 - (-1)^j}{p} \cos\left(\frac{2\pi}{p} j\right) = 0 \tag{11.42}$$

It should be noticed that for even values of j, the identity is already satisfied. For odd values of j, the above will be true if

$$\frac{2\pi}{p} j = \frac{\pi}{2} j \tag{11.43}$$

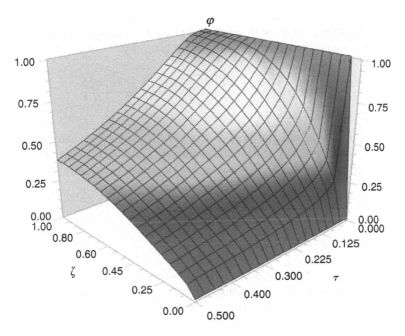

FIGURE 11.4 Dimensionless temperature (φ) against dimensionless time (τ) and position (ζ) in the heating bar (solution by separation of variables).

or

$$p = 4 \tag{11.44}$$

Applying this to Equation 11.41, the final form for the dimensionless temperature would be

$$\varphi(\tau,\zeta) = \frac{2}{\pi} \sum_{j=1}^{\infty} \frac{1 - (-1)^j}{j} \exp\left[-\left(\frac{\pi}{2}j\right)^2 \tau\right] \sin\left(\frac{\pi}{2}j\zeta\right) \tag{11.45}$$

Figure 11.4 illustrates the profile of dimensionless temperature (given above) against dimensionless time (τ) and space (ζ).

11.2.2.1 Comments

Despite the difference in the appearance of dimensionless temperature profiles in Figures 11.2 and 11.4, the results are equivalent. The reader may test it by verifying from Equations 11.6 and 11.27 that

$$\psi(\tau,\zeta) = 1 - \varphi(\tau,\zeta)$$

Equation 11.22 or 11.45 can be applied to compute the rate of heat transfer to the bar or the total amount of heat transferred to the bar during a given period. For instance, the rate of heat transfer to the bar is given by

$$\dot{Q} = A q_{z=0} = -A\lambda \frac{\partial T}{\partial z}\bigg|_{z=0} = -\frac{A\lambda(T_0 - T_b)}{L} \frac{\partial \psi}{\partial \zeta}\bigg|_{\zeta=0} \quad (11.46)$$

Here A is the cross-sectional area of the bar. Of course, the rate would be a function of time, and often called as instantaneous rate of heat transfer. The total heat transferred to the bar from between any two instants (t_1 and t_2) is given by

$$_1Q_2 = \int_{t_1}^{t_2} \dot{Q} \, dt = \frac{L^2}{D_T} \int_{\tau_1}^{\tau_2} \dot{Q} \, d\tau \quad (11.47)$$

The details of these deductions are left as an exercise.

Finally, it is important to notice that the above solutions are applicable to the case of a plate with infinite length and width, as illustrated in Figure 11.5. The whole plate is initially at temperature T_b and suddenly one face is set at temperature T_0 while the other remains insulated.

Owing to indefinite dimensions of the plate in all directions, except for its thickness (z), no variations of temperature in the x or y direction would occur. Therefore, the only two independent variables are t and z. In addition, assumptions 1, 2, 3, and 4 also remain valid, except for replacing the word rod by body.

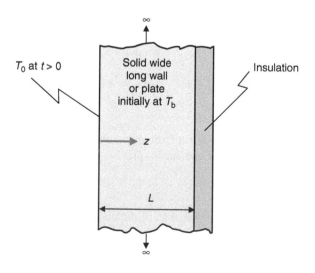

FIGURE 11.5 Indefinite insulated wall.

As seen, at least two methods could be applied to solve this problem. The method of similarity cannot be used because no coalescence between boundary conditions given by Equations 11.12 and 11.13, or even between the boundary conditions given by Equations 11.12 and 11.14, is possible.

11.3   DRYING OF A SPHERICAL PARTICLE

Drying is a very important branch of heat and mass transfer, with countless applications in several branches of industry.

Actually, drying is a very complex subject [1–4]. The solution presented here is a first approximation to the problem and a more realistic solution is shown in the next chapter.

Let us consider a wet, porous, spherical particle, as shown in Figure 11.6, initially with water concentration equal to ρ_{A0}.

At a given instant, the particle is exposed to relatively dry air which promotes its drying. The process involves several situations, as described below:

- During the first stage, liquid water migrates to the surface. Consequently, the surface is kept wet and the water concentration (ρ_{As}) is practically at equilibrium with the layer of a mixture of air and water above it. Usually, it is assumed that the air layer is saturated with water vapor and the temperature of this layer is similar to that of the wet bulb. This is called "first period of drying."
- After the completion of the first stage, the wet surface starts to recede toward the center leaving a layer of dry material. This is called the "second period of drying." This period can be modeled according to a similar procedure as shown in Chapter 6, or as an unreacted-core model [4]. Of course, except for some ionic interactions, the drying process involves no global chemical reaction.

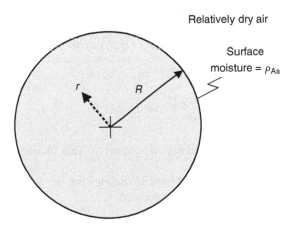

FIGURE 11.6 A wet porous sphere.

It is desired to obtain the water concentration profile inside the particle at any instant as well as the drying rate against time.

The following are assumed here:

1. Particle remains at the same temperature throughout the whole process. In other words, the particle is isothermal. This is possible for very small particles, or for conditions where Biot number remains below 0.1 (see Chapter 1). In addition, as drying involves a combination of heat and mass transfer, there will always be a temperature gradient inside the particle. In addition, other secondary effects—such as the Soret effect [5,6] describing the interference between mass and heat transfer—are neglected. This is possible if a low to moderate rate of mass transfer is assumed to allow thermal equilibrium to be restored, at least partially.
2. Porosity of the particle is uniform and the apparent or effective diffusivity (D_{eff}) of water through the porous structure remains constant. The D_{eff} can be computed by semiempirical relations [4,7].
3. Drying process occurs during the first period. Therefore, a constant concentration ρ_{As} is maintained at the particle surface. Mass transfer inside the particle would only occur if the water concentration at the surface is smaller than the initial concentration of water (ρ_{A0}) inside the particle. It is also assumed that the concentration ρ_{As} is known and is a constant. A more realistic approach for this situation will be presented in Chapter 12.
4. Overall velocity of fluid toward the particle surface is neglected.

Considering the above, Equation A.42 can be simplified to

$$\frac{\partial \rho_A}{\partial t} = D_{eff} \frac{1}{r^2} \frac{\partial}{\partial r}\left(r^2 \frac{\partial \rho_A}{\partial r}\right) \tag{11.48}$$

The boundary conditions are given by

$$\rho_A(0,r) = \rho_{A0}, \quad 0 < r < R \tag{11.49}$$

$$\rho_A(t,R) = \rho_{As}, \quad t > 0 \tag{11.50}$$

$$\left.\frac{\partial \rho_A}{\partial r}\right|_{r=0} = 0, \quad t > 0 \tag{11.51}$$

The last condition is a consequence of symmetry. This characterizes a second-type boundary condition.

Keeping in mind the discussed benefits drawn from the application of dimensionless variables, the following are proposed:

$$\varphi = \frac{\rho_A - \rho_{A0}}{\rho_{As} - \rho_{A0}} \tag{11.52}$$

$$\zeta = \frac{r}{R} \tag{11.53}$$

$$\tau = \gamma t \tag{11.54}$$

$$\gamma = \frac{D_{\text{eff}}}{R^2} \tag{11.55}$$

Notice that $\rho_A \le \rho_{A0}$, $\rho_{A0} \le \rho_{As}$, and $\rho_{As} \le \rho_A$. In addition, the first period of drying stops and the second period starts when $\rho_A = \rho_{As}$ at every position inside the particle. Therefore, during the first period or for the present case, φ would remain between 0 and 1.

In this way, the following can be written:

$$\frac{\partial \rho_A}{\partial t} = (\rho_{As} - \rho_{A0})\gamma \frac{\partial \varphi}{\partial \tau} \tag{11.56}$$

$$\frac{\partial \rho_A}{\partial r} = (\rho_{As} - \rho_{A0})\frac{1}{R} \frac{\partial \varphi}{\partial \zeta} \tag{11.57}$$

Using the above, Equation 11.48 becomes

$$\frac{\partial \varphi}{\partial \tau} = \frac{1}{\zeta^2} \frac{\partial}{\partial \zeta} \left(\zeta^2 \frac{\partial \varphi}{\partial \zeta} \right) \tag{11.58}$$

Let us now set an additional change of variable given by

$$\psi = \zeta \varphi \tag{11.59}$$

After this, it is possible to write

$$\frac{\partial \varphi}{\partial \tau} = \frac{1}{\zeta} \frac{\partial \psi}{\partial \tau} \tag{11.60}$$

$$\frac{\partial \varphi}{\partial \zeta} = \frac{1}{\zeta} \frac{\partial \psi}{\partial \zeta} - \frac{\psi}{\zeta^2} \tag{11.61}$$

Therefore, Equation 11.58 can be rewritten as

$$\frac{\partial \psi}{\partial \tau} = \frac{\partial^2 \psi}{\partial \zeta^2} \tag{11.62}$$

The details are left as an exercise to the reader.

Equation 11.62 is much simpler than Equation 11.48.

The boundary conditions given by Equations 11.49 and 11.50 become

$$\psi(0,\zeta) = 0, \quad 0 < \zeta < 1 \tag{11.63}$$

$$\psi(\tau,1) = 1, \quad \tau > 0 \tag{11.64}$$

If Equation 11.51 is transformed to the new variables, an apparent impossibility to achieve a proper definition for the boundary condition occurs. On the other hand, as φ is finite at the center of the sphere ($r = 0$, or $\zeta = 0$), Equation 11.59 allows one to write

$$\psi(\tau,0) = 0, \quad \tau > 0 \tag{11.65}$$

This condition replaces Equation 11.51. Although the present example departs from the intention of the present chapter, which is to deal with second-kind or -type boundary conditions, the present example is very important to situations dealing with spherical geometries.

11.3.1 Solution by Laplace Transform

As always, the transform can be applied at any independent variable (τ or ζ). However, one should look at the boundary conditions before making a hasty decision, which could demand unnecessary work or lead to unsurpassable difficulties.

Let the transform be defined by

$$\Phi(s,\zeta) = L\{\psi(\tau,\zeta)\} \tag{11.66}$$

Applying Equation 11.66 in Equation 11.62 and using the condition given by Equation 11.63, the following is obtained:

$$s\Phi = \frac{d^2\Phi}{d\zeta^2} \tag{11.67}$$

As seen, this is a homogeneous, second-order ordinary differential equation. According to Appendix B, the solution is

$$\Phi = A\sinh(\zeta\sqrt{s}) + B\cosh(\zeta\sqrt{s}) \tag{11.68}$$

The transforms of conditions given by Equations 11.64 and 11.65 are

$$\Phi(\zeta = 1) = \frac{1}{s} \tag{11.69}$$

$$\Phi(\zeta = 0) = 0 \tag{11.70}$$

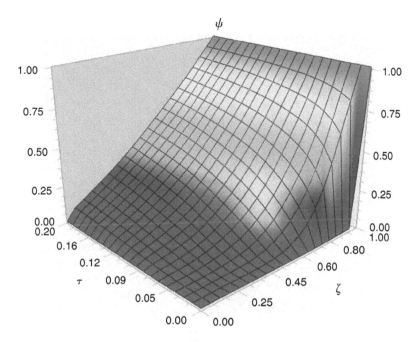

FIGURE 11.7 Dimensionless concentration (ψ) as a function of dimensionless time (τ) and dimensionless radius (ζ) of the drying sphere.

Using Equations 11.64 and 11.65 in Equation 11.68 leads to

$$\Phi = \frac{\sinh(\zeta\sqrt{s})}{s\sinh(\sqrt{s})} \tag{11.71}$$

The tables in Appendix D can be used to invert the above relation, and

$$\psi(\tau,\zeta) = \zeta + \frac{2}{\pi}\sum_{n=1}^{\infty}\frac{(-1)^n}{n}\exp\left(-n^2\pi^2\tau\right)\sin\left(n\pi\zeta\right) \tag{11.72}$$

This solution is illustrated in Figure 11.7.

Using Equation 11.59, the above equation can be written as

$$\varphi(\tau,\zeta) = 1 + \frac{2}{\pi\zeta}\sum_{n=1}^{\infty}\frac{(-1)^n}{n}\exp\left(-n^2\pi^2\tau\right)\sin\left(n\pi\zeta\right) \tag{11.73}$$

This is represented in Figure 11.8.

This verifies that all boundary conditions are satisfied. In particular, it should be noticed that the derivative of concentration (φ) tends to zero for positions approaching the center of the particle ($\zeta = 0$).

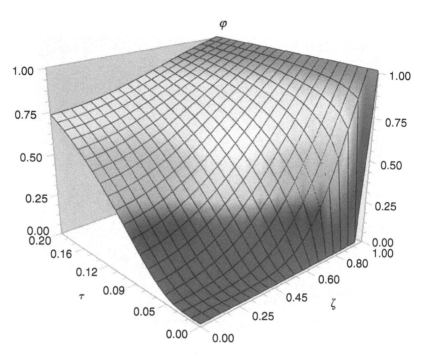

FIGURE 11.8 Dimensionless concentration (φ) as a function of dimensionless time (τ) and dimensionless radius (ζ) of the drying sphere.

The solution may also be written in terms of the variables: time and radius. Combining Equation 11.73 and the definitions given by Equations 11.52 through 11.54, the following results:

$$\frac{\rho_A - \rho_{A0}}{\rho_{As} - \rho_{A0}} = 1 + \frac{2R}{\pi r} \sum_{n=1}^{\infty} \frac{(-1)^n}{n} \exp\left(-n^2\pi^2 \frac{tD_{eff}}{R^2}\right) \sin\left(n\pi \frac{r}{R}\right) \qquad (11.74)$$

Equation 11.74 shows, for instance, that increases in D_{eff} of the porous matrix lead to an exponential decrease on the last term of the left-hand side, or a much faster drying rate. This can also be observed by Figure 11.3. First, one should remember that according to Equation 11.52 the particle is drier at positions where φ is greater. This occurs near the particle surface ($\zeta = 1$) or for longer periods (time or τ). According to Equation 11.54, τ is directly proportional to t and γ, while Equation 11.55 imposes γ proportional to D_{eff}. Therefore, for the same real time (t), increases in diffusivity lead to increases in τ. Hence, at any given position inside the particle, larger diffusivities lead to longer τ and larger φ, or lower moistures. A similar conclusion can be drawn from Figure 11.7.

The determination of D_{eff} is very complex [4,7]. However, relatively simple models lead to good approximations. One of the first and still most used correlations

[7] proposes that the D_{eff} of a gas through a solid is directly proportional to its porosity, or

$$D_{eff} = D_G \kappa^2 \tag{11.75}$$

Here D_G is the gas–gas diffusivity of gas. In the present case, it could be assumed as the diffusivity of steam in air. The parameter κ is the porosity of the particle, or the ratio between the total volume occupied by internal pores and the volume of the particle. This parameter is obtained by standard laboratory procedures.

It is also valuable to verify the condition at the center of the particle, or more specifically, the period taken for the drying process to start affecting the center. Let us exemplify this by using some numerical data related to the process. A typical value for porosity of biomass is 0.5. If we take the steam–air diffusivity as 1×10^{-5} m^2 s^{-1}, according to Equation 11.75 the D_{eff} would be around 2.5×10^{-6} m^2 s^{-1}. For a 2 cm diameter particle, the parameter γ would be around 6.25×10^{-3} s. From Figure 11.8, it is possible to see that the center of the particle is unaffected until τ is around 0.025 or, from Equation 11.54, not before approximately 4 s.

Equations 11.72 through 11.74 can be used to obtain important engineering functions, such as the drying rate of a particle during the first period. This rate is given by

$$F = \text{drying rate} = A_P \left. N_A \right|_{r=R} = -A_P D_{eff} \left. \frac{\partial \rho_A}{\partial r} \right|_{r=R}$$
$$= -\frac{A_P D_{eff}(\rho_{As} - \rho_{A0})}{R} \left. \frac{\partial \varphi}{\partial \zeta} \right|_{\zeta=1} \tag{11.76}$$

where A_P is the external area of the particle.

It is important to remember that the drying rate varies with time. Therefore, the period to reach a certain average concentration of water inside the particle can also be computed from the above equations.

11.4 HEATING A CYLINDER

The solid cylinder, as represented in Figure 11.9, is at uniform temperature T_0. At a given instant, its external surface is maintained at temperature T_1. It is desired to determine the temperature profile in the cylinder at any time as well as the rate of heat transfer to or from it.

The assumed conditions are as follows:

1. Length of the cylinder is much bigger than its radius. Therefore, end-effects are negligible and the temperature inside the cylinder is just a function of time and the radial coordinate.

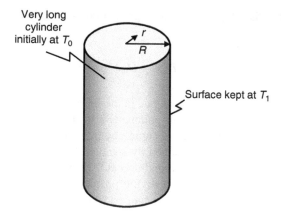

FIGURE 11.9 Scheme to illustrate the problem of transient heating of a cylinder.

2. No phase change is observed in the solid material and all properties of the cylinder material remain constant. Of course, depending on the gradients of temperature involved in the process, this might constitute a strong assumption. Hence, the solution achieved here is applicable for situations of small to moderate differences between T_0 and T_1.
3. No chemical reaction occurs in the cylinder material.

Since no velocity field is present here and no internal energy generation is observed in the cylinder, Equation A.35 can be written as

$$\rho C_p \frac{\partial T}{\partial t} = \lambda \frac{1}{r} \frac{\partial}{\partial r}\left(r \frac{\partial T}{\partial r}\right) \tag{11.77}$$

The following boundary conditions can be set

$$T(0,r) = T_0, \quad 0 \leq r \leq R \tag{11.78}$$

$$T(t,R) = T_1, \quad t > 0 \tag{11.79}$$

$$\left.\frac{\partial T}{\partial r}\right|_{r=0} = 0, \quad t > 0 \tag{11.80}$$

The last condition is due to symmetry.

To facilitate the solution as well as keeping in mind the generalization of dimensionless variables, the following new variables are proposed:

$$\varphi = \frac{T - T_0}{T_1 - T_0} \tag{11.81}$$

$$\zeta = \frac{r}{R} \tag{11.82}$$

$$\tau = t\frac{D_T}{R^2} = t\frac{\lambda}{\rho C_p R^2} \tag{11.83}$$

Of course, the above is just one possibility among several others.
    Using the above, Equation 11.77 becomes

$$\frac{\partial \varphi}{\partial \tau} = \frac{1}{\zeta}\frac{\partial \varphi}{\partial \zeta} + \frac{\partial^2 \varphi}{\partial \zeta^2} \tag{11.84}$$

The boundary conditions are written as follows:

$$\varphi(0,\zeta) = 0, \quad 0 < \zeta < 1 \tag{11.85}$$

$$\varphi(\tau,1) = 1, \quad \tau > 0 \tag{11.86}$$

$$\left.\frac{\partial \varphi}{\partial \zeta}\right|_{\zeta=0} = 0, \quad \tau > 0 \tag{11.87}$$

11.4.1 Solution by Laplace Transform

Let the Laplace transform regarding variable τ be described by

$$\Psi(s,\zeta) = L\{\varphi(\tau,\zeta)\} \tag{11.88}$$

Applying Equation 11.88 to Equation 11.84 and using the condition given by
Equation 11.89, it is possible to write

$$\zeta^2\frac{d^2\Psi}{d\zeta^2} + \zeta\frac{d\Psi}{d\zeta} - s\zeta^2\Psi = 0 \tag{11.89}$$

This second-order ordinary differential equation can be set as a standard Bessel
modified equation with the following change of variable:

$$y = \zeta\sqrt{s} \tag{11.90}$$

After some work, Equation 11.89 is written as

$$y^2\frac{d^2\Psi}{dy^2} + y\frac{d\Psi}{dy} - y^2\Psi = 0 \tag{11.91}$$

According to the explanation given in Appendix C, the general solution to the above equation is

$$\Psi = C_1 I_0(y) + C_2 K_0(y)$$

or

$$\Psi = C_1 I_0(\zeta\sqrt{s}) + C_2 K_0(\zeta\sqrt{s}) \tag{11.92}$$

The Laplace transforms of conditions using Equations 11.86 and 11.87 are

$$\Psi(s,1) = \frac{1}{s} \tag{11.93}$$

$$\frac{d\Psi}{d\zeta}\bigg|_{\zeta=0} = 0 \tag{11.94}$$

To apply the last condition, Equation 11.92 should be differentiated and one would obtain (Appendix C)

$$\frac{d\Psi}{d\zeta} = C_1 I_1\left(\zeta\sqrt{s}\right)\sqrt{s} + C_2 K_1\left(\zeta\sqrt{s}\right)\sqrt{s} \tag{11.95}$$

As can be verified, C_2 should be equal to zero because the Bessel functions present the following properties (Appendix C):

$$\lim_{y\to 0} K_1(y) = \infty \tag{11.96}$$

$$I_1(0) = 0 \tag{11.97}$$

Applying the above and the condition given by Equation 11.93 to Equation 11.95 allows determining the constants. Then, Equation 11.92 becomes

$$\Psi(s,\zeta) = \frac{I_0(\zeta\sqrt{s})}{s I_0(\sqrt{s})} \tag{11.98}$$

The inversion of the above transform can be found in the tables of Appendix D. Therefore

$$\varphi(\tau,\zeta) = 1 - 2\sum_{n=1}^{\infty} \frac{J_0(b_n\zeta)}{\beta_n J_1(b_n)} \exp\left(-b_n^2 \tau\right) \tag{11.99}$$

or

$$\frac{T - T_0}{T_1 - T_0} = 1 - 2\sum_{n=1}^{\infty} \frac{J_0\left(b_n\frac{r}{R}\right)}{\beta_n J_1(b_n)} \exp\left(-b_n^2 \frac{\lambda t}{\rho C_p R^2}\right) \tag{11.100}$$

In the above equations, the parameters b_n $(n = 1, 2, \ldots)$ are the positive roots of $J_0(b) = 0$.

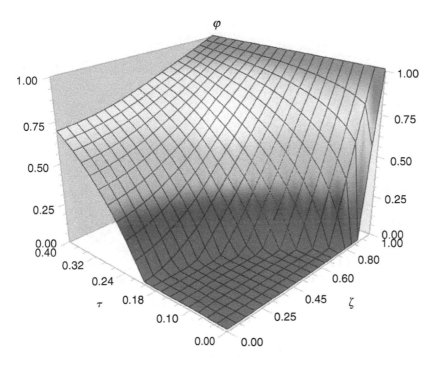

FIGURE 11.10 Dimensionless temperature (φ) as a function of dimensionless time (τ) and radius (ζ) for τ until 0.4.

Equation 11.99 is represented in Figure 11.10 for τ having values up to 0.4 and in Figure 11.11 for τ having values up to 1.0.

It is interesting to notice the following:

1. Conditions given by Equations 11.85 and 11.86 are satisfied.
2. By the tendency of lines of constant time (constant τ), Figures 11.10 and 11.11 also indicate that the condition given by Equation 11.87 is satisfied. However, this can be verified using Equation 11.99. Keeping in mind that (Appendix C)

$$\frac{\mathrm{d}J_0(x)}{\mathrm{d}x} = -J_1(x) \qquad (11.101)$$

and that $J_1(0) = 0$, it is possible to see from Equation 11.99 that the derivative of φ at $\zeta = 0$ would be equal to zero, therefore satisfying the condition given by Equation 11.87.
3. Equations 11.99 and 11.100 show that increases in the thermal conductivity of the cylinder contribute to the decrease of the terms inside the summation, consequently leading to a faster approaching of temperature

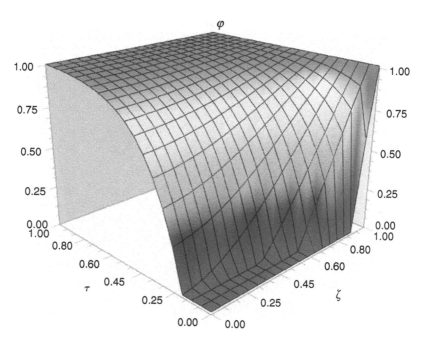

FIGURE 11.11 Dimensionless temperature (φ) as a function of dimensionless time (τ) and radius (ζ) for τ until 1.0.

to the final value of T_1. Actually, the thermal diffusivity (see definition given at Equation 10.95) is among the most important characteristic parameters to be considered in analyses of conductive processes.

4. It is also worthwhile to notice from Equation 11.83 that materials with lower density provide decreases in the thermal diffusivity and, therefore, constitute good insulators.

5. Figures 11.10 and 11.11 show how sensitive the temperature is regarding time or its dimensionless form (τ). For τ near 0.4, the temperature at the center of the cylinder ($\zeta = 0$) is progressing toward T_1 (or φ approaching 1). However, for τ near 1, the temperature at the center reaches T_1 (or $\varphi = 1$); therefore, the temperature throughout the cylinder is completely equalized.

11.5 INSULATED ROD WITH PRESCRIBED INITIAL TEMPERATURE PROFILE

Consider the insulated rod shown in Figure 11.12.

Unlike the situation presented in Section 10.2, here the initial temperature profile is described by a known function $f(z)$. In addition, it is imposed that for positions far from $z = 0$, the temperature is not affected.

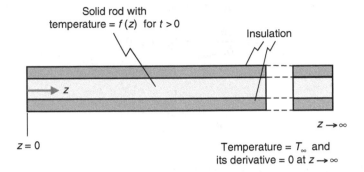

FIGURE 11.12 Insulated rod with prescribed initial temperature profile.

All assumptions listed in Section 10.2 are still valid, and therefore, the governing differential Equation 10.2 is repeated

$$\rho C_p \frac{\partial T}{\partial t} = \lambda \frac{\partial^2 T}{\partial z^2} \tag{11.102}$$

Nonetheless, the new boundary conditions are

$$T(0,z) = f(z), \quad z > 0 \tag{11.103}$$

$$T(t,\infty) = T_\infty, \quad t > 0 \tag{11.104}$$

$$\left. \frac{\partial T}{\partial z} \right|_{z \to \infty} = 0, \quad t > 0 \tag{11.105}$$

As seen, the last condition given by Equation 11.105 is a second-kind boundary condition.

Before applying any method, assume the following change of variable:

$$\theta = \frac{T - T_\infty}{T_\infty} \tag{11.106}$$

Therefore, Equation 11.102 can be rewritten as

$$\frac{\partial \theta}{\partial t} = D_T \frac{\partial^2 \theta}{\partial z^2} \tag{11.107}$$

where D_T is the thermal diffusivity and is defined by Equation 10.95.

The boundary conditions given by Equations 11.103 through 11.105 become

$$\theta(0,z) = \phi(z), \quad z > 0 \tag{11.108}$$

$$\theta(t,\infty) = 0, \quad t > 0 \tag{11.109}$$

$$\left.\frac{\partial\theta}{\partial z}\right|_{z\to\infty} = 0, \quad t > 0 \tag{11.110}$$

Here

$$\phi(z) = \frac{f(z) - T_\infty}{T_\infty} \tag{11.111}$$

11.5.1 Solution by Fourier Transform

Similar to the procedure in cases of Laplace transform, the Fourier transformed variable should be chosen. Let it be z, therefore

$$F\{\theta(t,z)\} = \Theta(t,s) \tag{11.112}$$

Appendix H describes several types of Fourier transforms, and with different levels of success, all of them might be applied to a given problem. Let us select the Fourier exponential transform and apply it to Equation 11.107. As variable t is not transformed, it is possible to write

$$F\left\{\frac{\partial\theta}{\partial t}\right\} = \frac{d\Theta}{dt} \tag{11.113}$$

The transform of second derivatives is given by Equation H.19. Hence

$$F\left\{\frac{\partial^2\theta}{\partial z^2}\right\} = -s^2\Theta(t,s) \tag{11.114}$$

With this, Equation 11.107 becomes

$$\frac{d\Theta}{dt} = -D_T s^2 \Theta \tag{11.115}$$

The solution to this simple separable differential equation is

$$\Theta(t,s) = C(s)\exp\left(-D_T s^2 t\right) \tag{11.116}$$

Note that C is a function of parameter s.

The Fourier transform of the condition given by Equation 11.108 provides the following:

$$\Theta(0,s) = F\{\phi(z)\} = \varphi(s) \tag{11.117}$$

Therefore

$$\Theta(t,s) = \varphi(s)\exp\left(-D_T s^2 t\right) \tag{11.118}$$

One should notice that once the function $f(z)$ is known, so is the function $\varphi(s)$. According to Equation H.13, the inversion of Equation 11.118 gives

$$\theta(t,z) = \frac{1}{2\pi}\int_{-\infty}^{\infty}\varphi(s)e^{-D_T s^2 t}e^{isz}\,ds \tag{11.119}$$

The important detail is that the above equation also satisfies the boundary condition given by Equation 11.110. Therefore, it is the solution. However, a more convenient form for computations can be written.

According to Equation H.12, the Fourier transform of the original function is

$$\varphi(s) = \int_{-\infty}^{\infty}f(w)e^{-isw}\,dw \tag{11.120}$$

Consequently, Equation 11.119 becomes

$$\theta(t,z) = \frac{1}{2\pi}\int_{-\infty}^{\infty}f(w)\left[\int_{-\infty}^{\infty}e^{-D_T s^2 t}e^{i(sz-sw)}ds\right]dw \tag{11.121}$$

At this point the Euler formula, given by Equation 11.122, is useful

$$\exp[i(sz - sw)] = \cos(sz - sw) + i\sin(sz - sw) \tag{11.122}$$

Since the imaginary part includes an odd function, the above integral is equated to zero. As the value of the integral from $-\infty$ to $+\infty$ of an even function is twice the value of the same integral from 0 to ∞, the final form of the solution is

$$\theta(t,z) = \frac{1}{\pi}\int_{-\infty}^{\infty}\phi(w)\left[\int_{0}^{\infty}e^{-D_T s^2 t}\cos(sz - sw)ds\right]dw \tag{11.123}$$

As a simple example, let $f(z)$ be a constant, or

$$f(z) = T_\infty + T_0 e^{-bz} \tag{11.124}$$

To be coherent with the physical situation, b is a positive real constant with units of 1/length. Therefore, from Equation 11.111

$$\phi(z) = \frac{T_0}{T_\infty} e^{-bz} \tag{11.125}$$

This satisfies the conditions given by Equations 11.109 and 11.110.

Using the general solution given by Equation 11.123, one obtains

$$\theta(t,z) = \frac{T_0}{\pi T_\infty} \int\limits_{-\infty}^{\infty} e^{-bw} \left[\int\limits_{0}^{\infty} e^{-D_T s^2 t} \cos(sz - sw)ds \right] dw \tag{11.126}$$

The inner integral is given by

$$\int\limits_{0}^{\infty} e^{-D_T s^2 t} \cos(sz - sw)ds = \frac{1}{2}\sqrt{\frac{\pi}{D_T t}} \exp\left[-\frac{(z-w)^2}{4 D_T t} \right] \tag{11.127}$$

Applying Equation 11.127 to Equation 11.126, the following can be found:

$$\theta(t,z) = \frac{T_0}{2T_\infty} \sqrt{\frac{1}{\pi D_T t}} \int\limits_{-\infty}^{\infty} e^{-bw} \exp\left[-\frac{(z-w)^2}{4 D_T t} \right] dw \tag{11.128}$$

Using a new variable, $y = z - w$, the following can be written:

$$\theta(t,z) = \frac{T_0}{2T_\infty} e^{-bz} \sqrt{\frac{1}{\pi D_T t}} \int\limits_{-\infty}^{\infty} \exp\left[by - \frac{y^2}{4 D_T t} \right] dy$$

$$= \frac{T_0}{T_\infty} e^{-bz} \exp(b^2 D_T t) \tag{11.129}$$

Figure 11.13 illustrates the above result when the following data are used: $T_0 = 600$ K, $T_\infty = 300$ K, $b = 0.1$ m^{-1}, and $D_T = 4.0 \times 10^{-6}$ m^2 s^{-1}. This thermal diffusivity is typical to that of steels.

Figure 11.14 shows the profile when just the diffusivity is changed to 7.4×10^{-7} m^2 s^{-1} (which is the case of bakelite).

Figures 11.13 and 11.14 show that:

- Boundary conditions given by Equations 11.109 and 11.110 are satisfied. In other words, the temperature tends to decline asymptotically with the distance from position $z = 0$.

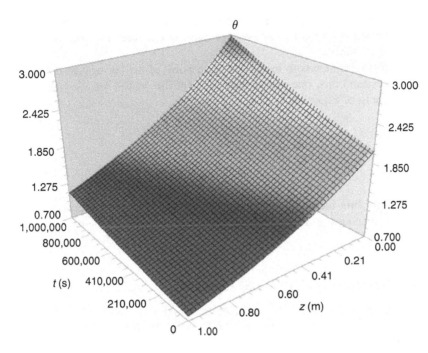

FIGURE 11.13 Dimensionless temperature (θ) profile as a function of time (t) and position (z) for the case when $T_0 = 600$ K, $T_\infty = 300$ K, $b = 0.1$ m^{-1}, and $D_T = 4.0 \times 10^{-6}$ m^2 s^{-1}.

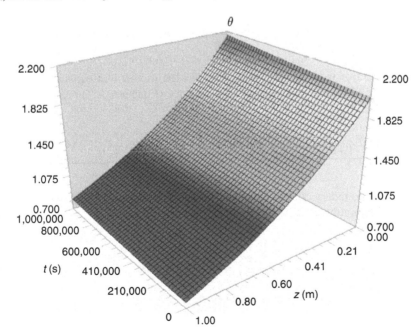

FIGURE 11.14 Dimensionless temperature (θ) profile as a function of time (t) and position (z) for the case when $T_0 = 600$ K, $T_\infty = 300$ K, $b = 0.1$ m^{-1}, and $D_T = 7.4 \times 10^{-7}$ m^2 s^{-1}.

- Exponential decline, imposed by Equation 11.124 or 11.125, can be followed at instant $t = 0$. The same curve is obtained, irrespective of the type of bar material.
- Temperature tends to increase at each point of the bar against time. To understand this, one may deduce the heat flux to the bar at $z = 0$. With the help of Equation 11.129, it is given by

$$q_z|_{z=0} = -\lambda \frac{\partial T}{\partial z}\bigg|_{z=0} = -\lambda T_\infty \frac{\partial \theta}{\partial z}\bigg|_{z=0}$$
$$= b\lambda T_0 \exp\left(b^2 D_T t\right) \tag{11.130}$$

As seen, the flux increases with time as well as for larger thermal conductivities. Higher values of thermal flux lead to larger increases of the bar temperature against time.

11.6 PLATE-AND-CONE VISCOMETER

A more rigorous solution to the problem presented in Section 1.11 is shown here because it does not need to assume the linear dependence of angular velocity regarding the radial coordinate.

Following the same assumptions made in Section 11.1, the continuity (Equation A.3) provides

$$\frac{\partial v_\phi}{\partial \phi} = 0 \tag{11.131}$$

Now, instead of working with shear stresses, the momentum equation for Newtonian fluid with constant viscosity and density (Equation A.21) can be readily simplified to give

$$\frac{\partial}{\partial r}\left(r^2 \frac{\partial v_\phi}{\partial r}\right) + \frac{1}{\sin\theta}\frac{\partial}{\partial\theta}\left(\frac{\partial v_\phi}{\partial\theta}\sin\theta\right) - \frac{v_\phi}{\sin^2\theta} = 0 \tag{11.132}$$

Hence, two boundary conditions should be set regarding the radial coordinate and two for the angular coordinate. From the imposed circumstances, the boundary conditions are

$$v_\phi\left(r, \frac{\pi}{2}\right) = 0, \quad 0 < r < R \tag{11.133}$$

$$v_\phi(r, \theta_1) = r\Omega \sin\theta_1, \quad 0 < r < R \tag{11.134}$$

$$v_\phi(0, \theta) = 0, \quad \theta_1 < \theta < (\theta_1 + \theta_0) \tag{11.135}$$

Just one more condition regarding the radial coordinate should be found. At this point, it is assumed that the shear stress between the fluid at the edge of the meniscus and the surrounding air is negligible. This is very reasonable and close to reality, leading to a very precise solution. Since v_r is zero, from Equation A.21a the following can be written:

$$\frac{\partial}{\partial r}\left(\frac{v_\phi}{r}\right)\bigg|_{r=R} = 0, \quad \theta_1 < \theta < (\theta_1 + \theta_0) \tag{11.136}$$

11.6.1 Solution by the Method of Variables Separation

According to Appendix G, it is assumed that it is possible to write

$$v_\phi = F(r)G(\theta) \tag{11.137}$$

Applying the above into Equation 11.132, the following two ordinary differential equations are obtained:

$$\frac{1}{F}\frac{d}{dr}\left(r^2\frac{dF}{dr}\right) = C \tag{11.138}$$

and

$$\frac{1}{G}\left[-\frac{1}{\sin\theta}\frac{d}{d\theta}\left(\frac{dG}{d\theta}\sin\theta\right) + \frac{G}{\sin^2\theta}\right] = C \tag{11.139}$$

where C is a constant.

Equation 11.138 can be written as

$$r^2\frac{d^2F}{dr^2} + 2r\frac{dF}{dr} - CF = 0 \tag{11.140}$$

This is a Cauchy equation, and according to Appendix B, with a solution

$$F(r) = C_1 r^{m_1} + C_2 r^{m_2} \tag{11.141}$$

where

$$m_1 = \frac{-1+\sqrt{1+4C}}{2} \quad \text{and} \quad m_2 = \frac{-1-\sqrt{1+4C}}{2} \tag{11.142a,b}$$

As seen, there are several possibilities for constant C:

1. A negative value would lead to imaginary solutions m_1 and m_2. Of course, our physical problem requires real solutions.
2. A zero value would frustrate the compatibility of solution for $F(r)$ with solution for $G(\theta)$. The reader is invited to try such a possibility and verify its impossibility.
3. A positive value would lead to a positive m_1 and a negative m_2.

Consequently, the last possibility is the only viable alternative. The form of solution given by Equation 11.141, combined with the fact that velocity is always finite, provides the conclusion that $C_2 = 0$. Therefore, the solution for $F(r)$ is

$$F(r) = C_1 r^{m_1} \tag{11.143}$$

On the other hand, the boundary condition given by Equation 11.135 leads to

$$F(0) = 0 \tag{11.144}$$

This is satisfied for any constant C_1.

The condition given by Equation 11.136 is now applied to give

$$C_1(m_1 - 1)R^{m_1-2} = 0 \tag{11.145}$$

Of course, C_1 cannot be equal to zero; otherwise, a trivial solution would be reached. The only possibility is $m_1 = 1$, which would lead to $C = 2$. With this, the solution for F becomes

$$F(r) = C_1 r \tag{11.146}$$

As seen, the educated guess of linear dependence of v_ϕ on the radius is correct.

The differential equation (Equation 11.139) can be solved by applying the following change of variable:

$$\eta = \cos\theta \tag{11.147}$$

With the value of C determined, it is possible to write

$$(1 - \eta^2)\frac{d^2G}{d\eta^2} - 2\eta\frac{dG}{d\eta} + \left(2 - \frac{1}{1-\eta^2}\right)G = 0 \tag{11.148}$$

According to the discussion in Appendix C, the above is a Legendre differential equation, with the following general solution:

$$G(\eta) = C_3 P_1^1(\eta) + C_4 Q_1^1(\eta) \tag{11.149}$$

$P_1^1(\eta)$ and $Q_1^1(\eta)$ are the associated Legendre functions of the first and second kinds.

The boundary condition given by Equation 11.133 applied to Equation 11.137 provides

$$G(0) = 0 \qquad (11.150)$$

Since

$$P_1^1(0) = 1 \text{ and } Q_1^1(0) = 0 \qquad (11.151\text{a,b})$$

the condition would be satisfied if $C_3 = 0$, for any constant C_4.
Finally, the general solution for the velocity is

$$v_\phi = C_6 r Q_1^1(\cos \theta) \qquad (11.152)$$

The remaining condition that has not been applied is Equation 11.134. Using this equation in the above provides the following:

$$v_\phi = \Omega r \sin \theta_1 \frac{Q_1^1(\cos \theta)}{Q_1^1(\cos \theta_1)} \qquad (11.153)$$

Similarly as in Section 1.11, the following modifications of variables are proposed to put the above result in a dimensionless form:

$$Y = \frac{v_\phi}{R\Omega \sin \theta_1} \qquad (11.154)$$

$$\zeta = \frac{r}{R} \qquad (11.155)$$

After this, Equation 11.153 becomes

$$Y = \zeta \frac{Q_1^1(\cos \theta)}{Q_1^1(\cos \theta_1)} \qquad (11.156)$$

Figure 11.15 shows the velocity profiles for the case where $\theta_0 = \pi/10$. It should be noticed despite the similarity with Figure 1.20. However, the present approach is more precise and rigorous.

11.7 HEATING A RECTANGULAR PLATE

The problem of steady-state heating of a plate is illustrated in Figure 11.16.

The solid is indefinite in the z direction and two side faces of it are kept at known temperature T_0, while the remaining two receive constant flux of energy by heat transfers. To take advantage of the geometric similarity, the origins of

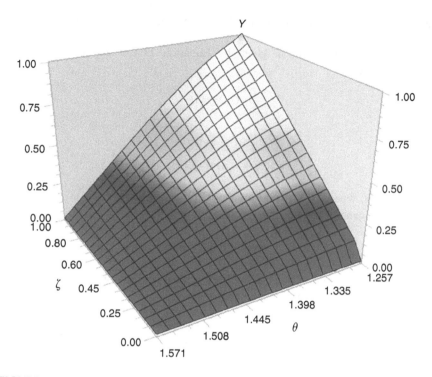

FIGURE 11.15 Dimensionless velocity (Y) profile as a function of dimensionless radial (ζ) and angle (θ) in the case of $\theta_0 = \pi/10$.

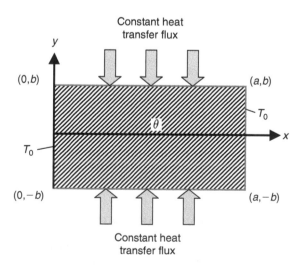

FIGURE 11.16 Scheme of the plate with constant temperature at two faces and receiving constant heat fluxes at the other two.

coordinates have been chosen at the center of the plate. Consequently, the derivatives of temperatures at $x = 0$ and $y = 0$ would be equal to zero.

It is desired to obtain the temperature profile throughout the plate. For this, the following are assumed:

1. Plate material is solid and, within the range of temperatures observed here, there is no phase change. Therefore, no velocity field in the plate is present.
2. There is no variation of temperature on the orthogonal coordinate (z) in the x–y plane. This condition is possible if the body is indefinite at that direction or if the plate is perfectly isolated in both main surfaces.
3. Despite variations of temperature, the plate material, its density, and its thermal conductivity remain approximately constant.
4. No chemical reactions or any other energy source term are involved.
5. Heat fluxes at the superior and inferior faces are equal and constant.

From the above assumptions, the energy balance shown by Equation A.31 (or Equation A.34) can be simplified to

$$\frac{\partial^2 T}{\partial x^2} + \frac{\partial^2 T}{\partial y^2} = 0 \qquad (11.157)$$

The boundary conditions are as follows:

$$T(0,y) = T_0, \quad -b < y < b \qquad (11.158)$$

$$T(a,y) = T_0, \quad -b < y < b \qquad (11.159)$$

$$\left.\frac{\partial T}{\partial y}\right|_{y=0} = 0, \quad 0 < x < a \qquad (11.160)$$

$$-\lambda \left.\frac{\partial T}{\partial y}\right|_{y=b} = q, \quad 0 < x < a \qquad (11.161)$$

Here q is a constant.

Of course, a similar condition given by Equation 11.161 could be written for $y = -b$. However, this has been replaced by Equation 11.160, which was derived by symmetry. The convenience of this will become clear ahead.

The dimensionless variables are as follows:

$$\theta = \frac{T - T_0}{T_0} \qquad (11.162)$$

$$\chi = \frac{x}{a} \qquad (11.163)$$

$$\xi = \frac{y}{b} \qquad (11.164)$$

Applying Equations 11.162 through 11.164 for the simpler case where $a = b$, Equation 11.157 is written as

$$\frac{\partial^2 \theta}{\partial \chi^2} + \frac{\partial^2 \theta}{\partial \xi^2} = 0 \qquad (11.165)$$

The boundary conditions become

$$\theta(0,\xi) = 0, \quad -1 < \xi < 1 \qquad (11.166)$$

$$\theta(1,\xi) = 0, \quad -1 < \xi < 1 \qquad (11.167)$$

$$\left. \frac{\partial \theta}{\partial \xi} \right|_{\xi=0} = 0, \quad 0 \le \chi \le 1 \qquad (11.168)$$

$$\left. \frac{\partial \theta}{\partial \xi} \right|_{\xi=1} = -H, \quad 0 < \chi < 1 \qquad (11.169)$$

Here

$$H = \frac{qb}{\lambda T_0} \qquad (11.170)$$

Thus we obtain the first- and second-kind boundary conditions.

11.7.1   SOLUTION BY SEPARATION OF VARIABLES

As introduced in Appendix G, this technique is based on the assumption that the following is allowed:

$$\theta(\chi,\xi) = F(\chi)G(\xi) \qquad (11.171)$$

Of course, this cannot be guaranteed beforehand and only trial would tell.
Employing the above, Equation 11.165 provides

$$\frac{F''(\chi)}{F(\chi)} = -\frac{G''(\xi)}{G(\xi)} = C \qquad (11.172)$$

The equality between groups involving a function of independent variables forces each side of the equation to be equal to a constant C. Consequently, two ordinary differential equations are generated. One of them is

$$F''(\chi) - CF(\chi) = 0 \qquad (11.173)$$

The general solution for such an equation presents the following possibilities:

1. If C is positive, or equal to γ^2 (where γ is a real parameter), according to Appendix B, the solution is

$$F(\chi) = C_1 e^{-\gamma\chi} + C_2 e^{\gamma\chi} \qquad (11.174)$$

However, from Equation 11.171 and conditions given by Equations 11.166 and 11.167, this alternative forces

$$F(0) = 0 \text{ and } F(1) = 0 \qquad (11.175a,b)$$

Using these two conditions, Equations 11.174 would be satisfied only if $C_1 = C_2 = 0$, i.e., the trivial solution for $F(\chi)$ and, consequently, to θ as well. Therefore, this is not an acceptable choice.

2. If C is equal to zero, the solution would be

$$F(\chi) = C_1 \chi + C_2 \qquad (11.176)$$

After the application of boundary conditions, the trivial solution is forced again.

3. If C is negative or equal to $-\gamma^2$ (where γ is a real parameter), the solution would be

$$F(\chi) = C_1 \sin(\gamma\chi) + C_2 \cos(\gamma\chi) \qquad (11.177)$$

Applying the condition given by Equation 11.166 leads to $C_2 = 0$ and the condition given by Equation 11.167 leads to the solution

$$0 = C_1 \sin(\gamma) \qquad (11.178)$$

If $\gamma = n\pi$ $(n = 0, 1, 2, \ldots)$, C_1 might be different from zero, therefore avoiding the trivial solution. The generalization of this solution is

$$F_n(\chi) = C_n \sin(n\pi\chi) \qquad (11.179)$$

The index n is used here to emphasize the multiple determinations of the solution. Since

$$C = -\gamma^2 = -n^2\pi^2 \qquad (11.180)$$

it is possible to obtain the solution for function G from Equation 11.172 as

$$G_n(\xi) = A_n e^{n\pi\xi} + B_n e^{-n\pi\xi} \qquad (11.181)$$

Applying the condition given by Equation 11.168 leads to $G'(0) = 0$. Hence, the above equation becomes

$$A_n = B_n \qquad (11.182)$$

With all these, Equation 11.181 can be written as

$$G_n(\xi) = A_n^* \cosh(n\pi\xi) \qquad (11.183)$$

Condition given by Equation 11.169 is not useful at this time to determine any constant.

Combining Equations 11.171, 11.179, and 11.183 the general solution becomes

$$\theta(\chi,\xi) = \sum_{n=1}^{\infty} K_n \sin(n\pi\chi)\cosh(n\pi\xi) \qquad (11.184)$$

Applying the condition given by Equation 11.169, one can write

$$1 = \sum_{n=1}^{\infty} \frac{n\pi}{H} K_n \sin(n\pi\chi)\sinh(n\pi) \qquad (11.185)$$

According to the equations given in Appendix G, the above can be seen as an odd half expansion of function 1. Thus

$$K_n = \frac{4H}{p n\pi \sinh(n\pi)} \int_0^{p/2} \sin(n\pi x)dx$$

$$= \frac{4H}{(n\pi)^2 p \sinh(n\pi)}\left[1 - \cos\left(\frac{n\pi p}{2}\right)\right], \quad n = 1, 2, 3, \dots \qquad (11.186)$$

Since the range of values for χ and ξ is between 0 and 1, period $p = 2$ should be set. Therefore

$$K_n = \frac{2H}{(n\pi)^2 \sinh(n\pi)}[1 - (-1)^n], \quad n = 1, 2, 3, \dots \qquad (11.187)$$

$$\theta(\chi,\xi) = \frac{2H}{\pi^2}\sum_{n=1}^{\infty} \frac{1 - (-1)^n}{n^2}\frac{\sin(n\pi\chi)\cosh(n\pi\xi)}{\sinh(n\pi)} \qquad (11.188)$$

This solution is represented in Figure 11.17 for H equal to 1.0.

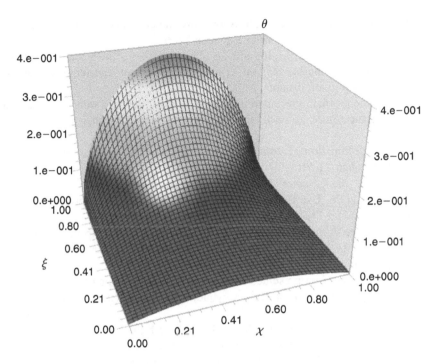

FIGURE 11.17 Dimensionless temperature (θ) profile in the plate as a function of dimensionless coordinates (χ and ξ).

From the above it is interesting to notice that

- According to Equations 11.170 and 11.188, it is clear that temperatures at any point of the plate are proportional to the heat flux. They are also inversely proportional to the thermal conductivity (λ) of the plate material. This is easy to understand because higher conductivity would dissipate heat through the plate, and the temperature at the heated faces would not reach values as high as for cases of lower conductivities.
- Constant heat fluxes force the temperatures at the faces receiving them to increase. Therefore, the temperatures at face $y = b$ (or $\xi = 1$) are larger than the rest of the plate. Of course, in practice there is a limit for the temperature because the assumptions 1 and 3 made must hold.

EXERCISES

1. Verify that the change in the definition of dimensionless temperature from Equations 11.6 through 11.27 would lead to the same differential equation as in Equation 11.11. In addition, show that by doing so the conditions given by Equations 11.12 and 11.13 would be replaced by the conditions given

by Equations 11.28 and 11.29. Show why the condition given by Equation 11.14 would remain similar to Equation 11.30.

2. Use Equations 11.22, 11.46, and 11.47 to compute the instantaneous rate of heat transfer to the bar as well as the amount of heat transferred to the bar between two given instants.

3. Repeat the problem presented in Section 11.2 for the case when, instead of a given temperature at $z = 0$, the heat flux at that position is given and is constant.

4. Show that Equation 11.74 satisfies boundary conditions given by Equations 11.49 through 11.51.

5. Determine the equation for the drying rate of a particle by developing Equation 11.76. Use Equation 11.74 or any previous forms given by Equations 11.72 and 11.73.

6. Integrate the drying rate obtained in the previous problem from $t = 0$ until a given instant t_f, in order to determine the total mass of water evaporated from the particle during that period.

7. Repeat the problem presented in Section 11.3 by replacing the sphere for a very long cylinder, or at least a cylinder with length much larger than its diameter.

8. From Equations 11.77 through 11.83, deduce Equations 11.84 and conditions given by Equations 11.85 through 11.87.

9. Using Equation 11.99 or 11.100, obtain an expression for the rate of heat transfer at the cylinder surface. Plot a graph for that rate against time.

10. Solve a similar situation as posed at Section 11.4 for the case of a sphere.

11. Solve the problem presented in Section 11.5 using Laplace transform.

12. Solve the differential equation (Equation 11.148) by any variation of method of weighted residuals (MWR). Find, at least the first approximation.

13. In the previous problem, find the second approximation for function G.

14. Use the last solution of function G combined with the solution for function F (Equation 11.146) into Equation 11.137. Verify if the solution satisfies the boundary conditions given by Equations 11.133 through 11.136.

15. Repeat problem 21 in chapter 10 assuming that a layer of impermeable rock is found at position or depth $z = L$.

16. Try to apply the possibility of $C = 0$ at Equation 11.140 and verify the consequences in obtaining compatible solutions for Equation 11.139 that could be used at Equation 11.137 to solve the boundary value problem posed at Section 11.6.

REFERENCES

1. Luikov, A.V., *Heat and Mass Transfer*, Mir, Moscow, 1980.
2. Kafarov, V., *Fundamentals of Mass Transfer*, Mir, Moscow, 1975.
3. King, C.J., *Separation Processes*, Tata McGraw Hill, New Delhi, 1974.
4. de Souza-Santos, M.L., *Solid Fuels Combustion and Gasification; Modeling, Simulation, and Equipment Operation*, Marcel Dekker, New York, 2004.

5. Bird, R.B., Stewart, W.E., and Lightfoot, E.N., *Transport Phenomena*, John Wiley & Sons, New York, 1960.
6. Slattery, J.C., *Momentum, Energy, and Mass Transfer in Continua*, Robert E. Krieger, New York, 1978.
7. Walker, P.L. Jr., Rusinko, F., and Austin, L.G., Gas Reactions of Carbon. In *Advances in Catalysis*, Vol. XI, Academic Press, New York, 1959.

12 Problems 223; Two Variables, 2nd Order, 3rd Kind Boundary Condition

12.1 INTRODUCTION

This chapter presents methods to solve problems with two independent variables involving second-order differential equations and third-kind boundary conditions. Therefore, partial differential equations are involved. Mathematically, this class of cases can be summarized as $f\left(\phi, \omega_1, \omega_2, \frac{\partial^2 \phi}{\partial \omega_i^2}, \frac{\partial^2 \phi}{\partial \omega_i \partial \omega_j}, \frac{\partial \phi}{\partial \omega_i}\right)$, third-kind boundary condition.

Again, most of the solutions presented here for mass transfer can be used for cases of heat transfer, and vice versa.

For several situations, the problems presented here are improvements to those presented in Chapters 11 and 12. Third-kind boundary conditions are more realistic descriptions for heat and mass transfers between solids and fluids.

12.2 CONVECTIVE HEATING OF AN INSULATED ROD OR SEMI-INFINITE BODY

This is a classical problem illustrated in Figure 12.1. Unlike the situation presented in Section 11.2, now the rod extremities exchange heat by convection with the environment, which is at temperature T_a. The initial temperature of the rod is T_b.

The solutions achieved here can be equally applied to the problem of a semi-infinite body or plate or wall, as illustrated in Figure 12.2. The whole wall is initially at temperature T_b and suddenly both faces are exposed to a fluid at temperature T_a.

This situation is very common in industrial applications during thermal treatments of plates.

To take advantage of the symmetry, the origin of the space coordinate is chosen to be at the middle of the bar.

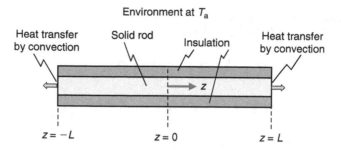

FIGURE 12.1 Insulated cylindrical rod exchanging heat by convection at both end faces.

To clearly pose the problem, let us list the assumptions:

1. Material of the solid body has uniform composition.
2. Thermal conductivity and density of the solid remain constant. Again, an approximation is valid for small-to-moderate temperature gradients within the body.
3. No phase change of the solid material of the rod is observed.
4. Insulation around the rod is perfect, i.e., no heat transfer is observed at surfaces covered with insulation material.
5. Temperature of the fluid (T_a), at positions far from the body surface, remains constant.
6. Convective heat transfer coefficient (α) between fluid and rod or plate surfaces is constant and equal at each heat exchanging side. As the

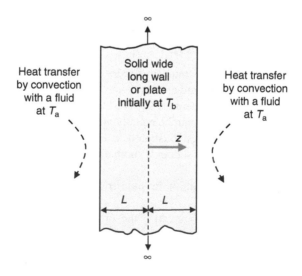

FIGURE 12.2 Semi-infinite flat plate exchanging heat by convection at both faces.

temperatures of the surfaces change, the physical properties of the fluid layer near them change. Since convective heat transfer coefficients depend on the properties of the fluid, such as density, viscosity, specific heat, etc., the present simplification is also valid if the variations of surface temperatures are not too high.

7. All heat transfer is assumed to take place by convection. This situation is valid as long as the radiative transfer is negligible. Such an assumption is reasonable for relatively small differences of temperatures between surfaces and environment.

Following the same arguments made in Sections 10.2 and 11.2, it is possible to arrive at

$$\rho C_p \frac{\partial T}{\partial t} = \lambda \frac{\partial^2 T}{\partial z^2} \tag{12.1}$$

Nevertheless, in the present case the boundary conditions are

$$T(0,z) = T_b, \quad -L \le z \le L \tag{12.2}$$

$$-\lambda \frac{\partial T}{\partial z}\bigg|_{z=L} = \alpha[T(t,L) - T_a], \quad t > 0 \tag{12.3}$$

$$-\lambda \frac{\partial T}{\partial z}\bigg|_{z=-L} = \alpha[T_a - T(t, -L)], \quad t > 0 \tag{12.4}$$

Of course, a symmetric temperature profile would develop and the last boundary condition can be replaced by

$$\frac{\partial T}{\partial z}\bigg|_{z=0} = 0, \quad t > 0 \tag{12.5}$$

Similar to the procedure used earlier, let us change variables to work with dimensionless variables or

$$\psi = \frac{T - T_a}{T_b - T_a} \tag{12.6}$$

$$\zeta = \frac{z}{L} \tag{12.7}$$

$$\tau = t \frac{\lambda}{\rho C_p L^2} = t \frac{D_T}{L^2} = \frac{t}{t_c} \tag{12.8}$$

The variable τ is also known as Fourier number.

After the transformations, it is possible to arrive at

$$\frac{\partial \psi}{\partial \tau} = \frac{\partial^2 \psi}{\partial \zeta^2} \tag{12.9}$$

The boundary conditions are now written as

$$\psi(0, \zeta) = 1, -1 < \zeta < 1 \tag{12.10}$$

$$\left. \frac{\partial \psi}{\partial \zeta} \right|_{\zeta=1} = -N_{Bi}\psi(\tau, 1), \quad \tau > 0 \tag{12.11}$$

$$\left. \frac{\partial \psi}{\partial \zeta} \right|_{\zeta=0} = 0, \quad \tau > 0 \tag{12.12}$$

Here

$$N_{Bi} = \frac{\alpha L}{\lambda} \tag{12.12a}$$

12.2.1   Solution by Laplace Transform

Let us apply the following Laplace transform:

$$\Psi(s, \zeta) = L\{\psi(\tau, \zeta)\} \tag{12.13}$$

Using the condition given by Equation 12.10 the transform of Equation 12.9 is

$$s\Psi - 1 = \frac{d^2 \Psi}{d\zeta^2} \tag{12.14}$$

This is a linear, second-order, nonhomogeneous ordinary differential equation and according to Appendix B, its general solution is

$$\Psi = C_1 \sinh(\zeta\sqrt{s}) + C_2 \cosh(\zeta\sqrt{s}) + \frac{1}{s} \tag{12.15}$$

The transform of conditions given by Equations 12.11 and 12.12 are

$$\left. \frac{d\Psi}{d\zeta} \right|_{\zeta=1} = -N_{Bi}\Psi(s, 1) \tag{12.16}$$

$$\left. \frac{d\Psi}{d\zeta} \right|_{\zeta=0} = 0 \tag{12.17}$$

Applying Equation 12.17 to Equation 12.15, one would conclude that $C_1 = 0$. Applying Equation 12.16 leads to

$$C_2 = -\frac{1}{s}\frac{N_{Bi}}{\sqrt{s}\sinh(\sqrt{s}) + N_{Bi}\cosh(\sqrt{s})} \qquad (12.18)$$

Using Equation 12.18, Equation 12.15 is written as

$$\Psi = \frac{1}{s} - \frac{1}{s}\frac{N_{Bi}\cosh(\zeta\sqrt{s})}{\sqrt{s}\sinh(\sqrt{s}) + N_{Bi}\cosh(\sqrt{s})} \qquad (12.19)$$

The inverse of the last term can be found in Table D.4 , or

$$L\left\{\frac{1}{s}\frac{N_{Bi}\cosh(\zeta\sqrt{s})}{\sqrt{s}\sinh(\sqrt{s}) + N_{Bi}\cosh(\sqrt{s})}\right\}$$

$$= 1 - \sum_{j=1}^{\infty}\frac{2\sin b_j}{b_j + \sin b_j\cos b_j}\cos(b_j\zeta)\exp\left(-b_j^2\tau\right) \qquad (12.20)$$

Here, parameters b_j are the roots to the following equation:

$$b_j\tan b_j = N_{Bi} \qquad (12.21)$$

Finally, it is possible to write

$$\psi(\tau,\zeta) = \sum_{j=1}^{\infty}\frac{2\sin b_j}{b_j + \sin b_j\cos b_j}\cos(b_j\zeta)\exp\left(-b_j^2\tau\right) \qquad (12.22)$$

Figures 12.3 through 12.5 represent the above equation for cases of $N_{Bi} = 0.1$, 1.0, and 10, respectively.

The following can be noticed from these figures:

- It is easy to verify, from either Equation 12.22 or the above figures, that the condition given by Equation 12.12 is satisfied. Of course, a similar profile would develop for the other side of the rod or plate. Hence, the solutions arrived here can be used for cases where the rod or plate is insulated at $z = 0$ and exchanges heat by convection on the exposed face to the environment or fluid.
- At the surfaces in contact with the fluid ($\zeta = 1$ or $\zeta = -1$) the boundary condition given by Equation 12.11 forces the profile to present a negative derivative. Its absolute value would increase with the Biot number. This can be seen by comparing Figures 12.3 and 12.4.
- Effect of thermal conductivity of the rod or plate material can be observed by considering a given instant (t). According to Equation 12.8, larger values of λ would lead to larger values of dimensionless

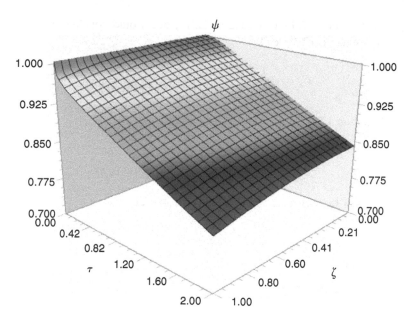

FIGURE 12.3 Dimensionless temperature (ψ) in an insulated rod or semi-infinite plate against dimensionless space (ζ) and time (τ) for $N_{Bi} = 0.1$.

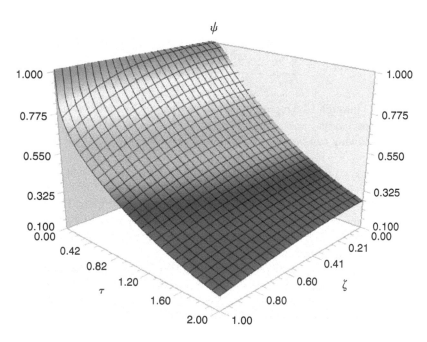

FIGURE 12.4 Dimensionless temperature (ψ) in an insulated rod or semi-infinite plate against dimensionless space (ζ) and time (τ) for $N_{Bi} = 1$.

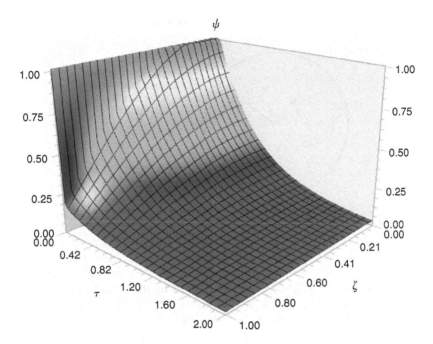

FIGURE 12.5 Dimensionless temperature (ψ) in an insulated rod or semi-infinite plate against dimensionless space (ζ) and time (τ) for $N_{Bi} = 10$.

time (τ), and therefore smaller values of dimensionless temperature (ψ) can be observed at the above figures. In other words, larger conductivities would force the temperature in the body to approach T_a at a faster rate.

- Of course, the contrary effect would be observed for thicker plates or longer rods (L).

12.3 DRYING OF A SPHERICAL PARTICLE

Similar to the situation described in Section 11.3, consider a spherical particle as shown in Figure 12.6.

The drying phases are described in Section 11.3. However now, the concentration of water at the particle surface is not known. Instead, this concentration is dictated by the rate of mass transfer between the sphere and the surrounding environment. Let us again determine the concentration of water inside the particle, which would provide conditions to compute the drying rate.

To clearly state the problem, the assumptions are listed as follows (justifications and clarifications of repeated assumptions are found in Section 11.3):

1. Isothermal particle.
2. Uniform porous particle with an apparent or effective constant diffusivity (D_{eff}) of water through the porous structure [1,2].

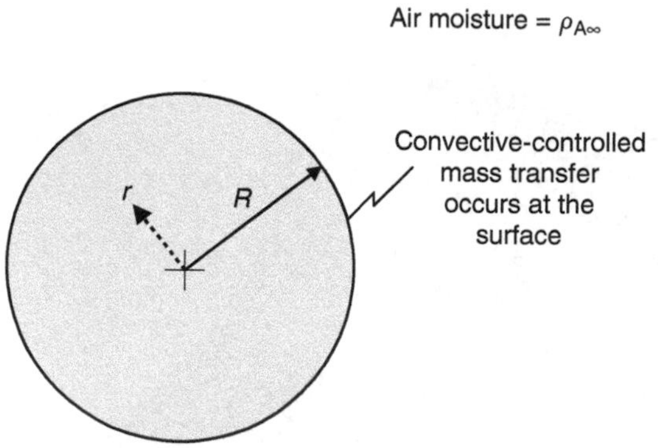

Air moisture = $\rho_{A\infty}$

FIGURE 12.6 A wet porous sphere.

3. No particular or constant concentration is assumed at the particle surface. Instead, the concentration there is dictated by the rate of mass transfer between the surface and the environment. This is a more realistic situation than the one posed in Chapter 11.
4. Water concentration in the environment (air or atmosphere) is constant and is denoted here by $\rho_{A\infty}$.
5. Mass transfer coefficient (β) between particle surface and environment is assumed constant. Usually, this coefficient is a function of temperature, pressure, and local concentration conditions. As the concentration changes, a variation may be expected for this coefficient. However, the variations are relatively small and the coefficient is assumed constant. Empirical and semiempirical correlations for mass transfer coefficients are available throughout the published literature [3–6].
6. Overall velocity of fluid toward the particle surface is neglected.

Likewise in Section 11.3, it is possible to simplify Equation A.42 to

$$\frac{\partial \rho_A}{\partial t} = D_{eff} \frac{1}{r^2} \frac{\partial}{\partial r}\left(r^2 \frac{\partial \rho_A}{\partial r}\right) \tag{12.23}$$

On the other hand, the boundary conditions for the present problem are

$$\rho_A(0,r) = \rho_{A0}, \quad 0 < r < R \tag{12.24}$$

$$-D_{AB}\frac{\partial \rho_A}{\partial r}\bigg|_{r=R} = \beta[\rho_A(t,R) - \rho_{A\infty}], \quad t > 0 \tag{12.25}$$

$$\frac{\partial \rho_A}{\partial r}\bigg|_{r=0} = 0, \quad t > 0 \tag{12.26}$$

One should notice that the only change from the problem presented in Section 11.3 is at the second boundary condition, which is a third-kind boundary condition.

Again, the same procedure as shown in Section 11.3 is used to reach at

$$\frac{\partial \psi}{\partial \tau} = \frac{\partial^2 \psi}{\partial \zeta^2} \tag{12.27}$$

Here

$$\psi = \zeta \varphi \tag{12.28}$$

$$\varphi = \frac{\rho_A - \rho_{A0}}{\rho_{A\infty} - \rho_{A0}} \tag{12.29}$$

$$\zeta = \frac{r}{R} \tag{12.30}$$

$$\tau = \gamma t \tag{12.31}$$

$$\gamma = \frac{D_{eff}}{R^2} \tag{12.32}$$

Notice that the present definition of dimensionless variable φ by Equation 12.29 differs from that set by Equation 11.52.

The boundary condition given by Equation 12.24 becomes

$$\psi(0,\zeta) = 0, \quad 0 < \zeta < 1 \tag{12.33}$$

The condition given in Equation 12.25 is rewritten as

$$\left. \frac{\partial \psi}{\partial \zeta} \right|_{\zeta=1} = -\psi(\tau,1)(N_{Sh} - 1) + N_{Sh}, \quad \tau > 0 \tag{12.34}$$

Here the Sherwood number is given by

$$N_{Sh} = \frac{\beta R}{D_{AB}} \tag{12.34a}$$

An apparent impossibility during the transformation of the boundary condition given by Equation 12.26 would occur. Instead, looking at Equation 12.28, and since φ always acquires a finite value at sphere center ($r=0$ or $\zeta=0$), it is possible to write

$$\psi(\tau,0) = 0, \quad \tau > 0 \tag{12.35}$$

Therefore, the above equation can replace the condition given by Equation 12.26, which avoids the indetermination problem.

12.3.1 Solution by Laplace Transform

Because of the reasoning described in Section 11.3.1 let the transform be

$$\Phi(s,\zeta) = L\{\psi(\tau,\zeta)\} \tag{12.36}$$

Applying the above equation in Equation 12.27 and using the condition given by Equation 12.33, the following can be obtained:

$$\frac{d^2\Phi}{d\zeta^2} = s\Phi \tag{12.37}$$

According to Appendix B, the general solution for this heterogeneous second-order linear differential equation is

$$\Phi = C_1 \sinh(\zeta\sqrt{s}) + C_2 \cosh(\zeta\sqrt{s}) \tag{12.38}$$

The conditions given by Equations 12.34 and 12.35 allow writing

$$\left.\frac{d\Phi}{d\zeta}\right|_{\zeta=1} = -\Phi(s,1)(N_{Sh} - 1) + \frac{N_{Sh}}{s} \tag{12.39}$$

$$\Phi(s,0) = 0 \tag{12.40}$$

After applying the above equations, Equation 12.38 becomes

$$\Phi = \frac{1}{s}\frac{N_{Sh}\sinh(\zeta\sqrt{s})}{\sqrt{s}\cosh(\sqrt{s}) + (N_{Sh} - 1)\sinh(\sqrt{s})} \tag{12.41}$$

Using Table D.4 the inverted function is

$$\psi(\tau,\zeta) = \zeta - 2\sum_{j=1}^{\infty}\frac{\sin b_j - b_j\cos b_j}{b_j - \sin b_j\cos b_j}\frac{\sin(b_j\zeta)}{b_j}\exp\left(-b_j^2\tau\right) \tag{12.42}$$

Alternatively, from Equation 12.28, the result can be written as

$$\varphi(\tau,\zeta) = 1 - \frac{2}{\zeta}\sum_{j=1}^{\infty}\frac{\sin b_j - b_j\cos b_j}{b_j - \sin b_j\cos b_j}\frac{\sin(b_j\zeta)}{b_j}\exp\left(-b_j^2\tau\right) \tag{12.43}$$

Here b_j are the roots of

$$N = 1 - b_j\cot b_j \tag{12.43a}$$

Equation 12.43 is represented in Figures 12.7 and 12.8 for the cases when $N_{Sh} = 4$ and $N_{Sh} = 100$, respectively.

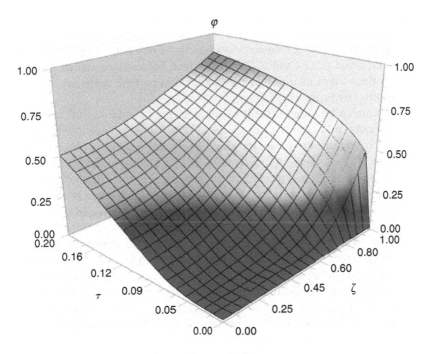

FIGURE 12.7 Dimensionless concentration (φ) of moisture inside a spherical particle as a function of dimensionless radius (ζ) and dimensionless time (τ) for the case of $N_{Sh} = 4$.

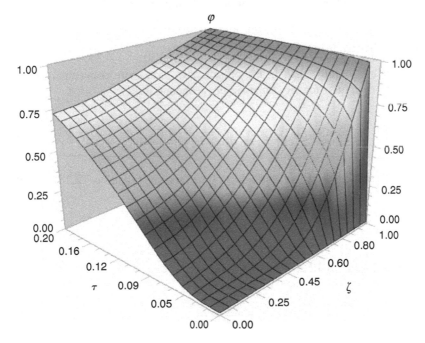

FIGURE 12.8 Dimensionless concentration (φ) of moisture inside a spherical particle as a function of dimensionless radius (ζ) and dimensionless time (τ) for the case of $N_{Sh} = 100$.

From the above figures, it is possible to notice the following:

- Equivalent on the dimensionless variables of boundary conditions given by Equations 12.24 and 12.26 are satisfied. For instance, one may verify that derivatives of moisture concentration at particle center are zero, as required by Equation 12.26.
- Although a bit more difficult to visualize, the boundary condition given by Equation 12.25 is also satisfied. This can be shown by Equation 12.43. An indication from the above figures is possible by observing the plane at $\zeta = 1$. There the derivatives of concentrations are always positive, indicating mass transfer from the particle to the surroundings.
- From the above two figures, it is possible to visualize that at any given position (r or ζ) inside the particle, the period of unaffected concentration is smaller for larger Sherwood numbers.
- Increases in the Sherwood number should provide higher drying rates. For instance, for dimensionless time (τ) equal to 0.2, the dimensionless concentration (φ) at the center ($\zeta = 0$) approaches 1 for higher Sherwood numbers. One should remember that the moisture in the particle would tend to the value at environment ($\rho_{A\infty}$), therefore φ should tend to 1.
- It can be observed that the concentration of water at the particle surface ($\zeta = 1$) is a function of time. The value starts at $\varphi = 0$ (or ρ_{A0}) and tends toward $\varphi = 1$ (or $\rho_{A\infty}$), which of course would be the situation where the particle would be completely dried. Of course, higher mass transfers (or N_{Sh}) would lead to a faster approach to the dry condition.

12.3.2 SOLUTION BY SEPARATION OF VARIABLES

It is possible to show that, in the present case, it is more convenient using the following definition for the dimensionless concentration:

$$\phi = \frac{\rho_A - \rho_{A\infty}}{\rho_{A0} - \rho_{A\infty}} \tag{12.44}$$

The equivalent to the new variable ψ (as given by Equation 12.28) would be called θ, and

$$\theta = \zeta\phi \tag{12.45}$$

With this, Equation 12.23 becomes

$$\frac{\partial\theta}{\partial\tau} = \frac{\partial^2\theta}{\partial\zeta^2} \tag{12.46}$$

The boundary conditions given by Equations 12.24 through 12.26 would be written as

$$\theta(0,\zeta) = \zeta, \quad 0 < \zeta < 1 \tag{12.47}$$

$$\left.\frac{\partial\theta}{\partial\zeta}\right|_{\zeta=1} = \theta(\tau,1)[1 - N_{Sh}], \quad \tau > 0 \tag{12.48}$$

$$\theta(\tau,0) = 0, \quad \tau > 0 \tag{12.49}$$

To apply the principle of the method of variable separation, one should write

$$\theta(\tau,\zeta) = F(\tau)G(\zeta) \tag{12.50}$$

Substituting Equation 12.50 in Equation 12.46 leads to

$$\frac{dF/d\tau}{F(\tau)} = \frac{d^2G/d\zeta^2}{G} = C \tag{12.51}$$

Here C is a constant.

The first differential equation is given by

$$\frac{dF/d\tau}{F(\tau)} = C \tag{12.52}$$

Its solution is

$$F(\tau) = C_1 \exp(C\tau) \tag{12.53}$$

However, the condition given by Equation 12.47 cannot be used to determine constants C and C_1. On the other hand, the general solution of an ordinary differential equation on variable G depends on whether C is a positive or a negative constant. For this, let us examine the situations regarding concentrations in the sphere and environment.

Since the concentration of water tends to $\rho_{A\infty}$, φ and ψ should decrease with time. Therefore, the solution given by Equation 12.53 would require C to be a negative constant. Let us force this by writing

$$C = -a^2 \tag{12.54}$$

Here parameter a is a real constant. Therefore

$$F(\tau) = C_1 \exp(-a^2\tau) \tag{12.55}$$

From the above, the solution of Equation 12.51 regarding function G is

$$G(\zeta) = C_2 \sin(a\zeta) + C_3 \cos(a\zeta) \tag{12.56}$$

Equation 12.50 combined with the condition given by Equation 12.49 would require

$$G(0) = 0 \tag{12.57}$$

Consequently, Equation 12.56 becomes

$$G(\zeta) = C_2 \sin(a\zeta) \tag{12.58}$$

In a more generalized way, this solution can be written as

$$G(\zeta) = \sum_{j=1}^{\infty} C_j \sin(a_j \zeta) \tag{12.59}$$

From Equation 12.50, it now possible to write

$$\theta(\tau,\zeta) = \sum_{j=1}^{\infty} C_j \exp(-a_j^2 \tau) \sin(a_j \zeta) \tag{12.60}$$

Without any loss of generality, it is possible to write

$$a_j = \frac{2\pi}{p} j \tag{12.61}$$

Here p is the period to be determined. With this, Equation 12.60 becomes

$$\theta(\tau,\zeta) = \sum_{j=1}^{\infty} C_j \exp\left[-\left(\frac{2\pi}{p} j\right)^2 \tau\right] \sin\left(\frac{2\pi}{p} j\zeta\right) \tag{12.62}$$

Now, condition given by Equation 12.47 can be used to provide

$$\zeta = \sum_{j=1}^{\infty} C_j \sin\left(\frac{2\pi}{p} j\zeta\right) \tag{12.63}$$

From Appendix G, the above can be seen as an odd half-range expansion of function ζ, and

$$C_j = \frac{4}{p} \int_0^{p/2} x \sin\left(\frac{2\pi}{p} jx\right) dx = -\frac{p}{\pi j} \cos(\pi j) = -\frac{p}{\pi j}(-1)^j \tag{12.64}$$

Finally

$$\theta(\tau,\zeta) = -\frac{p}{\pi} \sum_{j=1}^{\infty} \frac{(-1)^j}{j} \exp\left[-\left(\frac{2\pi}{p}j\right)^2 \tau\right] \sin\left(\frac{2\pi}{p}j\zeta\right) \qquad (12.65)$$

Therefore, Equation 12.65 satisfies conditions given by Equations 12.47 and 12.49. To satisfy the condition given by Equation 12.48 the following is required:

$$\sum_{j=1}^{\infty} \pi j \cos\left(\frac{2\pi}{p}j\right) - \frac{p}{\pi}(1 - N_{Sh}) \sin\left(\frac{2\pi}{p}j\right) = 0 \qquad (12.66)$$

Of course, the value of the representative period p depends on the Sherwood number. For instance, for the case of N_{Sh} equal to 2, the period becomes equal to 3.9.

The change of variables proposed by Equation 12.44 leads to a simpler form for Equation 12.66 when compared with the form that would be achieved if the condition given by Equation 12.34 were to be used.

In spite of applying different definitions for the dimensionless concentrations, as done when solving using the Laplace transform, both results are equivalent.

Figure 12.9 illustrates a case of results provided by Equation 12.65 when the Sherwood number equals to 2.

The reader is invited to obtain the rate of particle drying as a function of time.

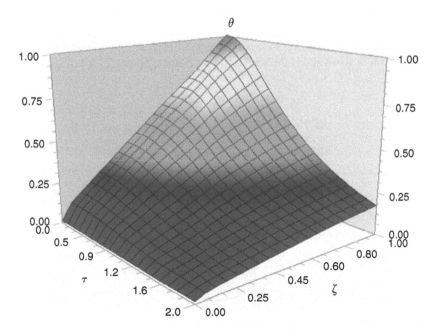

FIGURE 12.9 Dimensionless concentration (θ) of moisture inside a spherical particle as a function of dimensionless radius (ζ) and dimensionless time (τ) for the case of $N_{Sh} = 2$.

12.4   CONVECTIVE HEATING OF A CYLINDER

Convective heating of a cylinder is illustrated in Figure 12.10. This situation is similar to the situation solved in Section 11.4, however now the temperature at the surface is no longer known or imposed. Instead, it results from the heat transfer between this surface and the surrounding environment. Again, it is desired to determine the temperature profile in the cylinder at any time as well as the rate of heat transfer to or from it.

To clearly put the problem, consider the following simplifications:

1. Length of the cylinder is much bigger than its diameter, in a way that end-effects would be negligible. Therefore, the temperature inside the cylinder would be just a function of time and the radial coordinate.
2. Surrounding fluid, far from the cylinder surface, remains at a constant temperature T_a.
3. No phase change occurs in the solid material and all properties of the cylinder material remain constant. Of course, depending on the temperatures involved in the process, this might constitute strong assumptions. Therefore, the solution achieved here is applicable for situations with relatively small or moderate differences between T_0 and T_a.
4. No chemical reaction occurs in the cylinder material.
5. Heat transfer coefficient between the cylinder surface and the fluid remains constant. As the temperature of the surface changes with time, this would affect the physical properties of fluid layer near it. Since convective heat transfer coefficients depend on the properties of the fluid, such as density, viscosity, specific heat, etc., the present simplification is also valid if the variation of surface temperature is not too high.

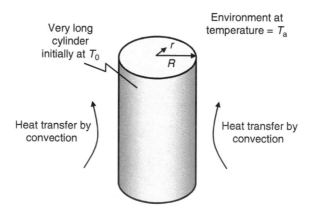

FIGURE 12.10 Infinite cylinder heated at surface by convective heat transfer with a surrounding fluid.

6. All heat transfer is assumed to take place by convection. This situation is valid as long as the radiative transfer is negligible. Such assumption is reasonable for moderate differences of temperatures (usually less than 100 K) between the cylinder surface and environment.

Applying the above assumptions allows Equation A.35 to be simplified and written as

$$\rho C_p \frac{\partial T}{\partial t} = \lambda \frac{1}{r} \frac{\partial}{\partial r} \left(r \frac{\partial T}{\partial r} \right) \tag{12.67}$$

The boundary conditions for this problem are

$$T(0,r) = T_0, \quad 0 \leq r \leq R \tag{12.68}$$

$$-\lambda \frac{\partial T}{\partial r}\bigg|_{r=R} = \alpha[T(t,R) - T_a], \quad t > 0 \tag{12.69}$$

$$\frac{\partial T}{\partial r}\bigg|_{r=0} = 0, \quad t > 0 \tag{12.70}$$

Notice that the boundary condition given by Equation 12.69 is a third-kind boundary condition and represents the heat transfer between the surface and environment at temperature T_a.

To facilitate the solution and the presentation of results, the following changes of variables are proposed:

$$\psi = \frac{T - T_0}{T_a - T_0} \tag{12.71}$$

$$\zeta = \frac{r}{R} \tag{12.72}$$

$$\tau = t \frac{D_T}{R^2} = t \frac{\lambda}{\rho C_p R^2} \tag{12.73}$$

The dimensionless variable τ is also known as the Fourier number. This leads to a similar equation as obtained in Section 11.4, or

$$\frac{\partial \psi}{\partial \tau} = \frac{1}{\zeta} \frac{\partial \psi}{\partial \zeta} + \frac{\partial^2 \psi}{\partial \zeta^2} \tag{12.74}$$

The boundary conditions become

$$\psi(0,\zeta) = 0, \quad 0 < \zeta < 1 \tag{12.75}$$

$$\frac{\partial \psi}{\partial \zeta}\bigg|_{\zeta=1} = -N_{Bi}[\psi(\tau,1) - 1], \quad \tau > 0 \tag{12.76}$$

$$\frac{\partial \psi}{\partial \zeta}\bigg|_{\zeta=0} = 0, \quad \tau > 0 \tag{12.77}$$

where the Biot number is given by

$$N_{Bi} = \frac{\alpha R}{\lambda} \tag{12.78}$$

12.4.1 SOLUTION BY LAPLACE TRANSFORM

Consider the Laplace transform on variable τ, or

$$\Psi(s,\zeta) = L\{\psi(\tau,\zeta)\} \tag{12.79}$$

From Equations 12.74 and 12.79, and using the condition given by Equation 12.75 it is possible to write

$$\zeta^2 \frac{d^2\Psi}{d\zeta^2} + \zeta \frac{d\Psi}{d\zeta} - s\zeta^2 \Psi = 0 \tag{12.80}$$

The following change of variable is proposed:

$$y = \zeta\sqrt{s} \tag{12.81}$$

With this, Equation 12.80 can be put in the exact form of a Bessel modified equation, or

$$y^2 \frac{d^2\Psi}{dy^2} + y \frac{d\Psi}{dy} - y^2 \Psi = 0 \tag{12.82}$$

As shown in Appendix C, the general solution of Equation 12.82 is

$$\Psi = C_1 I_0(y) + C_2 K_0(y)$$

or

$$\Psi = C_1 I_0(\zeta\sqrt{s}) + C_2 K_0(\zeta\sqrt{s}) \tag{12.83}$$

The transforms of conditions given by Equations 12.76 and 12.77 are

$$\frac{d\Psi}{d\zeta}\bigg|_{\zeta=1} = -N_{Bi}\left[\Psi(s,1) - \frac{1}{s}\right] \tag{12.84}$$

$$\frac{d\Psi}{d\zeta}\bigg|_{\zeta=0} = 0 \tag{12.85}$$

To apply the above conditions, Equation 12.83 should be differentiated, leading to

$$\frac{d\Psi}{d\zeta} = C_1 I_1\left(\zeta\sqrt{s}\right)\sqrt{s} + C_2 K_1\left(\zeta\sqrt{s}\right)\sqrt{s} \tag{12.86}$$

According to the condition given by Equation 12.84, C_2 should be equal to zero because (Appendix C)

$$\lim_{y \to 0} K_1(y) = \infty \tag{12.87}$$

and

$$I_1(0) = 0 \tag{12.88}$$

Using Equation 12.84, the solution can be written as

$$\Psi(s,\zeta) = \frac{N_{Bi}}{s}\frac{I_0(\zeta\sqrt{s})}{I_1(\sqrt{s})\sqrt{s} + N_{Bi}I_0(\sqrt{s})} \tag{12.89}$$

The inversion of such a complex function can be found in the tables in Appendix D and is written as

$$\psi(\tau,\zeta) = 1 - 2\sum_{j=1}^{\infty}\frac{N_{Bi}}{b_j^2 + N_{Bi}^2}\frac{J_0(b_j\zeta)}{J_0(b_j)}\exp\left(-b_j^2\tau\right) \tag{12.90}$$

Here, the parameters b_j are the real roots of the following equation:

$$\frac{b_j}{N_{Bi}} = \frac{J_0(b_j)}{J_1(b_j)} \tag{12.91}$$

Figure 12.11 represents Equation 12.90 for the case of the Biot number equal to 4 and Figure 12.12 for $N_{Bi} = 100$.

A brief examination of these figures shows the following:

- Cylinder is heated from the surface ($\zeta = 1$) to the center ($\zeta = 0$), hence the internal layers have their temperatures change as the "heating wave" advances from the surface to the center. The heating wave progresses faster for larger values of the Biot number.
- For relatively small N_{Bi} (Figure 12.11) and within the period (t or τ) covered by the graph, the cylinder surface would remain below the

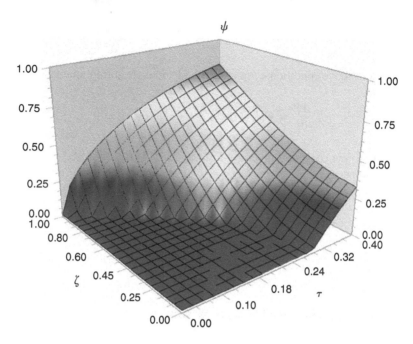

FIGURE 12.11 Dimensionless temperature (ψ) of a cylinder as a function of dimensionless radius (ζ) and dimensionless time (τ) for the case of $N_{Bi} = 4$.

ambiance temperature T_a. This contrasts with the case of large N_{Bi} (Figure 12.12), where almost the whole surface reached the ambiance temperature T_a, or $\psi = 1$ within the stipulated period.

• Heating rate is proportional to the derivative of temperature at the surface ($\zeta = 1$) and the figures illustrate how this rate decreases with time. This is inevitable due to the decrease of temperature difference between the surface and environment with time.

Using Equation 12.90 the reader is invited to deduce the formula for the computation of heating rate.

12.4.2 SOLUTION BY SEPARATION OF VARIABLES

According to Appendix G, the following is proposed:

$$\psi(\tau,\zeta) = F(\tau)\, G(\zeta) \tag{12.92}$$

When applied to Equation 12.74, it provides

$$\frac{1}{F}\frac{dF}{d\tau} = C_1 \tag{12.93}$$

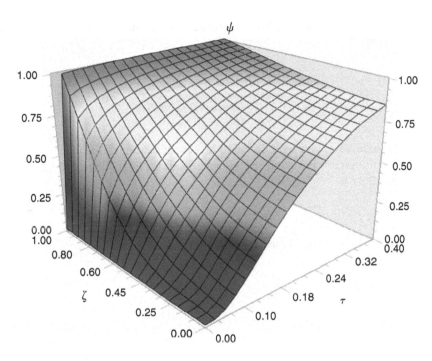

FIGURE 12.12 Dimensionless temperature (ψ) of a cylinder as a function of dimensionless radius (ζ) and dimensionless time (τ) for the case of $N_{Bi} = 100$.

$$\frac{1}{\zeta G}\frac{dG}{d\zeta} + \frac{1}{G}\frac{d^2 G}{d\zeta^2} = C_1 \qquad (12.94)$$

Here C_1 is a constant.

As before, the present definition of dimensionless temperature brings a problem because the boundary condition given by Equation 12.75 would force

$$F(0) = 0 \qquad (12.95)$$

which is impossible to conciliate with the solution given by Equation 12.92.

The same strategy as previously used is applied here, i.e., change the definition of function φ to avoid the zero value for boundary condition for F.

Let us try to set

$$\varphi = \frac{T - T_a}{T_0 - T_a} \qquad (12.96)$$

With this, one obtains the same differential equation as given by Equation 12.74 (just replacing ψ by φ). However, with the exception of the condition given by

Equation 12.77, the other two boundary conditions given by Equations 12.75 and 12.76 become

$$\varphi(0,\zeta) = 1, \quad 0 < \zeta < 1 \tag{12.97}$$

$$\left.\frac{\partial\varphi}{\partial\zeta}\right|_{\zeta=1} = -N_{\mathrm{Bi}}\varphi(\tau,1), \quad \tau > 0 \tag{12.98}$$

A similar relation as given by Equation 12.92 can be written for φ.

The solution of Equation 12.93 combined with the condition given by Equation 12.97 is

$$F(\tau) = e^{C_1\tau} \tag{12.99}$$

Equation 12.94 is written in the following form:

$$\zeta^2\frac{\mathrm{d}^2G}{\mathrm{d}\zeta^2} + \zeta\frac{\mathrm{d}G}{\mathrm{d}\zeta} - C_1\zeta^2G = 0 \tag{12.100}$$

This reminds us of a Bessel differential equation. Actually, as there are two boundary conditions to apply to the above equation, a degree of freedom allows setting an arbitrary value for C_1. If a positive or negative value for C_1 were chosen, a Bessel differential equation, or its modified version, would be generated. When the form given by Equation 12.96 is recollected, it is obvious that the dimensionless temperature φ should be a decreasing function of time. This happens for any situation of initial temperature T_0, i.e., if greater or smaller than the ambient value T_a. Therefore, the solution for Equation 12.99 should lead to a decreasing function $F(\tau)$. This forces us to choose a negative C_1, or

$$C_1 = -a^2 \tag{12.101}$$

Here, the parameter a acquires only real values.

According to Appendix C, the solution of Equation 12.100 would be

$$G(\zeta) = C_2J_0(a\zeta) + C_3Y_0(a\zeta) \tag{12.102}$$

Employing Equation 12.92, the condition given by Equation 12.77 leads to

$$\left.\frac{\mathrm{d}G}{\mathrm{d}\zeta}\right|_{\zeta=0} = 0 \tag{12.103}$$

The derivative of function G is (Appendix C)

$$\frac{\mathrm{d}G}{\mathrm{d}\zeta} = aC_2J_1(a\zeta) + aC_3Y_1(a\zeta) \tag{12.104}$$

According to Appendix C, for ζ approaching 0, $J_1(a\zeta)$ tends to zero and $Y_1(a\zeta)$ tends to minus infinity. Hence, $C_3 = 0$ and Equation 12.102 becomes

$$G(\zeta) = C_2 J_0(a\zeta) \tag{12.105}$$

Again, keeping in mind Equation 12.92, the boundary condition given by Equation 12.98 can be written as

$$\left.\frac{dG}{d\zeta}\right|_{\zeta=1} = -N_{\text{Bi}} G(1) \tag{12.106}$$

Since $C_3 = 0$, Equation 12.104 combined with Equations 12.105 and 12.106 leads to

$$\frac{aJ_1(a)}{J_0(a)} = -N_{\text{Bi}} \tag{12.107}$$

The roots of this characteristic equation provide the various values of parameter a. To indicate the generalization, they will be called a_j. Therefore, the solution will be a linear combination given by

$$\varphi(\tau,\zeta) = \sum_{j=1}^{\infty} C_j J_0(a_j\zeta) \exp\left(-a_j^2 \tau\right) \tag{12.108}$$

Now, the condition given by Equation 12.97 requires

$$\sum_{j=1}^{\infty} C_j J_0(a_j\zeta) = 1 \tag{12.109}$$

As shown in Appendix I, Bessel functions are orthogonal. Consequently, Equation 12.109 constitutes a generalized orthogonal expansion of function $f(x) = 1$. The coefficients C_j can be found by

$$C_j = \frac{2}{J_1^2(a_j)} \int_0^1 x J_0(a_j x)dx = \frac{2J_1(a_j)}{a_j J_1^2(a_j)} \tag{12.110}$$

Finally, it is possible to write

$$\varphi(\tau,\zeta) = \sum_{j=1}^{\infty} \frac{2}{a_j J_1(a_j)} J_0(a_j\zeta) \exp\left(-a_j^2 \tau\right) \tag{12.111}$$

Here a_j are the roots of the following relation:

$$\frac{a_j J_1(a_j)}{J_0(a_j)} = -N_{Bi} \tag{12.112}$$

The solutions given by Equations 12.108 or 12.111 are equivalent.

12.5 CONVECTIVE HEATING OF INSULATED ROD WITH PRESCRIBED TEMPERATURE AT ONE END

Consider a cylindrical solid rod as shown in Figure 12.13. Similar to the problems presented in Sections 10.2, 11.2, and 12.2, the rod is at uniform temperature T_b (in the temperature before heating). However, the situation here is a combination of prescribed temperature T_0 at one end with convective heat transfer at the other. It is desired to determine the temperature profile in the rod as a function of time. After this, it would be possible to determine the rate of heat transfer at each exposed surface.

As commented at the end of this section, the solutions achieved here can equally be applied to the problem of an infinite plate or wall.

The same basic assumptions listed in Section 12.2 are also considered here. Therefore, the following equation can be written:

$$q_r = -\lambda \frac{\partial T}{\partial r} = 0 \tag{12.113}$$

In addition, the same governing differential equation is obtained, or

$$\rho C_p \frac{\partial T}{\partial t} = \lambda \frac{\partial^2 T}{\partial z^2} \tag{12.114}$$

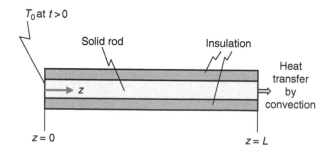

FIGURE 12.13 Insulated cylindrical rod with one face at prescribed temperature and exchanging heat by convection at the other face.

Nevertheless, in the present case the boundary conditions are

$$T(0,z) = T_b, \quad 0 < z \leq L \tag{12.115}$$

$$T(t,0) = T_0, \quad t > 0 \tag{12.116}$$

$$-\lambda \frac{\partial T}{\partial z}\bigg|_{z=L} = \alpha[T(t,L) - T_a], \quad t > 0 \tag{12.117}$$

Here, T_a is the ambiance temperature and the left-hand side of Equation 12.117 represents the heat flux transfer by conduction at interface ($z = L$), whereas the right-hand side represents the flux exchanged by convection between the bar and air (or any fluid) at the same interface. This is a third-kind boundary condition.

Similar to the procedure used before, let us change variables to work with dimensionless variables. The following are the proposed variables:

$$\psi = \frac{T - T_b}{T_0 - T_b} \tag{12.118}$$

$$\zeta = \frac{z}{L} \tag{12.119}$$

$$\tau = t\frac{\lambda}{\rho C_p L^2} = t\frac{D_T}{L^2} = \frac{t}{t_c} \tag{12.120}$$

After the transformations (see Section 11.2), it is possible to arrive at

$$\frac{\partial \psi}{\partial \tau} = \frac{\partial^2 \psi}{\partial \zeta^2} \tag{12.121}$$

The boundary conditions are written now as

$$\psi(0,\zeta) = 0, \quad 0 < \zeta \leq 1 \tag{12.122}$$

$$\psi(\tau,0) = 1, \quad \tau > 0 \tag{12.123}$$

$$\frac{\partial \psi}{\partial \zeta}\bigg|_{\zeta=1} = -N_{Bi}[\psi(\tau,1) - a], \quad \tau > 0 \tag{12.124}$$

where

$$a = \frac{T_a - T_b}{T_0 - T_b} \tag{12.125}$$

For the sake of an example, this problem is simplified by assuming that the rod was at the ambiance temperature before the process begins. Therefore $T_b = T_a$ and $a = 0$, which allows writing

$$\left.\frac{\partial \psi}{\partial \zeta}\right|_{\zeta=1} = -N_{Bi}\psi(\tau,1), \quad \tau > 0 \tag{12.126}$$

The Biot number is given by

$$N_{Bi} = \frac{\alpha L}{\lambda} \tag{12.127}$$

12.5.1   SOLUTION BY METHODS OF WEIGHTED RESIDUALS

As shown in Appendix E, trial functions may be set for either τ or ζ variables. On the other hand, it is worthwhile to remember that the residue is going to be integrated (after multiplied by the weighting function) on one of the variables. Therefore, it is more convenient to select the bounded variable, or the one with finite limits, to build the trial functions. As seen, time extends indefinitely, whereas ζ can vary only between zero and one. However, this choice does not guarantee better approximations against the alternative of functions on variable τ.

As shown in Appendix E, let us set the following approximate solution for Equation 12.121:

$$\psi_n = \exp\left(-N\frac{\zeta}{g(\tau)}\right) + \sum_{i=1}^{n} \zeta^{i+1}(1-\zeta)^2 f_i(\tau) \tag{12.128}$$

It is important to notice that the above form always satisfies the condition given by Equation 12.123. The condition given by Equation 12.122 would be satisfied as long as

$$g(0) = 0 \tag{12.129}$$

and

$$f_i(0) = 0 \tag{12.130}$$

On the other hand, the condition given by Equation 12.126 leads to

$$g(\tau) = 1, \quad \tau > 0 \tag{12.131}$$

From the two above relations, it becomes clear that $g(\tau)$ is a unit step-function. Of course, this is an approximation because no natural process follows true step-functions. Sudden jumps can be avoided by using Fourier series to represent near step variations. In this way, let us consider an odd representation of a step increase (see Appendix G) given by

$$g(\tau) = \sum_{j=1}^{\infty} \left[b_j \sin\left(\frac{2\pi}{p} j\tau\right) \right] \tag{12.132}$$

with

$$b_j = \frac{4}{p} \int_0^{p/2} \sin\left(\frac{2\pi}{p} jx\right) dx, \quad j = 1, 2, 3, \ldots \tag{12.133}$$

Therefore, Equation 11.121 can be written as

$$g(\tau) = \frac{2}{\pi} \sum_{j=1}^{\infty} \frac{1}{j} \left[1 - (-1)^j\right] \sin\left(\frac{2\pi}{p} j\tau\right) \tag{12.134}$$

Here, period p should be selected as twice the maximum achievable value of τ during the process. This would conform to the present mathematical representation of the process.

Now, applying the procedures described in Appendix E, it will be possible to find approximated solutions for the present problem.

12.5.1.1 First Approximation

The first approximation is

$$\psi_1 = \exp\left(-N\frac{\zeta}{g(\tau)}\right) + \zeta^2(1 - \zeta)^2 f_1(\tau) \tag{12.135}$$

Employing Equation 12.121, the residue is

$$\Lambda_1 = \frac{N\zeta}{g^2}\frac{dg}{d\tau}\exp\left(-N\frac{\zeta}{g}\right) + \zeta^2(1 - \zeta)^2 f_1' - \frac{N^2}{g^2}\exp\left(-N\frac{\zeta}{g}\right) - 4(1 - \zeta)^2 f_1$$
$$+ 8\zeta(1 - \zeta)f_1 - 2\zeta^2 f_1 \tag{12.136}$$

12.5.1.1.1 Collocation Method

For a simple initial approximation, let us apply the collocation method at point $\zeta = 1/2$. Therefore, the residual should be equal to zero at this point, or

$$f_1' + 8f_1 = \frac{16N}{g^2}e^{-\frac{N}{2g}}\left(N - \frac{g'}{2}\right) \tag{12.137}$$

Using variation of parameters (see Appendix B), the solution for Equation 12.137 is

$$f_1(\tau) = 16Ne^{-8\tau} \int_0^\tau \left(N - \frac{g'}{2} \right) \frac{1}{g^2} \exp\left(8\tau - \frac{N}{2g} \right) d\tau \qquad (12.138)$$

where

$$g'(\tau) = \frac{4}{p} \sum_{j=1}^\infty [1 - \cos(\pi j)] \cos\left(\frac{2\pi}{p} j\tau \right) \qquad (12.139)$$

Equation 12.138 satisfies the condition given by Equation 12.130. This function can now be used to obtain the first approximation, as given by Equation 12.135.

Figures 12.14 and 12.15 illustrate the results for this first approximation in cases when N_{Bi} is equal to 10 and 2, respectively.

Despite few plausible aspects, this first approximation is still a very crude representation of the temperature profile in the bar. An unlikely behavior is the very sharp increase of temperature at the very first instants ($\tau \approx 0$) of the

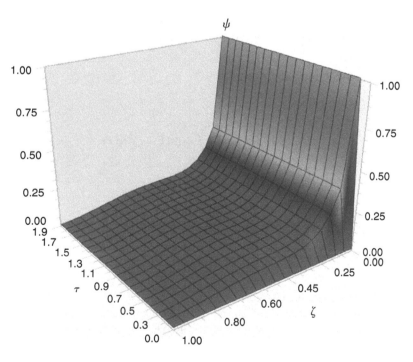

FIGURE 12.14 Dimensionless temperature (ψ) against dimensionless space (ζ) and time (τ) using the first approximation for $N_{Bi} = 10$.

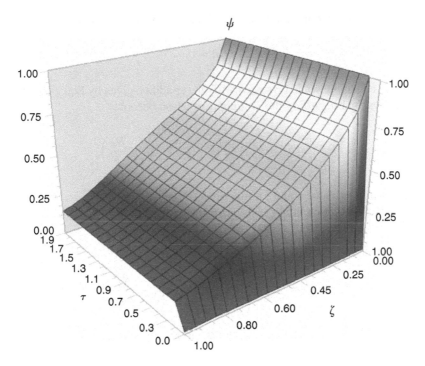

FIGURE 12.15 Dimensionless temperature (ψ) against dimensionless space (ζ) and time (τ) using the first approximation for $N_{Bi} = 2$.

conduction process. In addition, the temperature almost does not increase for instants after this ($\tau > 0$). Nonetheless, the profile is somewhat coherent regarding the dependence of the space variable. For instance, they show more uniform temperature profiles in the rod for lower values of Biot number. This is easy to understand from Equation 12.127, because N_{Bi} is inversely proportional to the conductivity of the rod.

Of course, the solution would improve if one applies a second approximation.

12.5.1.2 Alternative Form for the Approximation

An alternative form to Equation 12.128 for approximations is

$$\psi_n = \exp\left(-N\frac{\zeta}{g(\tau)}\right) + \sum_{i=1}^{n} \tau f_i(\zeta) \tag{12.140}$$

Notice that now, instead of τ, the functions (f_i) are found to depend on ζ.

To satisfy the condition given by Equation 12.122, it is required that

$$g(0) = 0 \tag{12.141}$$

To satisfy the condition given by 12.123, the following is necessary:

$$f_i(0) = 0 \qquad (12.142)$$

The other condition would be set to satisfy the condition given by Equation 12.126. The residual regarding the first approximation becomes

$$\Lambda_1 = \tau f_1'' - f_1 + \frac{N}{g^2}(N - \zeta g')\exp\left(-\frac{N\zeta}{g}\right) \qquad (12.143)$$

Here g is a function of τ and f a function of ζ.

The collocation method can be tried by setting the residual equal to zero at $\tau = \frac{p}{4} = \tau_c$. The length of period has been described earlier.

Let us call

$$\phi(\zeta) = \frac{N}{g(\tau_c)^2}[\zeta g'(\tau_c) - N]\exp\left(-\frac{N\zeta}{g(\tau_c)}\right) \qquad (12.144)$$

As required by the collocation method, the following is imposed:

$$\Lambda_1(\tau_c,\zeta) = 0 \qquad (12.145)$$

After this, the following differential equation can be obtained:

$$\tau_c f_1'' - f_1 = \phi(\zeta) \qquad (12.146)$$

Its solution is composed by the homogenous and particular solutions (see Appendix B), or

$$f_1(\zeta) = f_{1,h}(\zeta) + f_{1,p}(\zeta) \qquad (12.147)$$

The solution of the homogeneous part of Equation 12.147 is

$$f_{1,h} = C_1 \sinh\left(\frac{\zeta}{\sqrt{\tau_c}}\right) + C_2 \cosh\left(\frac{\zeta}{\sqrt{\tau_c}}\right) \qquad (12.148)$$

Following the method described in Appendix B, the particular solution is

$$f_{1,p}(\zeta) = -\sinh(\bar{\zeta})\int \frac{\cosh(\bar{\zeta})\phi(\zeta)}{W}\,d\zeta + \cosh(\bar{\zeta})\int \frac{\sinh(\bar{\zeta})\phi(\zeta)}{W}\,d\zeta \qquad (12.149)$$

where

$$\bar{\zeta} = \frac{\zeta}{\sqrt{\tau_c}} \tag{12.150}$$

$$W = \begin{vmatrix} \sinh(\bar{\zeta}) & \cosh(\bar{\zeta}) \\ \dfrac{1}{\sqrt{\tau_c}}\cosh(\bar{\zeta}) & \dfrac{1}{\sqrt{\tau_c}}\sinh(\bar{\zeta}) \end{vmatrix} = -\frac{1}{\sqrt{\tau_c}} \tag{12.151}$$

The limits for the integrals in Equation 12.149 can be chosen as long as one of them remains general. Bearing in mind the condition given by Equation 12.142, it would be convenient to choose the lower limit as zero. Therefore

$$f_{1,p}(\zeta) = \sqrt{\tau_c}\left[\sinh(\bar{\zeta})\int_0^\zeta \cosh(\bar{\zeta})\phi(\zeta)\,d\zeta - \cosh(\bar{\zeta})\int_0^\zeta \sinh(\bar{\zeta})\phi(\zeta)\,d\zeta\right] \tag{12.152}$$

This leads to

$$f_1(\zeta) = C_1\sinh(\bar{\zeta}) + C_2\cosh(\bar{\zeta})$$
$$+ \sqrt{\tau_c}\left[\sinh(\bar{\zeta})\int_0^\zeta \cosh(\bar{\zeta})\phi(\zeta)\,d\zeta - \cosh(\bar{\zeta})\int_0^\zeta \sinh(\bar{\zeta})\phi(\zeta)\,d\zeta\right] \tag{12.153}$$

The condition given by Equation 12.142 would be satisfied if $C_2 = 0$, hence

$$f_1(\zeta) = C_1\sinh(\bar{\zeta})$$
$$+ \sqrt{\tau_c}\left[\sinh(\bar{\zeta})\int_0^\zeta \cosh(\bar{\zeta})\phi(\zeta)\,d\zeta - \cosh(\bar{\zeta})\int_0^\zeta \sinh(\bar{\zeta})\phi(\zeta)\,d\zeta\right] \tag{12.154}$$

The first approximation would be

$$\psi_1 = \exp\left(-N\frac{\zeta}{g(\tau)}\right) + \tau f_1(\zeta) \tag{12.155}$$

The only task left is to find C_1 using boundary condition given by Equation 12.126. Applying the above equation, it is possible to find

$$C_1 = \frac{N\left[\frac{1}{g(\tau_c)} - 1\right]\exp\left(-\frac{N}{g(\tau_c)}\right) - \tau_c I_1\left[\cosh\left(\frac{1}{\sqrt{\tau_c}}\right) + N\sqrt{\tau_c}\sinh\left(\frac{1}{\sqrt{\tau_c}}\right)\right]}{\sqrt{\tau_c} + N\tau_c\sinh\left(\frac{1}{\sqrt{\tau_c}}\right)}$$
$$\frac{+\tau_c I_2\left[\sinh\left(\frac{1}{\sqrt{\tau_c}}\right) + N\sqrt{\tau_c}\cosh\left(\frac{1}{\sqrt{\tau_c}}\right)\right]}{}$$

(12.156)

Here I_1 and I_2 are not Bessel functions, but are defined as

$$I_1 = \int_0^1 \cosh(\bar{\zeta})\phi(\zeta)\,d\zeta \qquad (12.157)$$

and

$$I_2 = \int_0^1 \sinh(\bar{\zeta})\phi(\zeta)\,d\zeta \qquad (12.158)$$

Figure 12.16 illustrates Equation 12.155 for the case of $N_{Bi} = 2$.

This represents a significant improvement over the previous approximation, because a progressive increase of temperature against time is obtained. The reader is invited to improve this even further by obtaining the second approximation.

It is also interesting to compare the above figure with Figure 11.3. One should recall that in Section 11.2 the bar was isolated at its right-end. Therefore, the temperature tends to increase much faster than for the present case, where heat is exchanged with the environment at the same end.

A final comment should be made on the value chosen as collocation point τ_c.

As seen from Equation 12.120, the dimensionless variable τ is defined by a relation between the real time t and a characteristic time (t_c). This characteristic time is given by a relation between the square of bar length (L) and its thermal diffusivity (D_T). If one imagines the heat transfer process affecting the temperatures of points in the bar, t_c can be seen as the time taken to affect the entire bar, or positions from $z = 0$ to $z = L$. Therefore, another reasonable approach would be to select collocation values of dimensionless time as fractions of this characteristic time. For instance, the first choice could be $\tau_c = \frac{1}{2}$. In this way, other methods for choosing collocation points may follow, such as that hinted in Appendix E, among them being the roots of orthogonal polynomials.

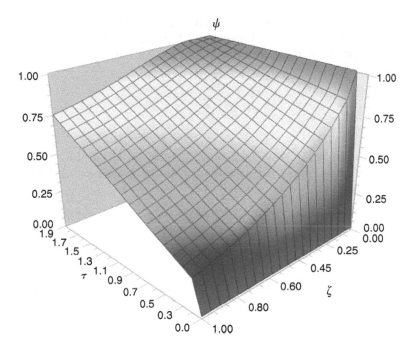

FIGURE 12.16 Dimensionless temperature (ψ) against dimensionless space (ζ) and time (τ) using the alternative first approximation for $N_{Bi} = 2$.

12.5.2 SOLUTION BY SEPARATION OF VARIABLES

Let us try the method of separation of variables, as described in Appendix G.
As seen in the appendix, the following is assumed:

$$\psi(\tau,\zeta) = F(\tau)\, G(\zeta) \tag{12.159}$$

Substituting Equation 12.159 in Equation 12.121, one gets

$$\frac{dF/d\tau}{F(\tau)} = \frac{d^2G/d\zeta^2}{G} = C \tag{12.160}$$

Here C is a constant. The solution of the first differential equation

$$\frac{dF/d\tau}{F(\tau)} = C \tag{12.161}$$

would be

$$F(\tau) = C_1 \exp(C\tau) \tag{12.162}$$

A similar problem as mentioned in Section 11.2 is now encountered, i.e., the boundary condition given by Equation 12.122 cannot be satisfied, unless the trivial solution is assumed, or $F = 0$. To circumvent this, the following change in the dimensionless variables is proposed:

$$\varphi = \frac{T - T_0}{T_b - T_0} \tag{12.163}$$

It is possible to verify that Equation 12.121 would be the same (of course, replacing ψ by φ). Now, the conditions given by Equations 12.122 through 12.124 become

$$\varphi(0,\zeta) = 1, \quad 0 < \zeta \leq 1 \tag{12.164}$$

$$\varphi(\tau,0) = 0, \quad \tau > 0 \tag{12.165}$$

$$\left.\frac{\partial \varphi}{\partial \zeta}\right|_{\zeta=1} = -N_{\mathrm{Bi}}[\varphi(\tau,1) - \varphi_a], \quad \tau > 0 \tag{12.166}$$

Here constant φ_a is defined as

$$\varphi_a = \frac{T_a - T_0}{T_b - T_0} \tag{12.167}$$

As seen, replacing ψ by φ, Equation 12.159 would result in Equation 12.161, its result given by Equation 12.162.

The general solution of the ordinary differential equation of variable G depends on whether C is a positive or negative constant. For this, let us examine the situation regarding temperatures T_b and T_0.

Depending on the relative values among T_b, T_0, and T_a, the temperature in the bar might increase or decrease at given points in the bar, but cannot increase or decrease indefinitely. In other words, no matter the situation, the temperature profile in the bar should converge to steady state. Hence, the influence of term or terms related to time should have a decreasing influence to the process, or an asymptotic approach to zero. Since φ is always positive, or better $0 \leq \varphi \leq 1$, the solution given by Equation 12.162 for function $F(\tau)$ would require C to be a negative constant, or

$$C = -a^2 \tag{12.168}$$

Here the constant a is a real number. Therefore

$$F(\tau) = C_1 \exp(-a^2 \tau) \tag{12.169}$$

According to Appendix B, the solution regarding function G in Equation 12.160 is

$$G(\zeta) = C_2 \sin(a\zeta) + C_3 \cos(a\zeta) \tag{12.170}$$

Equation 12.159 combined with the condition given by Equation 12.165 would demand

$$G(0) = 0 \tag{12.171}$$

Therefore, Equation 12.170 becomes

$$G(\zeta) = C_2 \sin(a\zeta) \tag{12.172}$$

In a more generalized way, this solution can be written as

$$G(\zeta) = \sum_{j-1}^{\infty} C_j \sin(a_j\zeta) \tag{12.173}$$

Applying Equations 12.173 and 12.169 to Equation 12.158, the equivalent equation for φ, it is now possible to write

$$\varphi(\tau,\zeta) = \sum_{j=1}^{\infty} C_j \exp(-a_j^2 \tau) \sin(a_j\zeta) \tag{12.174}$$

Without any loss of generality, parameters a_j can be written as

$$a_j = \frac{2\pi}{p} j \tag{12.175}$$

where p is the period to be determined. Hence, Equation 12.174 becomes

$$\varphi(\tau,\zeta) = \sum_{j=1}^{\infty} C_j \exp(-a_j^2 \tau) \sin\left(\frac{2\pi}{p} j\zeta\right) \tag{12.176}$$

The condition given by Equation 12.164 can be used to give

$$1 = \sum_{j=1}^{\infty} C_j \sin\left(\frac{2\pi}{p} j\zeta\right) \tag{12.177}$$

From Appendix G, the above can be seen as an odd half-range expansion of function 1, and

$$C_j = \frac{4}{p} \int\limits_{0}^{p/2} \sin\left(\frac{2\pi}{p}jx\right) dx = \frac{2}{\pi j}\left[1 - (-1)^j\right] \qquad (12.178)$$

Finally, it is possible to write

$$\varphi(\tau,\zeta) = \frac{2}{\pi} \sum_{j=1}^{\infty} \frac{1 - (-1)^j}{j} \exp\left(-\frac{4\pi^2}{p^2}j^2\tau\right) \sin\left(\frac{2\pi}{p}j\zeta\right) \qquad (12.179)$$

It is easy to see that Equation 12.179 satisfies conditions given by Equations 12.164 and 12.165.

The only parameter left to link with the original variables and conditions of this problem is the period p. This can be achieved through the application of the condition given by Equation 12.166 or by the determination of period p would complete the solution by forcing Equation 12.179 to comply with the remaining boundary condition given by Equation 12.166. Therefore, the following can be written:

$$\frac{2}{\pi} \sum_{j=1}^{\infty} \frac{1 - (-1)^j}{j} \exp\left(-\frac{4\pi^2}{p^2}j^2\tau\right)\left[\frac{2\pi}{p}j\cos\left(\frac{2\pi}{p}j\right) + N_{Bi}\sin\left(\frac{2\pi}{p}j\right)\right] = N_{Bi}\varphi_a$$

$$(12.180)$$

Consequently, the period p will be a function of time, or of its dimensionless representative τ. In a particular situation, or for $T_a = T_0$ (or $\varphi_a = 0$), the period will be independent of particular dimensionless time. Figure 12.17 illustrates the dimensionless (φ) temperature for the present particular case where Biot number was set as 2.

12.5.3 COMMENTS

As seen in Section 11.2.2, a period equal to four was obtained for the odd half-range expansion of unit function for ζ from 0 to 1. However, when applied to the present problem, the method of separation of variables leads to a parameter that should be adjusted for each situation of heat transfer. The graph presented in Figure 12.17 was developed for the case when $\varphi_a = 0$ and for a particular root of Equation 12.180. Nonetheless, this equation allows an infinite number of roots. One idea was to use the nearest root to $p = 4$. Routines to find roots of nonlinear equations are available in almost any package for mathematics. A guess of $p = 4$ leads to the root of approximately 3.9. This value is applied here.

Notwithstanding the use of different dimensionless variables and boundary conditions, Figure 12.17 can be compared with Figures 12.15 and 12.16, obtained by approximations. It is interesting to notice that despite a first approximation, Figure 12.15 (representing Equation 12.135) is not too far from the exact one.

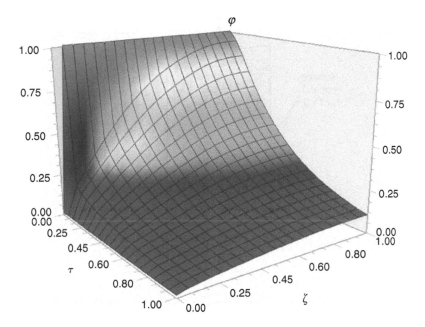

FIGURE 12.17 Exact solution for the dimensionless temperature (φ) against dimensionless space (ζ) and time (τ) for $N_{Bi} = 2$.

Finally, it is important to notice that the above solutions are applicable to the case of an indefinite wall or plate, as illustrated in Figure 12.18. The whole plate is initially at temperature T_b and suddenly one flat face is set at temperature T_0 while the other exchanges heat by convection with a fluid at temperature T_a.

Since the wall is indefinite on directions x and y, the temperature would vary only with time and direction z. It has been shown that this sort of problem is solvable by several methods; however, success is not always guaranteed due to the form of the differential equation or due to the boundary conditions. For instance, the method of similarity, as described in Appendix F, cannot be applied here because it is not possible to coalesce the boundary condition given by Equation 12.122 with Equation 12.123, even less with Equation 12.126.

12.6 CONVECTIVE HEATING OF A PLATE

Another example of steady-state heat transfer between a plate and surrounding fluid is presented in Figure 12.19. As seen, two side faces of the plate are kept at known temperature T_0, while the remaining two exchange heat by convection with a surrounding fluid.

In cases where convective heat transfers between a fluid and a plate, this problem can be seen as an improvement over the situation and conditions posed in Section 11.7.

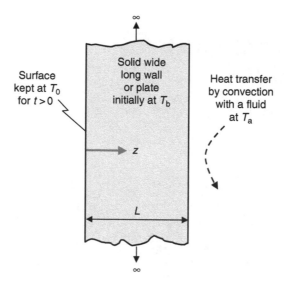

FIGURE 12.18 Semi-infinite plate or wall exchanging heat by convection in one face and the other kept at a prescribed temperature.

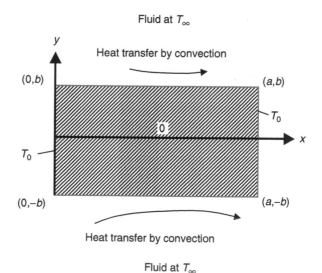

FIGURE 12.19 Scheme showing the plate heated at two surfaces due to contact with a fluid, while the other two are kept at constant temperature.

Again, to take advantage of the geometric similarity, the origins of coordinates have been chosen at the center of the plate.

One would desire to obtain the temperature profile throughout the plate, which would allow deducing the rate of heat transfer at any face. For this, the following are assumed:

1. Plate material is solid and, within the range of temperatures observed here, there is no phase change. Therefore, there is no velocity field in the plate.
2. No variation of temperature on the orthogonal coordinate (z) to the plane x–y. This condition is possible if the body is indefinite at this direction or if the plate is perfectly isolated in both main surfaces.
3. Despite different temperatures within the plate, its density and thermal conductivity remain constant, or approximately constant. Again, an approximation is valid to moderate the differences of temperature in the plate.
4. No chemical reactions or any other energy source term is involved.
5. Temperature of the fluid far from the plate remains constant and equal to T_∞.
6. Convective heat transfer coefficients at the upper and lower faces are equal and remain constant.

From the above, the energy balance shown by Equation A.31 (or Equation A.34) can be simplified to

$$\frac{\partial^2 T}{\partial x^2} + \frac{\partial^2 T}{\partial y^2} = 0 \qquad (12.181)$$

The boundary conditions are

$$T(0,y) = T_0, \quad -b < y < b \qquad (12.182)$$

$$T(a,y) = T_0, \quad -b < y < b \qquad (12.183)$$

$$\left.\frac{\partial T}{\partial y}\right|_{y=0} = 0, \quad 0 < x < a \qquad (12.184)$$

$$-\lambda \left.\frac{\partial T}{\partial y}\right|_{y=b} = \alpha[T(x,b) - T_\infty], \quad 0 < x < a \qquad (12.185)$$

Of course, a similar condition as given by Equation 12.185 could be written for $y = -b$. However, this has been replaced by Equation 12.184.

Consider the following new dimensionless variables:

$$\theta = \frac{T - T_0}{T_\infty - T_0} \qquad (12.186)$$

$$\chi = \frac{x}{a} \qquad (12.187)$$

$$\xi = \frac{y}{b} \qquad (12.188)$$

After this and for the simpler case where $a = b$, Equation 12.181 can be written as

$$\frac{\partial^2 \theta}{\partial \chi^2} + \frac{\partial^2 \theta}{\partial \xi^2} = 0 \qquad (12.189)$$

The boundary conditions become

$$\theta(0,\xi) = 0, \quad -1 < \xi < 1 \qquad (12.190)$$

$$\theta(1,\xi) = 0, \quad -1 < \xi < 1 \qquad (12.191)$$

$$\left.\frac{\partial \theta}{\partial \xi}\right|_{\xi=0} = 0, \quad 0 < \chi < 1 \qquad (12.192)$$

$$\left.\frac{\partial \theta}{\partial \xi}\right|_{\xi=1} = -N_{\mathrm{Bi}}[\theta(\chi,1) - 1], \quad 0 < \chi < 1 \qquad (12.193)$$

Here

$$N_{\mathrm{Bi}} = \frac{\alpha b}{\lambda} \qquad (12.194)$$

Therefore, one arrives at the first-, second-, and third-kind boundary conditions.

12.6.1 SOLUTION BY SEPARATION OF VARIABLES

According to Appendix G, it is assumed that

$$\theta(\chi,\xi) = F(\chi)\, G(\xi) \qquad (12.195)$$

Hence, Equation 12.189 leads to

$$\frac{F''(\chi)}{F(\chi)} = -\frac{G''(\xi)}{G(\xi)} = C \qquad (12.196)$$

The equality between groups involving the function of independent variables forces each side of the equation to be equal to a constant C. Hence, two ordinary differential equations are generated, and one of them is

$$F''(\chi) - C\, F(\chi) = 0 \qquad (12.197)$$

The general solution for such an equation presents the following possibilities:

1. According to Appendix B, if C is positive, or equal to γ^2 (where γ is a real parameter), the solution is

$$F(\chi) = C_1 e^{-\gamma\chi} + C_2 e^{\gamma\chi} \qquad (12.198)$$

However, Equation 12.195 and the conditions given by Equations 12.190 and 12.191 would demand that

$$F(0) = 0 \quad \text{and} \quad F(1) = 0 \qquad (12.198\text{a,b})$$

Employing these, Equation 12.198 would be satisfied only if $C_1 = C_2 = 0$, i.e., the trivial solution for $F(\chi)$ and, consequently, to θ as well. Therefore, this is not an acceptable choice.

2. If C is equal to zero, the solution of Equation 12.197 would be

$$F(\chi) = C_1 \chi + C_2 \qquad (12.199)$$

After the application of boundary conditions, the trivial solution is again obtained.

3. If C is negative or equal to $-\gamma^2$ (where γ is a real parameter), the solution would be

$$F(\chi) = C_1 \sin(\gamma\chi) + C_2 \cos(\gamma\chi) \qquad (12.200)$$

Application of the condition given by Equation 12.190 leads to $C_2 = 0$ and the condition given by Equation 12.191 to

$$0 = C_1 \sin(\gamma) \qquad (12.201)$$

If $\gamma = n\pi$ ($n = 0, 1, 2, \ldots$), C_1 might take any value other than zero, therefore avoiding the trivial solution. Thus, the general solution is

$$F_n(\chi) = C_n \sin(n\pi\chi) \qquad (12.202)$$

On the other hand, since

$$C = -\gamma^2 = -n^2\pi^2, \qquad (12.203)$$

it is possible to obtain the general solution for function G from Equation 12.196. According to Appendix B, this is given by

$$G_n(\xi) = A_n e^{n\pi\xi} + B_n e^{-n\pi\xi} \qquad (12.204)$$

The condition given by Equation 12.192 leads to $G'(0) = 0$, and the above equation would require that

$$A_n = B_n \tag{12.205}$$

Consequently, Equation 12.204 can be written as

$$G_n(\xi) = A_n^* \cosh(n\pi\xi) \tag{12.206}$$

Combining Equations 12.195, 12.202, and 12.206, it is possible to write a general solution as

$$\theta(\chi,\xi) = \sum_{n=1}^{\infty} K_n \sin(n\pi\chi) \cosh(n\pi\xi) \tag{12.207}$$

The following is achieved after applying the condition given by 12.193:

$$1 = \sum_{n=1}^{\infty} K_n \sin(n\pi\chi) \frac{n\pi \sinh(n\pi) + N_{Bi} \cosh(n\pi)}{N_{Bi}} \tag{12.208}$$

From Appendix G, the above can be seen as an odd half expansion of function 1. Thus

$$K_n = \frac{4N_{Bi}}{p[n\pi \sinh(n\pi) + N_{Bi} \cosh(b_n)]} \int_0^{p/2} \sin(n\pi x)\,dx$$
$$= \frac{4N_{Bi}}{n\pi p[n\pi \sinh(n\pi) + N_{Bi} \cosh(b_n)]} \left[1 - \cos\left(\frac{n\pi p}{2}\right)\right], \quad n = 1, 2, 3, \ldots \tag{12.209}$$

Since χ and ξ may acquire values only between 0 and 1, period $p = 2$ should be set. Therefore

$$K_n = \frac{2N_{Bi}}{n\pi[n\pi \sinh(n\pi) + N_{Bi} \cosh(b_n)]}[1 - (-1)^n], \quad n = 1, 2, 3, \ldots \tag{12.210}$$

Finally, the solution is

$$\theta(\chi,\xi) = \frac{2N_{Bi}}{\pi} \sum_{n=1}^{\infty} \frac{1 - (-1)^n}{n} \frac{\sin(n\pi\chi) \cosh(n\pi\xi)}{n\pi \sinh(n\pi) + N_{Bi} \cosh(n\pi)} \tag{12.211}$$

This solution is represented in Figures 12.20 and 12.21 for N_{Bi} equal to 10 and 100, respectively.

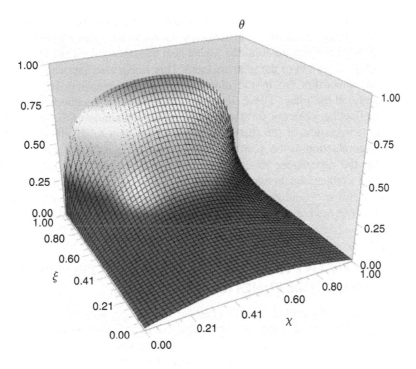

FIGURE 12.20 Temperature distribution in the plate for $N_{Bi} = 10$.

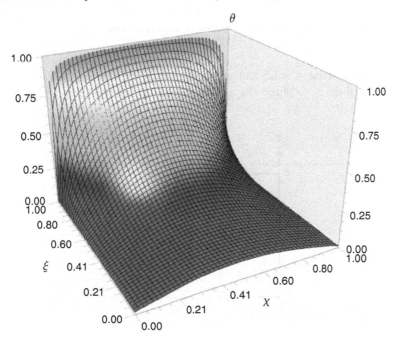

FIGURE 12.21 Temperature distribution in the plate for $N_{Bi} = 100$.

From the above it is interesting to notice how the Biot number influences the temperatures at face $y = b$ (or $\xi = 1$) to become closer to the environment (T_∞). Of course, this might be caused either by higher heat transfer coefficient (α) or lower thermal conductivity (λ) of the plate material. This would also happen in cases of larger plate dimension (b) in the direction (y) of heat transfers between the environment and the plate. These two last influences are understood to have an insulating effect between the surface ($y = b$) and the remaining of the plate. This allows the temperature of the surface to remain high with relatively lower heat transfer by conduction to the positions far from this surface.

Despite the differences in definitions of dimensionless temperatures, just a qualitative comparison between Figures 12.20 and 12.21 and Figure 11.17 might be useful to visualize the differences and similarities between constant heat flux and convective heat transfer. The reader is invited to obtain the heat flux at each face.

12.7 CONVECTIVE HEATING OF A PLATE WITH PRESCRIBED TEMPERATURE FUNCTION AT ONE FACE

An additional example of a steady-state heat transfer problem is illustrated in Figure 12.22. The temperatures at three side faces of the plate are maintained at T_0, while the temperature of the superior face follows a known function $f(x)$. Since this function is generic, not necessarily a constant or leading to constant derivative, the problem involves a third-kind boundary condition.

It is desired to obtain the temperature profile throughout the plate, which would allow deducing the heat transfer at any face. For this, the following are assumed:

1. Plate material is solid and, within the range of temperatures observed here, there is no phase change. Therefore, no velocity field is present.

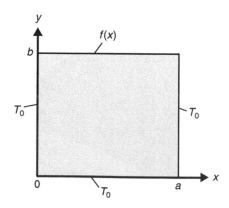

FIGURE 12.22 Rectangular plate with prescribed temperatures at the edges.

2. No variation of temperature on the orthogonal coordinate (z) to the plane x–y. This condition is possible if the body is indefinite at this direction or if the plate is perfectly isolated in both main surfaces.
3. Despite variations of temperature, the plate material, its density, and thermal conductivity remain constant, or approximately constant.
4. No chemical reactions or any other energy source term is involved.

From the above, the energy balance shown by Equation A.31 (or Equation A.34) can be simplified to

$$\frac{\partial^2 T}{\partial x^2} + \frac{\partial^2 T}{\partial y^2} = 0 \qquad (12.212)$$

The following are the new dimensionless variables:

$$\theta = \frac{T - T_0}{T_0} \qquad (12.213)$$

$$\chi = \frac{x}{a} \qquad (12.214)$$

$$\sigma = \frac{y}{b} \qquad (12.215)$$

For the simpler case where $a = b$, Equation 12.212 can be written as

$$\frac{\partial^2 \theta}{\partial \chi^2} + \frac{\partial^2 \theta}{\partial \sigma^2} = 0 \qquad (12.216)$$

As seen above, the particular solution requires two boundary conditions for each independent variable. From Figure 12.22, it is possible to set the following boundary conditions:

$$\theta(0,\sigma) = 0, \quad 0 < \sigma < 1 \qquad (12.217)$$

$$\theta(\chi,1) = \varphi(\chi), \quad 0 < \chi < 1 \qquad (12.218)$$

$$\theta(1,\sigma) = 0, \quad 0 < \sigma < 1 \qquad (12.219)$$

$$\theta(\chi,0) = 0, \quad 0 < \chi < 1 \qquad (12.220)$$

Here

$$\varphi(\chi) = \frac{f(x) - T_0}{T_0} \qquad (12.221)$$

Therefore, here there are first-kind boundary conditions for all situations, but the situation where a general function $f(x)$ is involved constitutes a third-kind boundary condition.

12.7.1 Solution by Separation of Variables

According to Appendix G, the following is assumed:

$$\theta(\chi,\sigma) = F(\chi)\, G(\sigma) \tag{12.222}$$

Consequently, Equation 12.216 leads to

$$\frac{F''(\chi)}{F(\chi)} = -\frac{G''(\sigma)}{G(\sigma)} = C \tag{12.223}$$

The equality between groups involving function of independent variables forces each side of the equation to be equal to a constant C. Hence, two ordinary differential equations are generated. One of them is

$$F''(\chi) - C\, F(\chi) = 0 \tag{12.224}$$

The general solution for such an equation (Appendix B) presents the following possibilities:

1. If C is positive, or equal to γ^2 (where γ is a real parameter), the solution is

$$F(\chi) = C_1 e^{-\gamma\chi} + C_2 e^{\gamma\chi} \tag{12.225}$$

   However, Equations 12.217, 12.219, and 12.222 would force:

$$F(0) = 0 \quad \text{and} \quad F(1) = 0 \tag{12.226a,b}$$

   Hence, Equation 12.225 would provide $C_1 = C_2 = 0$, which is the trivial solution for $F(\chi)$ and, consequently, for θ as well. Therefore, this is not an acceptable choice.
2. If C is equal to zero, the solution would be

$$F(\chi) = C_1\chi + C_2 \tag{12.227}$$

   After the application of boundary conditions, the trivial solution is obtained, again.
3. If C is negative or equal to $-\gamma^2$ (γ real), the solution would be

$$F(\chi) = C_1 \sin(\gamma\chi) + C_2 \cos(\gamma\chi) \tag{12.228}$$

Applying the condition given by Equation 12.226a leads to $C_2 = 0$ and the condition given by Equation 12.226b to

$$0 = C_1 \sin(\gamma) \tag{12.229}$$

If $\gamma = n\pi$ $(n = 1, 2, \ldots)$, C_1 might be different from zero, therefore avoiding the trivial solution. Thus,

$$F_n(\chi) = C_n \sin(n\pi\chi) \tag{12.230}$$

The index n is used here to emphasize the multiple determinations of the solution. Returning to Equation 12.223, the solution for function G would be

$$G_n(\sigma) = A_n e^{n\pi\sigma} + B_n e^{n\pi\sigma} \tag{12.231}$$

The condition given by Equation 12.220 leads to $G(0) = 0$, and in the above equation would require that

$$A_n = -B_n \tag{12.232}$$

Therefore, Equation 12.231 can be written as

$$G_n(\sigma) = A_n^* \sinh(n\pi\sigma) \tag{12.233}$$

At present, the condition given by Equation 12.218 does not allow any conclusion. Combining the solutions for functions F and G given by Equations 12.230 and 12.233, a particular form for the dimensionless temperature is obtained

$$\theta_n(\chi,\sigma) = K_n \sin(n\pi\chi) \sinh(n\pi\sigma) \tag{12.234}$$

The general solution would be a linear combination of these, or

$$\theta(\chi,\sigma) = \sum_{n=1}^{\infty} K_n \sin(n\pi\chi) \sinh(n\pi\sigma) \tag{12.235}$$

Applying the condition given by Equation 12.218 it is possible to write

$$\varphi(\chi) = \sum_{n=1}^{\infty} [K_n \sinh(n\pi)] \sin(n\pi\chi) \tag{12.236}$$

As seen in the Appendix G, this is a Fourier half-range expansion and

$$K_n = 2 \int_0^1 \frac{\varphi(\chi)}{\sinh(n\pi)} \sin(n\pi\chi) \, d\chi, \qquad n = 1, 2, \ldots \tag{12.237}$$

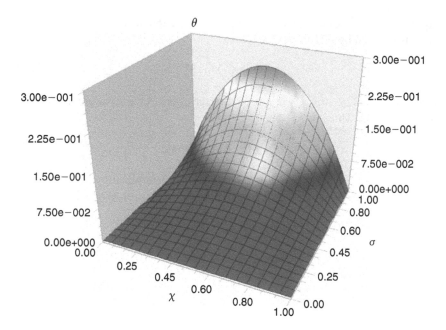

FIGURE 12.23 Dimensionless temperature (θ) profile in the plate for the case of temperature profile given by second-degree function at one edge.

Now, this solution can be applied to any continuous function $f(x)$, which prescribes the temperature at one face.

Consider an example where

$$f(x) = T_0[1 + x(a - x)] \tag{12.238}$$

For $a = 1$, the dimensionless temperature (θ) profile against dimensionless coordinates (χ and σ) is shown in Figure 12.23.

Figure 12.24 illustrates the case where

$$f(x) = T_0\left[1 + x^3(1 - x)\right] \tag{12.238a}$$

12.7.2 SOLUTION BY THE METHOD OF WEIGHTED RESIDUALS

The use of the method of weighted residuals (MWR) for the solution of boundary value problem defined by Equation 12.216 and conditions given by Equations 12.217 through 12.220 is illustrated here. In addition, consider the case where the temperature function $f(x)$ is given by Equation 12.238.

After defining function $f(x)$ as above, function $\varphi(\chi)$ (given by Equation 12.221) becomes

$$\varphi(\chi) = a^2\chi(1 - \chi) \tag{12.239}$$

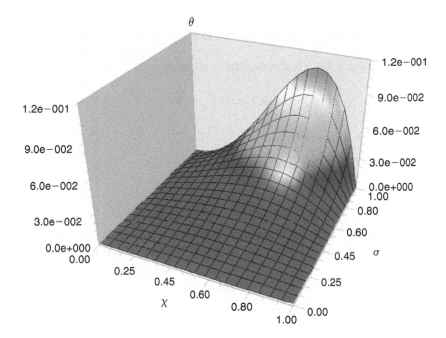

FIGURE 12.24 Dimensionless temperature (θ) profile in the plate for the case of temperature profile given by a fourth-degree function at one edge.

As seen in Appendix E, a good choice should:

1. Apply simple functions, such as polynomials. In the present case, the polynomials could be set using either χ or σ as an independent variable. The amount of work to be faced might depend on a particular choice. Bearing in mind condition given by Equation 12.218 or Equation 12.239, a polynomial on χ might be a reasonable selection.
2. Satisfy the boundary conditions.

A sensible choice for the trial functions could be

$$\theta_n = \sum_{j=1}^{n} \chi^j (1 - \chi)^j \psi_j(\sigma) \qquad (12.240)$$

The above function satisfies conditions given by Equations 12.217 and 12.219. To simplify the discussion, let us take the case of a square plate, or $a = b = 1$. Therefore, if

$$\psi_1(1) = 1, \qquad (12.241)$$

the condition defined by Equation 12.218 would be reproduced for the first approximation or $n = 1$. This would also facilitate treatment during subsequent approximations.

Using Equation 12.216, the residual is given by

$$\Lambda(\chi,\psi_n) = (\theta_n)_{\chi\chi} + (\theta_n)_{\sigma\sigma} \tag{12.242}$$

12.7.2.1 First Approximation

From Equation 12.240, the first approximation is

$$\theta_1 = \chi(1 - \chi)\psi_1(\sigma) \tag{12.243}$$

Hence, boundary conditions given by Equations 12.217 and 12.219 are fulfilled and the condition given by Equation 12.220 would be satisfied if

$$\psi_1(0) = 0 \tag{12.244}$$

From Equations 12.242 and 12.243, the residual becomes

$$\Lambda_1 = -2\psi_1 + \chi(1 - \chi)\psi_1'' \tag{12.245}$$

To exemplify this, let us apply Galerkin's method. Thus, the weighting function (see Appendix E) would be

$$\frac{\partial\theta_1}{\partial\psi_1} = \chi(1 - \chi) \tag{12.246}$$

In addition, the method requires that

$$\int_0^1 \Lambda_1\chi(1 - \chi)\,d\chi = 0 \tag{12.247}$$

Using Equation 12.245, the following differential equation is obtained:

$$\psi_1'' - 10\psi_1 = 0 \tag{12.248}$$

According to Appendix B, the general solution of Equation 12.248 is

$$\psi_1(\sigma) = C_1 \sinh\left(\sigma\sqrt{10}\right) + C_2 \cosh\left(\sigma\sqrt{10}\right) \tag{12.249}$$

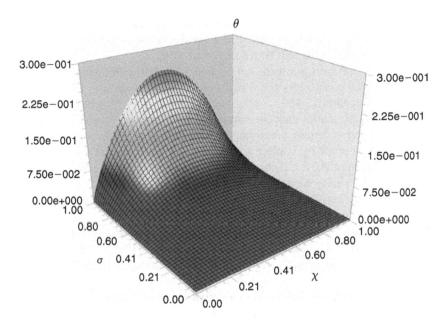

FIGURE 12.25 Dimensionless temperature (θ) profile in the plate for the case of temperature profile given by second-degree function at one edge, obtained using MWR (first approximation).

Applying conditions given by Equations 12.241 and 12.244, one gets $C_2 = 0$ and the following equation can be written:

$$\psi_1(\sigma) = \frac{\sinh\left(\sigma\sqrt{10}\right)}{\sinh\left(\sqrt{10}\right)} \tag{12.250}$$

Consequently, the first approximation for the temperature profile in the plate is

$$\theta_1(\chi,\sigma) = \chi(1-\chi)\frac{\sinh\left(\sigma\sqrt{10}\right)}{\sinh\left(\sqrt{10}\right)} \tag{12.251}$$

This is illustrated in Figure 12.25.

It is possible to verify that it is already an excellent approximation by comparing Figure 12.25 with Figure 12.23, which represents the exact solution.

EXERCISES

1. Show the details to arrive at Equation 12.9.
2. Solve Equation 12.14 to arrive at Equation 12.15.

3. Using the equation describing the temperature profile in Section 12.2, obtain a formula to compute the rate of heat transfer between the rod and environment. Of course, this will be a function of time.

4. Repeat the problem presented in Section 12.2, but with distinct convective heat transfer coefficients at each side of the rod or plate.

5. Deduce Equation 12.41.

6. From the solution for the concentration profile given in Section 12.3, deduce the formula to compute the drying rate of particles as a function of time. Use the solution either by Laplace transform or by separation of variables.

7. Using Equation 12.87, deduce the formula for computation of the heating rate of the cylinder as a function of time (or τ).

8. Develop the details for solving the part related to function G in Equation 12.51 to reach at Equation 12.58.

9. Work the details to solve Equation 12.56 to arrive at Equation 12.58.

10. From the solution of temperature profile in Section 12.5, obtain formulas to compute the rates of heat transfer to or from the rod at each end as functions of time.

11. Repeat the solution for the first approximation in Section 12.5.1.2 but by applying the method of momentum and the method of collocation.

12. Develop a second approximation for the problem set in Section 12.5. Use the same approach as employed in Section 12.5.1.2. Use collocation values for time as fractions of the characteristic time $\frac{\lambda}{\rho C_p L^2}$, or τ_c equal to $1/3$ and $2/3$.

13. Repeat the previous problem using the roots of an orthogonal polynomial as collocation points. Apply, for instance, the Legendre polynomials, as shown in Appendix E.

14. Solve the problem of Section 12.6 using MWR. Apply any method and compare the first approximation to the exact solution presented there.

15. Try to solve the problem in Section 12.6 using Laplace transform.

16. Using the result for temperature profile in a plate, as given in Section 12.6, obtain the expressions for the heat fluxes at faces $x=0$ and $y=b$.

17. Obtain the temperature profile inside a plate where one face is exposed to a fluid while the other is perfectly insulated, as shown in Figure 12.26. Deduce the rate of heat transfer between plate surface and fluid as well. Apply the same assumptions used in Sections 12.2 and 12.5.

18. Using the solution given by Equation 12.236, deduce the formulas to compute rates of heat transfers at each face of the plate.

19. Solve the problem described in Section 12.7, however, setting a known continuous function $g(x)$ at the edge $y=0$. In other words solve the problem, after the same change of variables suggested there, with the following boundary conditions:

$$\theta(0,\sigma) = 0, \quad 0 < \sigma < 1$$
$$\theta(\chi,1) = \varphi(\chi), \quad 0 < \chi < 1$$
$$\theta(1,\sigma) = 0, \quad 0 < \sigma < 1$$
$$\theta(\chi,0) = \omega(\chi), \quad 0 < \chi < 1$$

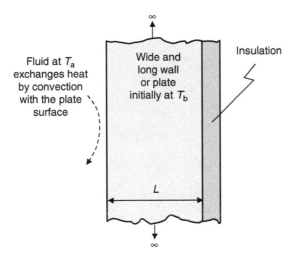

FIGURE 12.26 Indefinite wall exposed to convection at one face and insulated at the other face.

where

$$\omega(\chi) = \frac{g(x) - T_0}{T_0}$$

REFERENCES

1. Walker, P.L. Jr., Rusinko, F., and Austin, L.G., Gas Reactions of Carbon. In *Advances in Catalysis*, Vol. XI, Academic Press, New York, 1959.
2. de Souza-Santos, M.L., *Solid Fuels Combustion and Gasification: Modeling, Simulation, and Equipment Operation*, Marcel Dekker, New York, 2004.
3. Mikhailov, M.D. and Ozisik, M.N., *Unified Analysis and Solutions of Heat and Mass Diffusion*, Dover, New York, 1994.
4. Luikov, A.V., *Heat and Mass Transfer*, Mir, Moscow, 1980.
5. Kafarov, V., *Fundamentals of Mass Transfer*, Mir, Moscow, 1975.
6. King, C.J., *Separation Processes*, Tata McGraw Hill, New Delhi, 1974.

13 Problems 311; Three Variables, 1st Order, 1st Kind Boundary Condition

13.1 INTRODUCTION

This chapter presents methods to solve problems with three independent variables involving first-order differential equations and first-kind boundary conditions. Therefore, partial differential equations are involved. Mathematically, this class of cases can be summarized as $f\left(\phi,\omega_1,\omega_2,\omega_3,\frac{\partial\phi}{\partial\omega_i}\right)$, first-kind boundary condition.

13.2 HEATING A FLOWING LIQUID

This problem is a generalization of the situation presented in Section 7.3. To verify this, let us refer to Figure 13.1 showing the x–y region into which a fluid is injected at a given direction. The same situation is shown in Figure 13.2 in a three-dimensional (3-D) view. The fluid is at temperature T_0 and its thickness is indefinite at a direction orthogonal (z) to the x–y plane. At the distributing vertical planes $(0,y)$ and $(x,0)$, let us consider electrical grids that work to instantly maintain the injected fluid at temperature T_1. It is desired to find the temperature of the fluid at any position.

The present problem is applicable to two directions, whereas at Section 7.3, just one direction was considered.

The following assumptions are made here:

1. Space occupied by the fluid is not limited with regard to direction z (orthogonal to x and y). In other words, the heating grids are very wide in the z direction that allows border effects to be neglected in this direction.
2. In each direction of x and y, the flow can be approximated to plug-flow. In other words, the terms related to convective momentum transfers are much higher than those related to viscous transfers.

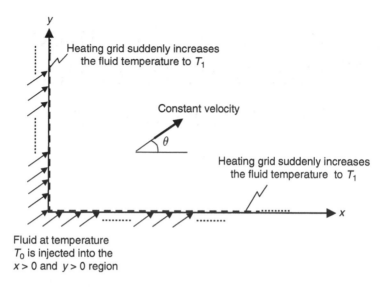

FIGURE 13.1 Scheme for the problem of heating fluid at two-dimensional (2-D) flow.

3. Fluid is initially flowing at temperature T_0. At a given instant, the grids start working and instantly elevate the temperature of the incoming fluid to T_1. Of course, this implies an instantaneous jump in the temperature, which is physically impossible. Despite this, such situations can be approximated in practice by a high heat transfer between the grid and

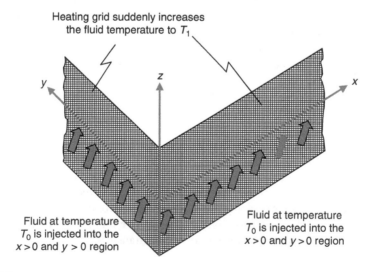

FIGURE 13.2 A three-dimensional (3-D) scheme for the problem of heating fluid at 2-D flow.

the flowing fluid. Such sharp step functions are useful in studies of process and control performance.
4. Fluid is Newtonian and its properties (for instance, density) can be taken as constants. This is assumed despite the heating or temperature variations. Obviously, this might be a strong approximation if large variations of temperature are to be expected. Such an assumption would be invalid for cases of gases.
5. No energy generation due to friction is observed.
6. For a first approximation, convection transfer of energy is considered much higher than that achieved by conduction.

The equations of momentum transfer do not provide any useful information here because the fluid maintains a constant velocity. As the fluid flows parallel to the x–y plane, component v_z does not exist.

Keeping in mind assumptions 5 and 6, the energy balance, or Equation A.34, can be written as

$$\rho C_p \left(\frac{\partial T}{\partial t} + v_x \frac{\partial T}{\partial x} + v_y \frac{\partial T}{\partial y} \right) = \lambda \left(\frac{\partial^2 T}{\partial x^2} + \frac{\partial^2 T}{\partial y^2} \right) \tag{13.1}$$

Due to assumption 6, it is possible to write

$$\frac{\partial T}{\partial t} + v_x \frac{\partial T}{\partial x} + v_y \frac{\partial T}{\partial y} = 0 \tag{13.2}$$

where v_x and v_y are constants.

The boundary conditions are

$$T(0,x,y) = T_0, \quad x > 0, \quad y > 0 \tag{13.3}$$

$$T(t,0,y) = T_1, \quad t > 0, \quad y \geq 0 \tag{13.4}$$

$$T(t,x,0) = T_1, \quad t > 0, \quad x \geq 0 \tag{13.5}$$

The following form for dimensionless temperature is proposed:

$$\psi = \frac{T - T_0}{T_1 - T_0} \tag{13.6}$$

Applying the above, Equation 13.2 can be written as

$$\frac{\partial \psi}{\partial t} + v_x \frac{\partial \psi}{\partial x} + v_y \frac{\partial \psi}{\partial y} = 0 \tag{13.7}$$

The boundary conditions become

$$\psi(0,x,y) = 0, \quad x > 0, \quad y > 0 \tag{13.8}$$

$$\psi(t,0,y) = 1, \quad t > 0, \quad y \geq 0 \tag{13.9}$$

$$\psi(t,x,0) = 1, \quad t > 0, \quad x \geq 0 \tag{13.10}$$

13.2.1 SOLUTION BY LAPLACE TRANSFORM

In view of the condition given by Equation 13.8, the Laplace transform regarding time seems the most convenient choice. Hence

$$\Psi(s,x,y) = L\{\psi(t,x,y)\} \tag{13.11}$$

From Equation 13.11 and using the condition given by Equation 13.8, Equation 13.7 becomes

$$s\Psi + v_x \frac{\partial \Psi}{\partial x} + v_y \frac{\partial \Psi}{\partial y} = 0 \tag{13.12}$$

The new boundary conditions are

$$\Psi(0,y) = \frac{1}{s}, \quad y \geq 0 \tag{13.13}$$

$$\Psi(x,0) = \frac{1}{s}, \quad x \geq 0 \tag{13.14}$$

The solution to the problem posed by Equation 13.12 and that of boundary conditions given by Equations 13.13 and 13.14 is possible by several methods. For a while, let us try a second application of the Laplace transform. Now, variable x is the one used for the transform operation, or

$$\Xi(s,r;y) = L\{\Psi(s;x,y)\} \tag{13.15}$$

Employing the boundary condition given by Equation 13.13, the transform of Equation 13.12 is given by

$$s\Xi + v_x\left(r\Xi - \frac{1}{s}\right) + v_y \frac{d\Xi}{dy} = 0 \tag{13.16}$$

The remaining boundary condition given by Equation 13.14 becomes

$$\Xi(0) = \frac{1}{sr} \tag{13.17}$$

The differential equation (Equation 13.16) is linear and its solution may be sought through the method of parameter variation, as shown in Appendix B. For this, the following is proposed:

$$\Xi(y) = \sigma(y)\,\omega(y) \tag{13.18}$$

Functions σ and ω should now be determined.

Applying Equation 13.18 into Equation 13.16, one gets

$$s\sigma\omega + v_x\left(r\sigma\omega - \frac{1}{s}\right) + v_y\omega\frac{d\sigma}{dy} + v_y\sigma\frac{d\omega}{dy} = 0 \tag{13.19}$$

This can be put in the following form:

$$\omega\left(s\sigma + v_x r\sigma + v_y\frac{d\sigma}{dy}\right) - v_x\frac{1}{s} + v_y\sigma\frac{d\omega}{dy} = 0 \tag{13.20}$$

As shown in Appendix B, with no loss of generality, it is possible to set

$$s\sigma + v_x r\sigma + v_y\frac{d\sigma}{dy} = 0 \tag{13.21}$$

This has a particular solution given by

$$\sigma = \exp\left[-(s + rv_x)\frac{1}{v_y}y\right] \tag{13.22}$$

Using Equation 13.22 in Equation 13.20, the following remains:

$$\frac{d\omega}{dy} = \frac{v_x}{v_y}\frac{1}{s}\exp\left[(s + rv_x)\frac{y}{v_y}\right] \tag{13.23}$$

The general solution of this equation is

$$\omega = \frac{v_x}{s(s + rv_x)}\exp\left[(s + rv_x)\frac{y}{v_y}\right] + C_1 \tag{13.24}$$

After using these results in Equation 13.18 it is possible to obtain

$$\Xi = \frac{v_x}{s(s + rv_x)} + C_1\exp\left[-(s + rv_x)\frac{y}{v_y}\right] \tag{13.25}$$

Using the condition given by Equation 13.17, it is possible to write

$$\Xi = \frac{v_x}{s(s + rv_x)} + \left[\frac{1}{sr} - \frac{v_x}{s(s + rv_x)}\right] \exp\left[-(s + rv_x)\frac{y}{v_y}\right] \tag{13.26}$$

The first inversion is related to Equation 13.15 or to the parameter r and variable x, and the inversion of the first term in the right-hand side of Equation 13.26 becomes (see Appendix D)

$$L^{-1}\left\{\frac{v_x}{s(s + rv_x)}\right\} = \frac{1}{s}\exp\left(-\frac{s}{v_x}x\right) \tag{13.27}$$

The inversion of the first part of the second term is

$$L^{-1}\left\{\frac{1}{sr}\exp\left[-(s + rv_x)\frac{y}{v_y}\right]\right\} = L^{-1}\left\{\frac{1}{s}\exp\left(-s\frac{y}{v_y}\right)\frac{1}{r}\exp\left(-\frac{v_x}{v_y}yr\right)\right\}$$

$$= \frac{1}{s}\exp\left(-s\frac{y}{v_y}\right)L^{-1}\left\{\frac{1}{r}\exp\left(-\frac{v_x}{v_y}yr\right)\right\}$$

$$= \frac{1}{s}\exp\left(-s\frac{y}{v_y}\right)u\left(x - \frac{v_x}{v_y}y\right) \tag{13.28}$$

The inversion of the final part of the second term becomes

$$L^{-1}\left\{\frac{v_x}{s(s + rv_x)}\exp\left[-(s + rv_x)\frac{y}{v_y}\right]\right\} = L^{-1}\left\{\frac{1/s}{\frac{s}{v_x} + r}\exp\left(-s\frac{y}{v_y}\right)\exp\left(-\frac{v_x}{v_y}yr\right)\right\}$$

$$= \frac{1}{s}\exp\left(-s\frac{y}{v_y}\right)L^{-1}\left\{\frac{1}{\frac{s}{v_x} + r}\exp\left(-\frac{v_x}{v_y}yr\right)\right\}$$

$$= \frac{1}{s}\exp\left(-s\frac{y}{v_y}\right)\exp\left[-\frac{s}{v_x}\left(x - \frac{v_x}{v_y}y\right)\right]u\left(x - \frac{v_x}{v_y}y\right)$$

$$= \frac{1}{s}\exp\left(-s\frac{x}{v_x}\right)u\left(x - \frac{v_x}{v_y}y\right) \tag{13.29}$$

As seen, the last two terms are equal and they cancel each other in the final form. Thus

$$\Psi(s;x,y) = \frac{1}{s}\exp\left(-s\frac{x}{v_x}\right) \tag{13.30}$$

The next inversion is related to the transform in Equation 13.11, or is concerning parameter s and time. Consequently, from the above equation the following is obtained:

$$\psi(t, x, y) = u\left(t - \frac{x}{v_x}\right) \tag{13.31}$$

The independence of the above on variable y may seem a contradiction. However, if the total velocity is kept constant in modulus and direction, a relation between the modulus of components v_x and v_y exists in the form $v_x = av_y$, where parameter a is constant. Hence, the heating wave front reaches the position x (measured in the x-axis) at the same time it reaches the position y (measured by the y-coordinate) and they are related by the following:

$$\frac{x}{v_x} = \frac{y}{v_y} \tag{13.32}$$

Therefore, Equation 13.31 can also be written as

$$\psi(t,x,y) = u\left(t - \frac{y}{v_y}\right) \tag{13.33}$$

The above relations would be easier understood by adopting the following notation:

$$t_x = \frac{x}{v_x} \tag{13.34}$$

This can be understood as the necessary time for the heating wave front—traveling at speed equal to v_x—to reach a position x. Similarly, for the y-coordinates we write

$$t_y = \frac{y}{v_y} \tag{13.35}$$

Hence, Equation 13.31 becomes

$$\psi(t,x,y) = u(t - t_x) = u(t - t_y) \tag{13.36}$$

Figure 13.3 illustrates the above solution for regions where the heating wave front has already passed by, or $t > t_x$ or $t > t_y$, or the dimensionless temperature ψ becomes equal to 1, or $T = T_1$. The remaining regions would be at the original temperature T_0.

Figure 13.4 illustrates a 3-D view for a case when $v_x = 0.5$ m s^{-1} and $v_y = 1.0$ m s^{-1} at instant $t = 0.5$ s.

13.2.1.1 Completely Dimensionless Variables

It is interesting to show that the present problem can also be put under complete dimensionless variables, which would simplify the treatment and provide wider generalization of the solution.

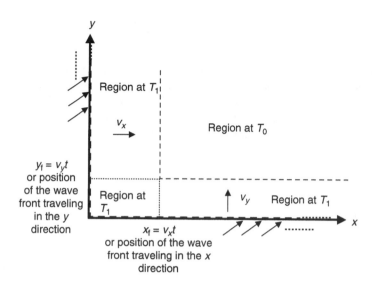

FIGURE 13.3 Scheme showing the temperature at various regions of the x–y space.

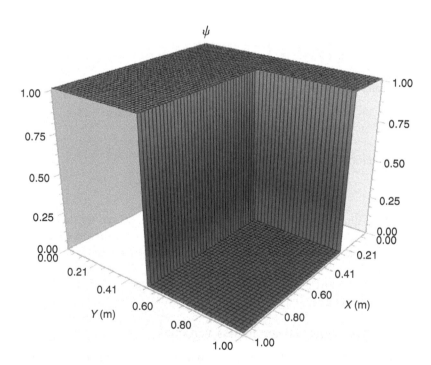

FIGURE 13.4 Profile of dimensionless temperature in the x–y space for the case of $v_x = 0.5$ m s^{-1} and $v_y = 1.0$ m s^{-1} at instant $t = 0.5$ s.

The dimensionless form for the dependent variable, shown in Equation 13.6, is maintained. In addition, the following are proposed for the dependent variables:

$$\chi = \frac{x}{L} \tag{13.37}$$

$$\xi = \frac{v_x}{v_y}\frac{y}{L} \tag{13.38}$$

$$\tau = t\frac{v_x}{L} \tag{13.39}$$

L is a given representative length. One might set it, for instance, as $L = 1$ m.
Using Equations 13.37 through 13.39, Equation 13.2 can be written as

$$\frac{\partial \psi}{\partial \tau} + \frac{\partial \psi}{\partial \chi} + \frac{\partial \psi}{\partial \xi} = 0 \tag{13.40}$$

The boundary conditions given by Equations 13.3 through 13.5 become

$$\psi(0,\chi,\xi) = 0, \quad \chi > 0, \quad \xi > 0 \tag{13.41}$$

$$\psi(\tau,0,\xi) = 1, \quad \tau > 0, \quad \xi \geq 0 \tag{13.42}$$

$$\psi(\tau,\chi,0) = 1, \quad \tau > 0, \quad \chi \geq 0 \tag{13.43}$$

The reader is now invited to reconstruct the solution on this frame of variables.

13.2.2 Heating Rate

Usually, the rate of heating imposed on the fluid per unit of grid area at $y = 0$ is given by

$$q_x(x = 0) = -\lambda \frac{\partial T}{\partial x}\Big|_{x=0} \tag{13.44}$$

Of course, a similar relation can be written for the grid at $x = 0$.

However, owing to the approximation of instantaneous jump on temperature values at the grids, their derivatives would be infinite. Therefore, the approach based on the local derivative cannot be applied here. Instead an energy balance would be applicable and described by

$$\dot{Q}_x = F_x(h_{x>0} - h_{x<0}) \tag{13.45}$$

Here F is the mass flow of liquid passing through the grid at $y = 0$. In the case of an incompressible fluid, as most of the liquids, the above can be approximated by

$$\dot{Q}_x = \rho C_p A_{y=0} v \sin \theta \ (T_1 - T_0) = \rho C_p \ A_{y=0} \ v_x (T_1 - T_0) \qquad (13.46)$$

Here $A_{y=0}$ is the area of the grid at $y=0$ and all properties refer to the fluid at the grid.

Of course, the total heating rate would be the sum of heating rates delivered by the two grids.

13.3 TWO-DIMENSIONAL REACTING FLOW

A similar problem to the one shown in the previous section is considered. However here, instead of being heated, the fluid reacts. In other words, the incoming fluid is composed of a pure substance B. When entering the region $x > 0$ and $y > 0$, there is a grid formed by a network of porous tubes with very small diameter. At a given instant, the fluid with pure component A is injected into the tubes, which reacts with B. The situation is illustrated in Figures 13.5 and 13.6.

As always, let us list the following assumptions:

1. Space is not limited regarding direction z (orthogonal to x and y).
2. Plug-flow regime is approximated in each direction. Despite an approximation, where the terms involving viscosity are assumed much smaller than the convective ones, such treatment allows interesting and useful solutions, mainly in the field of control engineering.
3. Distributing grids do not interfere in the flow.
4. Fluid is initially composed of a single chemical species (B). At a given instant, the grids start delivering a fluid A, which reacts with B. The

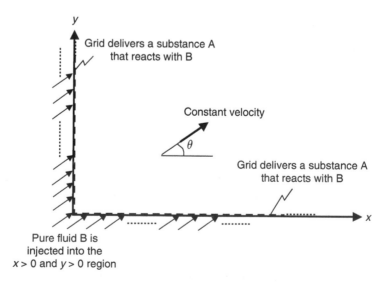

FIGURE 13.5 Scheme for the problem of reacting fluid at 2-D flow.

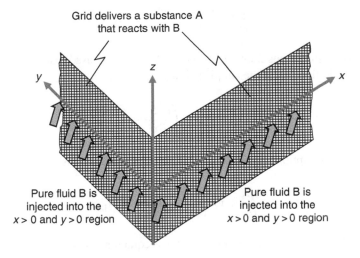

Grid delivers a substance A
that reacts with B

Pure fluid B is
injected into the
$x > 0$ and $y > 0$ region

Pure fluid B is
injected into the
$x > 0$ and $y > 0$ region

FIGURE 13.6 A 3-D scheme for the problem of reacting fluid at 2-D flow.

delivering rate is such that the concentration of A jumps instantaneously for a given value at the grid. Of course, rigorously speaking, this is impossible because any delivery of a fluid starting from zero flow would reach a desired level only after some time. However, the present approximation can provide several interesting informations to model controlling systems.

5. Reaction is neither endothermic nor exothermic and can be represented by

$$A + B \rightarrow C \qquad (13.47)$$

Of course, this is a very special condition. However, the solution might be useful for cases of slightly endothermic or exothermic reactions.

6. Reaction is a zero order and irreversible.

7. Any mixture of components A, B, and C results in a Newtonian fluid and its properties (for instance, density, viscosity, diffusivities, etc.) can be taken as constants. This is assumed despite the change in composition.

8. No energy generation due to friction is observed.

9. For a first approximation, convection transfer of mass is considered much higher than those achieved by diffusion processes.

The equations of momentum transfer are not necessary here because the fluid keeps a constant velocity. As the fluid flows parallel to the $x–y$ plane, component v_z does not exist.

Keeping in mind assumption 1, the mass balance or Equation A.37 can be written as

$$\frac{\partial \rho_A}{\partial t} + \frac{\partial N_{Ax}}{\partial x} + \frac{\partial N_{Ay}}{\partial y} = R_{M,A} \qquad (13.48)$$

In view of assumptions 5 and 7, Equation A.40 can be used instead of Equation 13.48. In addition, it can be simplified to give

$$\frac{\partial \rho_A}{\partial t} + v_x \frac{\partial \rho_A}{\partial x} + v_y \frac{\partial \rho_A}{\partial y} = D_{AB} \left(\frac{\partial^2 \rho_A}{\partial x^2} + \frac{\partial^2 \rho_A}{\partial y^2} \right) + R_{M,A} \qquad (13.49)$$

Moreover, assumption 9 allows neglecting the diffusivity terms, thereby leading to

$$\frac{\partial \rho_A}{\partial t} + v_x \frac{\partial \rho_A}{\partial x} + v_y \frac{\partial \rho_A}{\partial y} = R_{M,A} \qquad (13.50)$$

After application of assumption 6, the above equation becomes

$$\frac{\partial \rho_A}{\partial t} + v_x \frac{\partial \rho_A}{\partial x} + v_y \frac{\partial \rho_A}{\partial y} = -k \qquad (13.51)$$

Here, because of assumptions 5 and 6, k is a constant. Notice that the units of k in SI system would be kg m^{-3} s^{-1}.

The boundary conditions are as follows:

$$\rho_A(0,x,y) = \rho_{A0}, \quad x \geq 0, \quad y \geq 0 \qquad (13.52)$$

$$\rho_A(t,0,y) = \rho_{A1}, \quad t > 0, \quad y \geq 0 \qquad (13.53)$$

$$\rho_A(t,x,0) = \rho_{A1}, \quad t > 0, \quad x \geq 0 \qquad (13.54)$$

Here, ρ_{A0} is the concentration (kg m^{-3}) of A in the fluid before starting its injection through the grids. Of course, in the present situation since only B is flowing before the injection of A, its concentration is zero. Nonetheless, for the sake of generality, this is not assumed here and the solution might be applicable to more general situations than that described above. For instance, instead of B, a mixture of component A and a neutral substance will flow and B is delivered at the grid. In this case, ρ_{A0} would be the concentration of A before the injection of B. The value of ρ_{A1} is the concentration of A as soon it is mixed with the injected reactant B at the grid positions. At present, to maintain the formalism as in Sections 13.3 and 14.3, ρ_{A0} can be dealt with as a reference concentration, which may be even set as 1 kg m^{-3}.

As always, it is interesting to use dimensionless variables. The following are proposed:

$$\psi = \frac{\rho_A - \rho_{A0}}{\rho_{A1} - \rho_{A0}} \qquad (13.55)$$

$$\chi = \frac{ax}{v_x} \qquad (13.56)$$

$$\xi = \frac{ay}{v_y} \qquad (13.57)$$

$$\tau = at \qquad (13.58)$$

Here

$$a = \frac{k}{\rho_{A1} - \rho_{A0}} \qquad (13.59)$$

Using Equations 13.55 through 13.59, Equation 13.51 becomes

$$\frac{\partial \psi}{\partial \tau} + \frac{\partial \psi}{\partial \chi} + \frac{\partial \psi}{\partial \xi} = -1 \qquad (13.60)$$

The boundary conditions given by Equations 13.52 through 13.54 become

$$\psi(0,\chi,\xi) = 0, \quad \chi > 0, \quad \xi > 0 \qquad (13.61)$$

$$\psi(\tau,0,\xi) = 1, \quad \tau > 0, \quad \xi \geq 0 \qquad (13.62)$$

$$\psi(\tau,\chi,0) = 1, \quad \tau > 0, \quad \chi \geq 0 \qquad (13.63)$$

This demonstrates how the correct choice of dimensionless forms might lead to an elegantly simple form for a boundary value problem.

13.3.1 Solution by Laplace Transform

In view of the sudden modification of the dependent variable value, Laplace transform seems an attractive method to be applied here. Consider the following:

$$\Psi(s;\chi,\xi) = L\{\psi(\tau,\chi,\xi)\} \qquad (13.64)$$

After the application at Equation 13.60 combined with the condition given by Equation 13.61, one gets

$$s\Psi + \frac{\partial \Psi}{\partial \chi} + \frac{\partial \Psi}{\partial \xi} = -\frac{1}{s} \qquad (13.65)$$

The transforms of boundary conditions are

$$\Psi(0,\xi) = \frac{1}{s}, \quad \xi \geq 0 \qquad (13.66)$$

$$\Psi(\chi,0) = \frac{1}{s}, \quad \chi \geq 0 \qquad (13.67)$$

As shown in the last section, Equation 13.65 might be solved by several methods, one among them is the Laplace transform. In this way, let a second transform be given by

$$\Xi(s,r;\xi) = L\{\Psi(s;\chi,\xi)\} \tag{13.68}$$

If the condition given by Equation 13.66 is applied to Equation 13.65, it is possible to write

$$(s+r)\Xi + \frac{d\Xi}{d\xi} = \frac{1}{s} - \frac{1}{sr} \tag{13.69}$$

The remaining boundary condition is

$$\Xi(0) = \frac{1}{sr} \tag{13.70}$$

As before, Equation 13.69 is linear and the solution can be sought by using the method of parameter variation, as shown in Appendix B. Therefore, the following is written:

$$\Xi(\xi) = \sigma(\xi)\omega(\xi) \tag{13.71}$$

Using Equation 13.71 in Equation 13.69, one gets

$$\omega\left[(s+r)\sigma + \frac{d\sigma}{d\xi}\right] + \sigma\frac{d\omega}{d\xi} = \frac{1}{s} - \frac{1}{sr} \tag{13.72}$$

As shown in Appendix B, it is possible to set

$$\frac{d\sigma}{d\xi} = -(s+r)\sigma \tag{13.73}$$

A particular solution of this is

$$\sigma = e^{-(s+r)\xi} \tag{13.74}$$

Applying this into Equation 13.72, one gets

$$\frac{d\omega}{d\xi} = \left(1 - \frac{1}{r}\right)\frac{1}{s}e^{(s+r)\xi} \tag{13.75}$$

Its general solution is given by

$$\omega = \frac{r-1}{sr(s+r)}e^{(s+r)\xi} + C_1 \tag{13.76}$$

Applying all the above, Equation 13.71 becomes

$$\Xi = \frac{r-1}{sr(s+r)} + C_1 e^{-(s+r)\xi} \tag{13.77}$$

Employing the condition given by Equation 13.70 it is possible to write

$$\Xi = \frac{r-1}{sr(s+r)} + \frac{s+1}{sr(s+r)} e^{-(s+r)\xi} \tag{13.78}$$

The first inversions are related to the transform given by Equation 13.68. According to Appendix D, the first term on the right-hand side provides

$$L^{-1}\left\{\frac{r-1}{sr(s+r)}\right\} = L^{-1}\left\{\frac{1}{s(s+r)}\right\} - L^{-1}\left\{\frac{1}{sr(s+r)}\right\}$$

$$= \frac{1}{s} e^{-sx} - \frac{1}{s^2}(1 - e^{-sx}) \tag{13.79}$$

The inversion of the other term is

$$L^{-1}\left\{\frac{s+1}{sr(s+r)} e^{-(s+r)\xi}\right\} = \frac{s+1}{s^2} e^{-s\xi} L^{-1}\left\{\frac{s}{r(s+r)} e^{-r\xi}\right\}$$

$$= \frac{s+1}{s^2}\left(e^{-s\xi} - e^{-sx}\right)u(\chi - \xi) \tag{13.80}$$

After collecting all terms, one gets

$$\Psi(s;\chi,\xi) = \frac{1}{s} e^{-sx} - \frac{1}{s^2}(1 - e^{-sx}) + \frac{s+1}{s^2}\left(e^{-s\xi} - e^{-sx}\right)u(\chi - \xi) \tag{13.81}$$

The final inversion would lead to

$$\psi(\tau,\chi,\xi) = -\tau + (\tau - \chi + 1)u(\tau - \chi)u(\xi - \chi)$$
$$+ (\tau - \xi + 1)u(\tau - \xi)u(\chi - \xi) \tag{13.82}$$

Figure 13.7 illustrates the dimensionless concentration profile for $\tau = 0.5$.

Obviously, the affected concentration of the front will expand with time. The following additional comments are valuable:

1. Effect on the concentration because of variations in fluid velocity can be verified. For instance, increases in the injection velocity of the fluid would lead to faster advance of the front. This leaves higher concentrations at a position behind the front or at positions already affected by the injection of reactant through the grids.

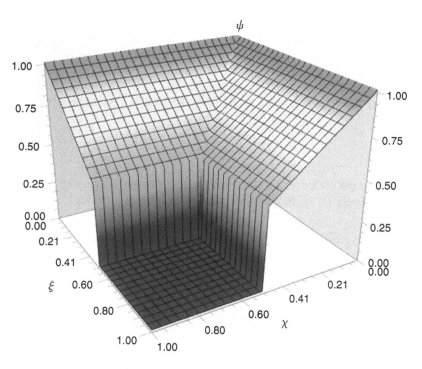

FIGURE 13.7 Profile of dimensionless concentration (ψ) against dimensionless coordinates (χ and ξ) for a particular case of dimensionless time $\tau = 0.5$.

2. According to Equations 13.56 through 13.59, given a real position (x,y), doubling the reaction rate (or parameter a) would double the values of χ, ξ, and τ. Figure 13.7 would show an expansion of the front, and previous values of concentration—at the real position (x,y)—would decrease. This is obvious, since increases in the rate of reaction lead to decreases in the concentration of reacting component A.

13.3.2 RATE OF REACTANT CONSUMPTION

The reaction is irreversible and, sooner or later, all the injected component A would be consumed in the space behind the grids. The rate of reactant A consumption would be computed by the mass or molar flux of this component passing through the grid. Rigorously, this is given by Equation A.53. Its component in the x direction is

$$\mathbf{N}_{Ax} = -D_{AB}\rho\frac{\partial w_A}{\partial x} + w_A(\mathbf{N}_{Ax} + \mathbf{N}_{Bx}) \qquad (13.83)$$

Using Equation A.55

$$\mathbf{N}_{Ax} = -D_{AB}\rho\frac{\partial w_A}{\partial x} + w_A\,\rho v_x \qquad (13.84)$$

The first term in the right-hand side is the diffusive component and the last term is the convective component. However, for the present case, the diffusion is assumed much lower than convection. Finally, the rate of consumption of component A passing through the grid at $y = 0$ is given by

$$F_{Ax} = A_{y=0} N_{Ax} = A_{y=0} \rho_A v_x \qquad (13.85)$$

The consumption of A per unit of time (kg s^{-1}) would add the similar amount passing through the grid at $x = 0$.

EXERCISES

1. In Section 13.2, instead of dealing with a liquid, assume the flowing fluid as an ideal gas. One should notice that now the velocity is no longer constant, but depends on the temperature. Set the governing equations and solve the problem for such a situation using any method.
2. Work the details to arrive at Equation 13.82.
3. Solve the problem present in Section 13.3 for the case of an irreversible zero-order reaction.
4. Solve the previous problem by any method.
5. Solve the problem presented in Section 13.3 for the case of a first-order, reversible reaction.
6. Write all differential equations and boundary conditions for the problem presented in Section 13.3 for the case of a first-order, irreversible, but exothermic reaction.

14 Problems 312; Three Variables, 1st Order, 2nd Kind Boundary Condition

14.1 INTRODUCTION

This chapter presents methods to solve problems with three independent variables involving first-order differential equations and second-kind boundary conditions. Therefore, partial differential equations are involved. Mathematically, this class of cases can be summarized as $f\left(\phi, \omega_1, \omega_2, \omega_3, \frac{\partial \phi}{\partial \omega_i}\right)$, second-kind boundary condition.

14.2 HEATING A FLOWING LIQUID

This problem is an improvement such that it provides a more feasible situation than that presented in Section 13.2. For this, let us refer to Figures 14.1 and 14.2.

It is similar to Figure 13.1; however, instead of an instantaneous jump in temperature, the grids deliver a constant heat flux. This is a more realistic picture than before and can be accomplished by, for instance, grids made of electrical resistances.

It is desired to find the temperature of the fluid at any position as well as the rate of energy delivered to the fluid.

For the sake of clarity, let us list the assumptions:

1. Space occupied by the fluid is not limited regarding direction z (orthogonal to x and y).
2. In each direction x and y, the flow can be approximated to plug-flow. In other words, the terms related to convective momentum transfers are much higher than those related to viscous transfers.
3. Distributing grids do not interfere in the flow.
4. Fluid is initially flowing at temperature T_0. At a given instant, the grids impose a preset value of heat flux exchanged with the incoming fluid.

411

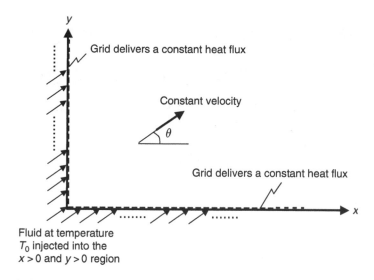

FIGURE 14.1 Scheme for the problem of heating fluid at two-dimensional (2-D) flow.

5. Liquid is Newtonian and its properties (for instance, density and thermal conductivity) can be taken as constants. This is assumed despite the heating, or temperature variations.
6. No energy dissipation due to viscous flow is observed.
7. For a first approximation, convection transfer of energy is considered much higher than those achieved by conduction.

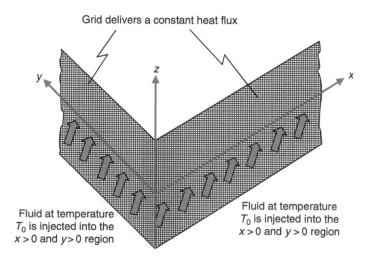

FIGURE 14.2 A three-dimensional (3-D) scheme for the problem of heating fluid at 2-D flow.

14.2.1   BASIC EQUATIONS AND BOUNDARY CONDITIONS

Similarly to Section 13.2, Equation A.34 can be written as

$$\rho C_p \left(\frac{\partial T}{\partial t} + v_x \frac{\partial T}{\partial x} + v_y \frac{\partial T}{\partial y} \right) = \lambda \left(\frac{\partial^2 T}{\partial x^2} + \frac{\partial^2 T}{\partial y^2} \right) \tag{14.1}$$

Because of assumption 7, it is possible to write

$$\frac{\partial T}{\partial t} + v_x \frac{\partial T}{\partial x} + v_y \frac{\partial T}{\partial y} = 0 \tag{14.2}$$

where v_x and v_y are constants.

The boundary conditions are

$$T(0,x,y) = T_0, \quad x > 0, \quad y > 0 \tag{14.3}$$

$$-\lambda \frac{\partial T}{\partial x} \bigg|_{x=0} = q_x = a, \quad t > 0, \quad y \geq 0 \tag{14.4}$$

$$-\lambda \frac{\partial T}{\partial y} \bigg|_{y=0} = q_y = a, \quad t > 0, \quad x \geq 0 \tag{14.5}$$

Therefore, the grids at $x = 0$ and $y = 0$ impose the same constant heat flux equal to a.

The present problem can also be put under complete dimensionless variables. The following are suggested:

$$\psi = \frac{T - T_0}{T_0} \tag{14.6}$$

$$\chi = \frac{a}{\lambda T_0} x \tag{14.7}$$

$$\xi = \frac{v_x a}{v_y \lambda T_0} y \tag{14.8}$$

$$\tau = \frac{a v_x}{\lambda T_0} t \tag{14.9}$$

Using Equations 14.6 through 14.9, Equation 14.2 becomes

$$\frac{\partial \psi}{\partial \tau} + \frac{\partial \psi}{\partial \chi} + \frac{\partial \psi}{\partial \xi} = 0 \tag{14.10}$$

The new forms for the boundary conditions are

$$\psi(0,\chi,\xi) = 0, \quad \chi > 0, \quad \xi > 0 \tag{14.11}$$

$$\frac{\partial \psi}{\partial \chi}\bigg|_{\chi=0} = -1, \quad \tau > 0, \quad \xi \geq 0 \tag{14.12}$$

$$\frac{\partial \psi}{\partial \xi}\bigg|_{\xi=0} = -\frac{v_y}{v_x} = -b, \quad \tau > 0, \quad \chi \geq 0 \tag{14.13}$$

14.2.2 Solution by Laplace Transform

In view of the above, a suggestion would be to apply the following transform:

$$\Psi(s,\chi,\xi) = L\{\psi(\tau,\chi,\xi)\} \tag{14.14}$$

Applying the condition given by Equation 14.11, Equation 14.10 becomes

$$s\Psi + \frac{\partial \Psi}{\partial \chi} + \frac{\partial \Psi}{\partial \xi} = 0 \tag{14.15}$$

The transforms of the remaining boundary conditions are

$$\frac{\partial \Psi}{\partial \chi}\bigg|_{\chi=0} = -\frac{1}{s}, \quad \xi \geq 0 \tag{14.16}$$

$$\frac{\partial \Psi}{\partial \xi}\bigg|_{\xi=0} = -\frac{b}{s}, \quad \chi \geq 0 \tag{14.17}$$

The solution for the partial differential equation (Equation 14.15), under boundary conditions given by Equations 14.16 and 14.17, is possible by several methods. One among them being a second application of Laplace Transform, or

$$\Xi(s,r,\xi) = L\{\Psi(s,\chi,\xi)\} \tag{14.18}$$

After this, Equation 14.15 becomes

$$(s+r)\Xi - f(s;\xi) + \frac{d\Xi}{d\xi} = 0 \tag{14.19}$$

Here f is a function of ξ and represents $\Psi(s;\chi=0,\xi)$. The semicolon is set to remind that unlike the variable ξ, s is just a parameter involved in function f. Of course, for now, this function is unknown.

The transform of the remaining boundary condition is

$$\frac{d\Xi}{d\xi}\bigg|_{\xi=0} = -\frac{b}{sr} \tag{14.20}$$

Equation 14.19 is linear and the solution is obtainable through the method of parameter variation, as shown in Appendix B. Therefore, the following is written:

$$\Xi(\xi) = \sigma(\xi)\omega(\xi) \tag{14.21}$$

Using Equation 14.21, Equation 14.19 becomes

$$(s + r)\sigma\omega - f(s;\xi) + \omega\frac{d\sigma}{d\xi} + \sigma\frac{d\omega}{d\xi} = 0 \tag{14.22}$$

From Equation 14.22, we may write

$$\omega\left[(s + r)\omega + \frac{d\sigma}{d\xi}\right] - f(s;\xi) + \sigma\frac{d\omega}{d\xi} = 0 \tag{14.23}$$

As shown in Appendix B, it is possible to set

$$(s + r)\sigma + \frac{d\sigma}{d\xi} = 0 \tag{14.24}$$

A particular solution is

$$\sigma = e^{-(s+r)\xi} \tag{14.25}$$

Substituting Equation 14.25 into Equation 14.23, one gets

$$\frac{d\omega}{d\xi} = f(s;\xi)e^{(s+r)\xi} \tag{14.26}$$

Hence, the general solution is given by

$$\omega = \int_0^{\xi} f(s;\xi)e^{(s+r)\xi}d\xi + C_1 \tag{14.27}$$

The finite integral is chosen here to facilitate the application of the boundary condition given by Equation 14.20, as shown below.

The following remains after substituting the results into Equation 14.21:

$$\Xi = e^{-(s+r)\xi}\int_0^{\xi} f(s;\xi)e^{(s+r)\xi}d\xi + C_1e^{-(s+r)\xi} \tag{14.28}$$

The application of the condition given by Equation 14.20 requires the derivative of the former expression, which is given by

$$\frac{d\Xi}{d\xi} = -(s+r)e^{-(s+r)\xi}\int_0^\xi f(s;\xi)e^{(s+r)\xi}d\xi + f(s;\xi) - (s+r)C_1e^{-(s+r)\xi} \tag{14.29}$$

Applying Equation 14.20, it is possible to deduce the constant as

$$C_1 = \frac{1}{s+r}\left[f(s;0) + \frac{b}{sr}\right] \tag{14.30}$$

Hence

$$\Xi = e^{-(s+r)\xi}\int_0^\xi f(s;\xi)e^{(s+r)\xi}d\xi + \frac{1}{s+r}\left[f(s;0) + \frac{b}{sr}\right]e^{-(s+r)\xi} \tag{14.31}$$

Likewise in Chapter 13, the first inversion would be performed involving parameter r and variable χ. Such an inversion is not trivial. Fortunately, in the present case, it is easy to imagine that the condition $\Psi(\chi = 0, \xi)$ should not be a function of the variable ξ. This can be understood because the conditions at $\chi = 0$ or at the vertical grid may vary with time (or τ) but remain the same for the whole field of the indefinite variable ξ. If so, the new unknown function $f(s;\xi)$ will be just a relation involving parameter s, which will be called $g(s)$. After this, the above equation becomes

$$\Xi = g(s)\frac{1}{s+r}\left[1 - e^{-(s+r)\xi}\right] + \frac{1}{s+r}\left[g(s) + \frac{b}{sr}\right]e^{-(s+r)\xi}$$

$$= g(s)\frac{1}{r+s} + \frac{b}{s}e^{-s\xi}\frac{1}{r(s+r)}e^{-r\xi} \tag{14.32}$$

The inversion results in

$$\Psi = g(s)e^{-s\chi} + \frac{b}{s^2}\left(e^{-s\xi} - e^{-s\chi}\right)u(\chi - \xi) \tag{14.33}$$

The application of the condition in Equation 14.16 requires the derivative of the above, which is given by

$$\frac{\partial\Psi}{\partial\chi} = -sg(s)e^{-s\chi} + \frac{b}{s^2}\left(se^{-s\chi}\right)u(\chi - \xi) \tag{14.34}$$

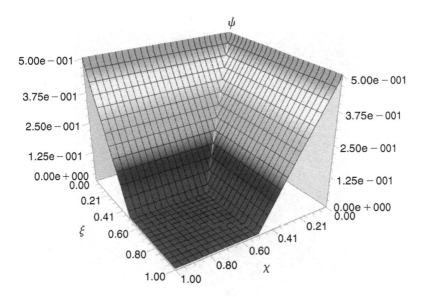

FIGURE 14.3 Dimensionless temperature (ψ) profile against dimensionless coordinates χ and ξ at $\tau = 0.5$ in the case of $b = 1$.

From the above equation, we can arrive at the following:

$$g(s) = \frac{b}{s^2} \qquad (14.35)$$

Therefore

$$\Psi = \frac{b}{s^2} e^{-s\chi} + \frac{b}{s^2} \left(e^{-s\xi} - e^{-s\chi} \right) u(\chi - \xi) \qquad (14.36)$$

Finally, the last inversion would give the function ψ as

$$\psi = b(\tau - \chi)u(\tau - \chi) + b(\tau - \xi)u(\tau - \xi)u(\chi - \xi) - (\tau - \chi)u(\tau - \chi)u(\chi - \xi) \qquad (14.37)$$

Figure 14.3 illustrates a situation for the temperature profiles in two-dimensional (2-D) space at $\tau = 0.5$ for $b = 1$.

It is possible to observe an unaffected region by the wave fronts coming from the grids at $\chi = 0$ and $\xi = 0$. Thus, this region remains at the original temperature T_0 (or $\psi = 0$). Of course, with the advance of τ (or time), this region will advance to higher values of χ (or x) and ξ (or y).

Figure 14.4 presents the same situation for $b = 2$. Since $b = v_y/v_x$ and according to Equation 14.8, for a given value of dimensionless coordinate ξ, higher values of parameter b would represent either increases on the heat flux (parameter a) or

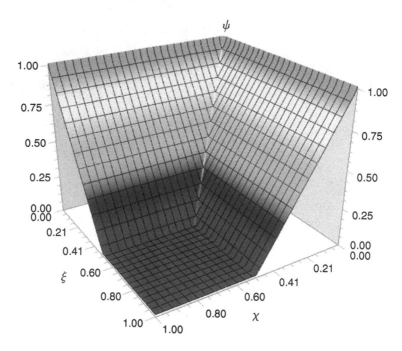

FIGURE 14.4 Dimensionless temperature (ψ) profile against dimensionless coordinates χ and ξ at $\tau = 0.5$ in the case of $b = 2$.

decreases on the thermal conductivity of the fluid. Both effects tend to augment the temperatures between the grids and wave fronts.

Obviously, the rate of heat transfer is obtained by just multiplying the already-imposed flux (or parameter a) and the area of the grids.

14.3 TWO-DIMENSIONAL REACTING FLOW

This is a more realistic and feasible situation than that presented in Section 13.3. Again, the incoming fluid is composed of pure substance B, which receives a flow of pure component. Both the components react. However, instead of the situation given at Section 13.3, a constant mass rate of species A diffuses through the porous grid into the fluid. The situation is illustrated by Figures 14.5 and 14.6.

As always, let us list the following assumptions:

1. Space is not limited regarding direction z (orthogonal to x and y).
2. In each direction x and y, the flow can be approximated to plug-flow. In other words, the terms related to convective momentum transfers are much higher than those related to viscous transfers.
3. Distributing grids are composed of networks of small-diameter porous tubes that do not interfere in the flow.

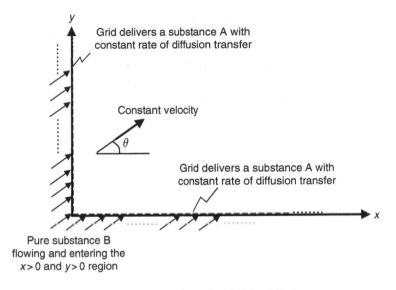

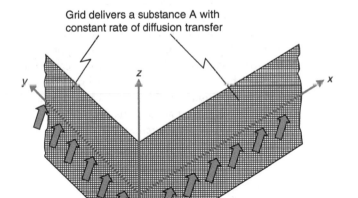

FIGURE 14.5 Scheme for the problem of reacting fluid at 2-D flow.

4. Fluid is initially composed of a single chemical species B. At a given instant, the grids start delivering fluid A, which reacts with B. The delivering occurs just by diffusion of A from the porous grid into the fluid. In addition, the delivering rate is constant.
5. Reaction is neither endothermic nor exothermic and can be represented by

$$A + B \rightarrow C \tag{14.38}$$

FIGURE 14.6 A 3-D scheme for the problem of reacting fluid at 2-D flow.

6. Reaction is zero order and irreversible.
7. Any mixture of components A, B, and C results in a Newtonian fluid and its properties (for instance, density, viscosity, diffusivities, etc.) can be taken as constants. This is assumed despite the change in composition.
8. Negligible dissipation of energy due to viscous flow.
9. For a first approximation, convection transfer of mass is considered much higher than those achieved by diffusion processes.

The procedure to set the governing equation is the same as shown in Section 13.3, hence leading to

$$\frac{\partial \rho_A}{\partial t} + v_x \frac{\partial \rho_A}{\partial x} + v_y \frac{\partial \rho_A}{\partial y} = -k \tag{14.39}$$

However, the boundary conditions are

$$\rho_A(0,x,y) = \rho_{A0}, \quad x \geq 0, \quad y \geq 0 \tag{14.40}$$

$$-D_{AB} \frac{\partial \rho_A}{\partial x}\bigg|_{x=0} = N_{A0}, \quad t > 0, \quad y \geq 0 \tag{14.41}$$

$$-D_{AB} \frac{\partial \rho_A}{\partial y}\bigg|_{y=0} = N_{A0}, \quad t > 0, \quad x \geq 0 \tag{14.42}$$

Of course, here the value of diffusivity remains constant and actually represents the diffusivity of component A into a mixture of components A and B, and reaction product C. In addition, and according to assumption 4, the mass flux N_{A0} at the grid is a constant.

Since only component B flows before the injection of component A, ρ_{A0} is zero. The notation is maintained to allow a more general solution. For instance, if instead of pure component B, a mixture of component A and a neutral substance passes through the grid at which component B is delivered, ρ_{A0} would be the concentration of component A before the injection of component B. In the present case, to maintain the formalism as in Section 13.3, ρ_{A0} can be understood as a reference concentration, which may be even set as 1 kg m^{-3}.

The conditions given by Equations 14.41 and 14.42 increase the concentration of species A in the flowing fluid at the grids.

Let us propose the following dimensionless variables:

$$\psi = \frac{\rho_A - \rho_{A0}}{\rho_{A0}} \tag{14.43}$$

$$\chi = \frac{ax}{v_x} \tag{14.44}$$

$$\xi = \frac{ay}{v_y} \tag{14.45}$$

$$\tau = at \tag{14.46}$$

Here

$$a = \frac{k}{\rho_{A0}} \tag{14.47}$$

The unit for the above constant parameter a would be s^{-1}.

Applying Equations 14.43 through 14.47, Equation 14.39 becomes

$$\frac{\partial \psi}{\partial \tau} + \frac{\partial \psi}{\partial \chi} + \frac{\partial \psi}{\partial \xi} = -1 \tag{14.48}$$

The boundary conditions given by Equations 14.40 through 14.42 are

$$\psi(0,\chi,\xi) = 0, \quad \chi > 0, \quad \xi > 0 \tag{14.49}$$

$$\left. \frac{\partial \psi}{\partial \chi} \right|_{\chi=0} = -b_1, \quad \tau > 0, \quad \xi \geq 0 \tag{14.50}$$

$$\left. \frac{\partial \psi}{\partial \xi} \right|_{\xi=0} = -b_2, \quad \tau > 0, \quad \chi \geq 0 \tag{14.51}$$

Here

$$b_1 = \frac{N_{A0}v_x}{D_{AB}a\rho_{A0}} \quad \text{and} \quad b_2 = \frac{N_{A0}v_y}{D_{AB}a\rho_{A0}} \tag{14.52}$$

For the sake of an example, the problem is solved for the particular case where $b_1 = b_2 = b$, or when $v_x = v_y$.

14.3.1 SOLUTION BY LAPLACE TRANSFORM

In view of the condition given by Equation 14.49, the transform on variable τ seems attractive, and

$$\Psi(s;\chi,\xi) = L\{\psi(\tau,\chi,\xi)\} \tag{14.53}$$

Transforming Equation 14.48 and combining it with condition given by Equation 14.49 leads to

$$s\Psi + \frac{\partial \Psi}{\partial \chi} + \frac{\partial \Psi}{\partial \xi} = -\frac{1}{s} \tag{14.54}$$

The transforms of the remaining boundary conditions are

$$\left. \frac{\partial \Psi}{\partial \chi} \right|_{\chi=0} = -\frac{b}{s}, \quad \xi \geq 0 \tag{14.55}$$

$$\frac{\partial \Psi}{\partial \xi}\Big|_{\xi=0} = -\frac{b}{s}, \quad \chi \geq 0 \tag{14.56}$$

Equation 14.54 might be solved by several methods, one among them is using the Laplace transform. Let us consider

$$\Xi(s,r;\xi) = L\{\Psi(s;\chi,\xi)\} \tag{14.57}$$

Applying Equation 14.57 in Equation 14.54, we obtain

$$(s+r)\Xi - f(s;\xi) + \frac{d\Xi}{d\xi} = -\frac{1}{sr} \tag{14.58}$$

Again, f is a function of ξ and represents $\Psi(s;\chi=0,\xi)$. The semicolon is set to remind that unlike the variable ξ, s is just a parameter involved in function f. Of course, for now this function is unknown.

The remaining boundary condition given by Equation 14.56 becomes

$$\frac{d\Xi}{d\xi}\Big|_{\xi=0} = -\frac{b}{sr} \tag{14.59}$$

Equation 14.58 is linear and the solution can be achieved through the method of parameter variation, as shown in Appendix B. Therefore, the following is set:

$$\Xi(\xi) = \sigma(\xi)\omega(\xi) \tag{14.60}$$

Applying Equation 14.60 to Equation 14.58 allows writing

$$\omega\left[(s+r)\sigma + \frac{d\sigma}{d\xi}\right] - f(s;\xi) + \sigma\frac{d\omega}{d\xi} = -\frac{1}{sr} \tag{14.61}$$

Following the same steps as in Section 14.2, it is possible to arrive at

$$\Xi = e^{-(s+r)\xi}\int_0^\xi f(s;\xi)e^{(s+r)\xi}d\xi - \frac{1}{rs(s+r)} + C_1 e^{-(s+r)\xi} \tag{14.62}$$

The application of the condition given by Equation 14.59 requires the derivative of the former expression, which is given by

$$\frac{d\Xi}{dy} = -(s+r)e^{-(s+r)\xi}\int_0^\xi f(s;\xi)e^{(s+r)\xi}d\xi + f(s;\xi) - (s+r)C_1 e^{-(s+r)\xi} \tag{14.63}$$

Through Equation 14.59, the constant is determined, or

$$C_1 = \frac{1}{s+r}\left[f(s;0) + \frac{b}{sr}\right]$$
(14.64)

Consequently

$$\Xi = e^{-(s+r)\xi}\int_0^\xi f(s;\xi)e^{(s+r)\xi}d\xi - \frac{a}{rs(s+r)} + \frac{1}{s+r}\left[f(s;0) + \frac{b}{sr}\right]e^{-(s+r)\xi}$$
(14.65)

As before, let us consider the inversion involving parameter r and variable χ. Nonetheless, similarly to the justification presented at the previous section, the condition $\Psi(\chi=0,\xi)$ should not be a function of variable ξ. Hence, the new unknown function $f(s;\xi)$ will be just a relation involving parameter s, which will be called $g(s)$. After this the above equation becomes

$$\Xi = \frac{g(s)}{r+s} + \frac{b}{sr(s+r)}e^{-(s+r)\xi} - \frac{1}{sr(r+s)}$$
(14.66)

The inversion results into

$$\Psi = g(s)e^{-\chi s} + \frac{b}{s^2}\left(e^{-s\xi} - e^{-s\chi}\right)u(\chi-\xi) - \frac{1}{s^2}(1 - e^{-s\chi})$$
(14.67)

To employ the condition given by Equation 14.55 one requires the derivative of the above, which is given by

$$\frac{\partial\Psi}{\partial\chi} = -sg(s)e^{-s\chi} + \frac{b}{s}e^{-s\chi}u(\chi-\xi) - \frac{1}{s}e^{-s\chi}$$
(14.68)

Applying Equation 14.68, the condition provides

$$g(s) = \frac{1}{s^2}(b-1)$$
(14.69)

Therefore

$$\Psi = \frac{b}{s^2}e^{-\chi s} + \frac{b}{s^2}\left(e^{-s\xi} - e^{-s\chi}\right)u(\chi-\xi) - \frac{1}{s^2}$$
(14.70)

Finally, the last inversion allows the solution to be written as

$$\psi = b(\tau-\chi)u(\tau-\chi) + b[(\tau-\xi)\,u(\tau-\xi) - (\tau-\chi)\,u(\tau-\chi)]\,u(\chi-\xi) - \tau$$
(14.71)

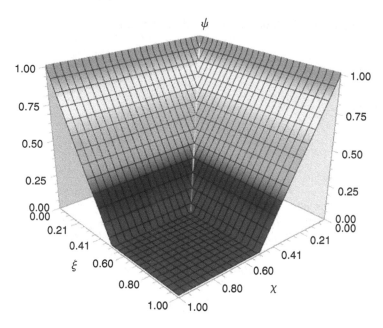

FIGURE 14.7 Dimensionless concentration (ψ) profile against dimensionless coordinates χ and ξ at $\tau = 1.0$ in the case of $b = 2$.

Figure 14.7 shows the dimensionless concentration profile at $\tau = 1.0$ for the case where $b = 2$, while Figure 14.8 shows the dimensionless concentration profile at the same instant for $b = 4$.

According to conditions given by Equations 14.50 and 14.51, the rate of component A injection is proportional to that of parameter b, given by Equation 14.52 ($b = b_1 = b_2$). Thus, increases in the flux N_{A0} or in the velocity v (here $v = v_x = v_y$) would increase this parameter. The same effect would be produced by decreases in the diffusivity of A in the mixture as well as by lower reaction rates (k or a). In any of these alternatives, the concentration of A near the grid would increase.

As once component A is injected into the volume it will eventually react with component B, the rate of its consumption is obtained by multiplying with N_{A0} and the area of the grids.

EXERCISES

1. Solve the problem presented in Section 14.3, when v_x is different from v_y.
2. Try to solve the problem presented in Section 14.2 by a method other than Laplace transform.
3. Solve the problem presented in Section 14.3 for the case of a first-order reaction, with the sink term at Equation 14.39 as

$$R_{M,A} = -k\rho_A$$

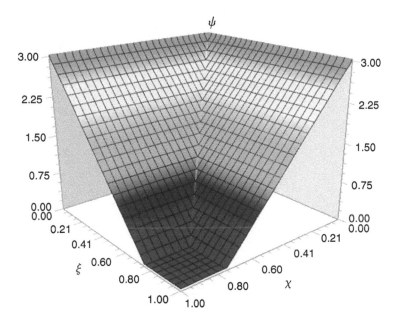

FIGURE 14.8 Dimensionless concentration (ψ) profile against dimensionless coordinates χ and ξ at $\tau = 1.0$ in the case of $b = 4$.

4. Rework the problem presented in Section 14.3 for the case of a first-order reaction, where the sink term is given by

$$R_{M,A} = -k\rho_A\rho_B$$

15 Problems 313; Three Variables, 1st Order, 3rd Kind Boundary Condition

15.1 INTRODUCTION

This chapter presents methods to solve problems with three independent variables involving first-order differential equations and third-kind boundary conditions. Therefore, partial differential equations are involved. Mathematically, this class of cases can be summarized as $f\left(\phi,\omega_1,\omega_2,\omega_3,\frac{\partial\phi}{\partial\omega_i}\right)$, third-kind boundary condition.

15.2 HEATING A FLOWING LIQUID

This problem is an improvement over the situation presented in Sections 13.2 and 14.2. For this, let us refer to Figures 15.1 and 15.2. As seen they are similar to Figures 14.1 and 14.2; however here, the heat flux is given by the rate of convective transfer between the grid and the fluid. This is a more realistic and feasible situation than the one presented in the previous chapters.

Again, it is desired to determine the temperature of the fluid at any position as well as the rate of heat transfer to the fluid.

For the sake of clarity, let us list the assumptions:

1. Space is not limited regarding direction z (orthogonal to x and y).
2. In each direction of x and y, the flow can be approximated to plug-flow. In other words, the terms related to convective momentum transfers are much higher than those related to viscous transfers.
3. Distributing grids do not interfere with the flow.
4. Fluid is initially flowing at temperature T_0. At a given instant, the porous grids start heating the incoming fluid. As described below, the rate of heat exchange between each grid and passing fluid is given by a combination of conductive and convective heat transfers.

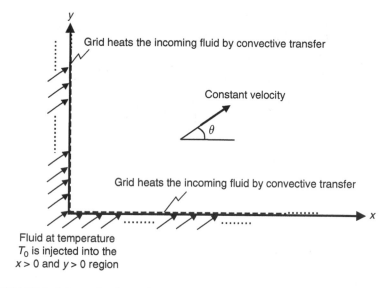

FIGURE 15.1 Scheme for the problem of heating fluid at two-dimensional (2-D) flow.

5. Fluid is Newtonian and its properties (for instance, density and thermal conductivity) can be taken as constants. This is assumed despite the heating or temperature variations. Of course, this might be a strong approximation if large variations of temperature are to be expected. This assumption would not be possible in cases of gases.

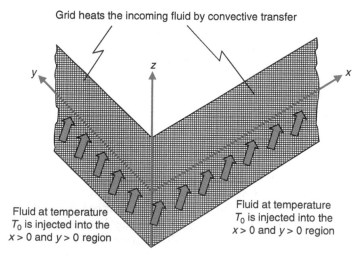

FIGURE 15.2 A three-dimensional (3-D) scheme for the problem of heating fluid at 2-D flow.

6. Dissipation of energy is negligible due to viscous flow.
7. For a first approximation, convective transfer of energy is considered much higher than those achieved by conduction.
8. Coefficient of heat transfer by convection is the same for both grids and remains constant.

Similarly to Section 14.2, Equation A.34 can be written as

$$\rho C_p \left(\frac{\partial T}{\partial t} + v_x \frac{\partial T}{\partial x} + v_y \frac{\partial T}{\partial y} \right) = \lambda \left(\frac{\partial^2 T}{\partial x^2} + \frac{\partial^2 T}{\partial y^2} \right) \tag{15.1}$$

From assumption 7, it is possible to write

$$\frac{\partial T}{\partial t} + v_x \frac{\partial T}{\partial x} + v_y \frac{\partial T}{\partial y} = 0 \tag{15.2}$$

where v_x and v_y are constants.
The boundary conditions are

$$T(0,x,y) = T_0, \quad x > 0, \quad y > 0 \tag{15.3}$$

$$-\lambda \frac{\partial T}{\partial x}\bigg|_{x=0} = q_{x=0} + \alpha[T(t,0,y) - T_0], \quad t > 0, \quad y \geq 0 \tag{15.4}$$

$$-\lambda \frac{\partial T}{\partial y}\bigg|_{y=0} = q_{y=0} + \alpha[T(t,x,0) - T_0], \quad t > 0, \quad x \geq 0 \tag{15.5}$$

The grids are at $x = 0$ and $y = 0$, and as seen earlier there is a relation that provides the rate of heat transfer by two terms. The first term is related to conduction and the second to convection between the grids and the neighboring fluid.
As usual, let us apply the following variable transformation:

$$\psi = \frac{T - T_0}{T_0} \tag{15.6}$$

$$\chi = \frac{\alpha}{\lambda} x \tag{15.7}$$

$$\xi = \frac{\alpha v_x}{\lambda v_y} y \tag{15.8}$$

$$\tau = t \frac{\alpha v_x}{\lambda} \tag{15.9}$$

Notice that, because of assumptions 5 and 6, the parameters α and λ are considered as constants.

By applying the above changes in Equation 15.2, it is possible to write

$$\frac{\partial \psi}{\partial \tau} + \frac{\partial \psi}{\partial \chi} + \frac{\partial \psi}{\partial \xi} = 0 \tag{15.10}$$

The boundary conditions are written as

$$\psi(0,\chi,\xi) = 0, \quad \chi > 0, \quad \xi > 0 \tag{15.11}$$

$$\left.\frac{\partial \psi}{\partial \chi}\right|_{\chi=0} = -b_\chi - c_\chi\, \psi(\chi = 0), \quad \tau > 0, \quad \xi \geq 0 \tag{15.12}$$

$$\left.\frac{\partial \psi}{\partial \xi}\right|_{\xi=0} = -b_\xi - c_\xi\, \psi(\xi = 0), \quad \tau > 0, \quad \chi \geq 0 \tag{15.13}$$

where

$$b_\chi = \frac{q_{x=0}}{\alpha T_0}, \quad b_\xi = \frac{q_{y=0}v_y}{\alpha T_0 v_x} \tag{15.14}$$

$$c_\chi = 1 \tag{15.15}$$

$$c_\xi = \frac{v_y}{v_x} \tag{15.16}$$

As observed, the conditions given by Equations 15.12 and 15.49 are third-kind boundary conditions.

15.2.1   SOLUTION BY LAPLACE TRANSFORM

In view of the condition given by Equation 15.11, the Laplace transform on variable τ seems attractive, and

$$\Psi(s;\chi,\xi) = L\{\psi(\tau,\chi,\xi)\} \tag{15.17}$$

After applying this in Equation 15.10, which when combined with condition given by Equation 15.11, provides

$$s\Psi + \frac{\partial \Psi}{\partial \chi} + \frac{\partial \Psi}{\partial \xi} = 0 \tag{15.18}$$

The transforms of the remaining boundary conditions are

$$\left.\frac{\partial \Psi}{\partial \chi}\right|_{\chi=0} = -\frac{b_\chi}{s} - c_\chi\, \Psi(s;0,\xi), \quad \xi \geq 0 \tag{15.19}$$

$$\left.\frac{\partial \Psi}{\partial \xi}\right|_{\xi=0} = -\frac{b_\xi}{s} - c_\xi \, \Psi(s;\chi,0), \quad \chi \geq 0 \tag{15.20}$$

As before, Equation 15.18 can be solved by several methods, one among them being the Laplace transform. Let the second transform be as follows:

$$\Xi(s,r;\xi) = L\{\Psi(s;\chi,\xi)\} \tag{15.21}$$

Applying Equation 15.21 to Equation 15.18, one gets

$$(s + r)\Xi - f(s;\xi) + \frac{d\Xi}{d\xi} = 0 \tag{15.22}$$

Again, f is a function of ξ and represents $\Psi(\chi=0,\xi)$. The semicolon is set to remind that unlike the variable ξ, s is just a parameter involved in function f. Of course, for now, this function is unknown.

The transform of the condition given by Equation 15.20 is

$$\left.\frac{d\Xi}{d\xi}\right|_{\xi=0} = -\frac{b_\xi}{rs} - c_\xi \Xi(s,r;0) \tag{15.23}$$

Equation 15.22 is linear and the solution is obtainable by the method of parameter variation, as shown in Appendix B. Therefore, the following is set:

$$\Xi(\xi) = \sigma(\xi)\omega(\xi) \tag{15.24}$$

Applying Equation 15.24 to 15.22, allows writing

$$\omega\left[(s + r)\sigma + \frac{d\sigma}{d\xi}\right] - f(s;\xi) + \sigma\frac{d\omega}{d\xi} = 0 \tag{15.25}$$

Without any loss of generality, the term inside the bracket can be set equal to zero, which leads to

$$\sigma = \exp[-(s + r)\xi] \tag{15.26}$$

Consequently, Equation 15.25 yields

$$\omega = \int_0^\xi f(s;\xi)e^{(s+r)\xi}d\xi + C_1 \tag{15.27}$$

Using the above solutions into Equation 15.24, one gets

$$\Xi = e^{-(s+r)\xi} \int_0^\xi f(s;\xi) \, e^{(s+r)\xi} \, d\xi + C_1 e^{-(s+r)\xi} \tag{15.28}$$

The application of the condition given by Equation 15.23 requires the derivative of the former expression, which is given by

$$\frac{d\Xi}{d\xi} = -(s+r)e^{-(s+r)\xi} \int_0^\xi f(s;\xi)e^{(s+r)\xi} \, d\xi + f(s;\xi) - (s+r)C_1 e^{-(s+r)\xi} \tag{15.29}$$

The following is obtained after applying the condition given by Equation 15.23:

$$C_1 = \frac{1}{r+s-c_\xi} \left[f(s;0) + \frac{b_\xi}{sr} \right] \tag{15.30}$$

Substituting Equation 15.30 in Equation 15.28 gives

$$\Xi = e^{-(s+r)\xi} \int_0^\xi f(s;\xi)e^{(s+r)\xi} \, d\xi + \frac{e^{-(s+r)\xi}}{r+s-c_\xi} \left[f(s;0) + \frac{b_\xi}{sr} \right] \tag{15.31}$$

As described before, the first inversion should be performed involving parameter r and variable χ. Similarly, it is assumed that the condition $\Psi(\chi=0,\xi)$ is not a function of variable ξ. This is tested below and if found true the new unknown function $f(s;\xi)$ is just a relation involving parameter s, which will be called $g(s)$. After this, the above equation becomes

$$\Xi = \frac{g(s)}{r+s} - \frac{g(s)e^{-(s+r)\xi}}{r+s} + \frac{e^{-(s+r)\xi}}{r+s-c_\xi} \left[g(s) + \frac{b_\xi}{sr} \right] \tag{15.32}$$

The inversion regarding parameter r results into

$$\Psi = g(s)e^{-s\chi} - g(s)e^{-s\chi}u(\chi - \xi)$$

$$+ g(s)e^{-(s-c_\xi)\chi - c_\xi\xi}u(\chi - \xi) + b_\xi \frac{e^{-s\xi} - e^{-(s-c_\xi)\chi - c_\xi\xi}}{s(s-c_\xi)}u(\chi - \xi) \tag{15.33}$$

Before inverting the above, some information regarding function $g(s)$ might be obtained by applying the condition given by Equation 15.19. This provides the following:

$$g(s) = \frac{b_\chi}{s(s-c_\chi)} \tag{15.34}$$

Therefore, the assumption of $f(s;\xi)$ equal to $g(s)$, or independence of f on ξ, leads to a possible relation. Using the above, Equation 15.33 is rewritten as

$$
\Psi = \frac{b_\chi e^{-s\chi}}{s(s - c_\chi)} - \frac{b_\chi e^{-s\chi}}{s(s - c_\chi)} u(\chi - \xi) + \frac{b_\chi}{s(s - c_\chi)} e^{-(s - c_\xi)\chi - c_\xi\xi} u(\chi - \xi)
$$
$$
+ b_\xi \frac{e^{-s\xi} - e^{-(s - c_\xi)\chi - c_\xi\xi}}{s(s - c_\xi)} u(\chi - \xi) \tag{15.35}
$$

The inverse regarding parameter s is

$$
\psi = - b_\chi \frac{1 - e^{c_\chi(\tau - \chi)}}{c_\chi} u(\tau - \chi) + b_\chi \frac{1 - e^{c_\chi(\tau - \chi)}}{c_\chi} u(\tau - \chi)u(\chi - \xi)
$$
$$
- b_\chi \frac{1 - e^{c_\chi(\tau - \chi)}}{c_\chi} e^{c_\xi(\chi - \xi)} u(\tau - \chi)u(\chi - \xi)
$$
$$
- b_\xi \frac{1 - e^{c_\xi(\tau - \xi)}}{c_\xi} u(\tau - \xi)u(\chi - \xi) + b_\xi \frac{e^{c_\xi(\chi - \xi)} - e^{c_\xi(\tau - \xi)}}{c_\xi} u(\tau - \chi)u(\chi - \xi)
$$

The above equation can be put in the following form:

$$
\psi = - b_\chi \frac{1 - e^{c_\chi(\tau - \chi)}}{c_\chi} u(\tau - \chi)
$$
$$
+ \left[b_\chi \frac{1 - e^{c_\chi(\tau - \chi)}}{c_\chi} \left[1 - e^{c_\xi(\chi - \xi)} \right] + b_\xi \frac{e^{c_\xi(\chi - \xi)} - e^{c_\xi(\tau - \xi)}}{c_\xi} \right] u(\tau - \chi)u(\chi - \xi)
$$
$$
- b_\xi \frac{1 - e^{c_\xi(\tau - \xi)}}{c_\xi} u(\tau - \xi)u(\chi - \xi) \tag{15.36}
$$

Figure 15.3 illustrates the case where $b_\chi = b_\xi = 2$ and $c_\chi = c_\xi = 1$ at dimensionless instant $\tau = 1$ and Figure 15.4 illustrates the case where $b_\chi = b_\xi = 4$ and $c_\chi = c_\xi = 1$ at the same instant. Figure 15.5 illustrates the case where $b_\chi = 2$, $b_\xi = 4$, and $c_\chi = c_\xi = 1$ at dimensionless instant $\tau = 1$ and Figure 15.6 illustrates the case where $b_\chi = b_\xi = 2$ and $c_\chi = 1$, $c_\xi = 2$ at the same instant.

As before, a wave front of heated fluid travels from the grids toward the rest of the region. Owing to the simplification of plug-flow, a sharp modification or change in the derivative characterization at the wave front is observed. Of course, the real situation is a more smooth variation, and this can be obtained by including thermal conductivity terms to the governing equations. As these are second-order derivatives, the effect is to anticipate the change and introduce a gradual variation of concentration at the wave front.

Despite the present approximation, the treatment provides some interesting features of the process. For instance, the rate of heat transfer by conduction from the grid located at the $\chi = 0$ plane is represented by increases in parameters b_χ,

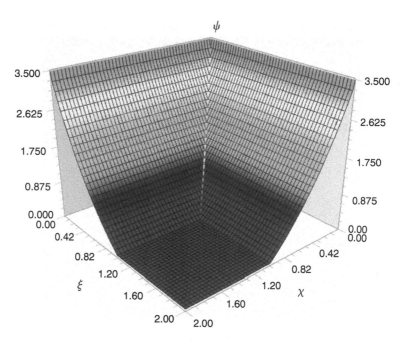

FIGURE 15.3 Dimensionless temperature (ψ) against dimensionless coordinates (χ, ξ) for the case where $b_\chi = b_\xi = 2$ and $c_\chi = c_\xi = 1$ at dimensionless instant $\tau = 1$.

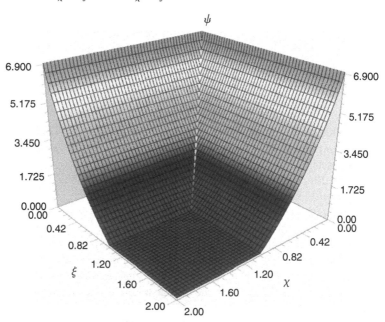

FIGURE 15.4 Dimensionless temperature (ψ) against dimensionless coordinates (χ, ξ) for the case where $b_\chi = b_\xi = 4$ and $c_\chi = c_\xi = 1$ at dimensionless instant $\tau = 1$.

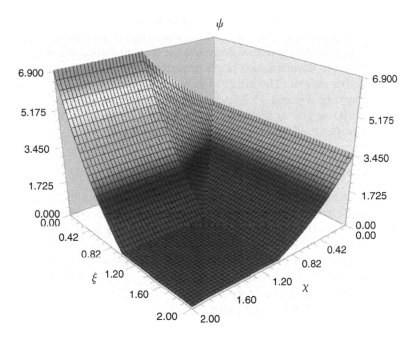

FIGURE 15.5 Dimensionless temperature (ψ) against dimensionless coordinates (χ,ξ) for the case where $b_\chi = 2$, $b_\xi = 4$, and $c_\chi = c_\xi = 1$ at dimensionless instant $\tau = 1$.

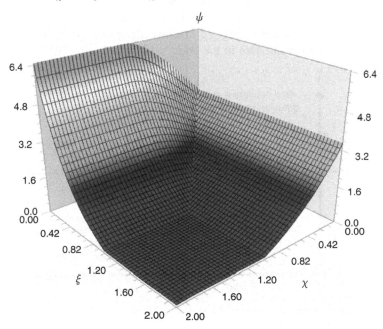

FIGURE 15.6 Dimensionless temperature (ψ) against dimensionless coordinates (χ,ξ) for the case where $b_\chi = b_\xi = 2$ and $c_\chi = 1$, $c_\xi = 2$ at dimensionless instant $\tau = 1$.

whereas for the one at $\xi = 0$ by b_ξ. The rates of heat transfers by convection are represented by parameters c_χ and c_ξ. Therefore, increases in these parameters would provoke higher temperature derivatives at these positions and ahead of the grid. In addition, to maintain such larger derivatives, the temperature at the grids should be higher as well. The effect of increases in conduction transfer can be seen by comparing Figures 15.3 and 15.4. The effect of conduction transfer increases at just one grid as shown in Figure 15.5, while the effect of convective transfer can be observed by comparing Figures 15.3 and 15.6.

15.3 TWO-DIMENSIONAL REACTING FLOW

Let us consider a similar problem as shown in Section 15.2, but instead of being heated the fluid reacts. In other words, the incoming fluid is composed of pure substance B. When entering the region $x > 0$ and $y > 0$, the grid delivers another fluid A that reacts with B. However, instead of the situation described in Section 14.3, the rate of mass transfer of A by diffusion through the porous grids equals its rate of mass transfer by convection into the fluid. This realistic situation is illustrated in Figures 15.7 and 15.8.

As always, let us list the assumptions:

1. Space is not limited regarding direction z (orthogonal to x and y).
2. In each direction of x and y, the flow can be approximated to plug-flow. In other words, the terms related to convective momentum transfers are much higher than those related to viscous transfers.
3. Distributing grids are comprised of networks of small-diameter porous tubes and do not interfere in the flow.

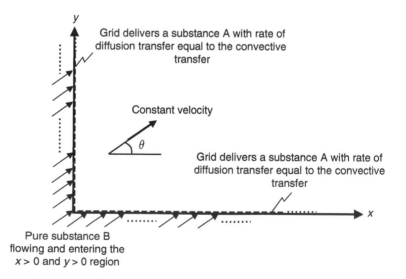

FIGURE 15.7 Scheme for the problem of reacting fluid at 2-D flow.

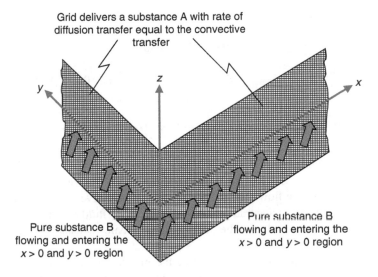

Grid delivers a substance A with rate of
diffusion transfer equal to the convective
transfer

Pure substance B
flowing and entering the
$x > 0$ and $y > 0$ region

Pure substance B
flowing and entering the
$x > 0$ and $y > 0$ region

FIGURE 15.8 A 3-D scheme for the problem of reacting fluid at 2-D flow.

4. Fluid is initially composed of a single chemical species B. At a given instant, the grids start delivering a fluid of pure component A, which reacts with B. The mass transfer of A is given by diffusion and convection. This will become clear during the mathematical treatment below.

5. Reaction is neither endothermic nor exothermic and can be represented by

$$A + B \rightarrow C \qquad (15.37)$$

6. Reaction is zero order and irreversible.

7. Any mixture of components A, B, and C results in a Newtonian fluid and its properties (for instance, density, viscosity, diffusivities, etc.) can be taken as constants. This is assumed despite the change in composition.

8. No energy dissipation due to viscous flow is observed.

9. For a first approximation, convection transfer of mass is considered much higher than those achieved by diffusion processes.

The procedure is the same as shown in Sections 13.3 and 14.3. Hence, one is entitled to write the governing equation as

$$\frac{\partial \rho_A}{\partial t} + v_x \frac{\partial \rho_A}{\partial x} + v_y \frac{\partial \rho_A}{\partial y} = R_{M,A} = -k \qquad (15.38)$$

where k is the reaction rate constant.

However here, the boundary conditions are

$$\rho_A(0,x,y) = \rho_{A0} \tag{15.39}$$

$$-D_{AB}\frac{\partial \rho_A}{\partial x}\bigg|_{x=0} = N_{A,x=0} + \beta[\rho_A(x=0) - \rho_{A0}], \quad t > 0, \quad y \geq 0 \tag{15.40}$$

$$-D_{AB}\frac{\partial \rho_A}{\partial y}\bigg|_{y=0} = N_{A,y=0} + \beta[\rho_A(y=0) - \rho_{A0}], \quad t > 0, \quad x \geq 0 \tag{15.41}$$

Notice that the mass transfer at the grid is given by two terms. The first is similar to that described in Section 14.3 and is related to diffusive mass transfer of component A into the flow. The second refers to convective mass transfer.

Again, ρ_{A0} is the concentration (kg m^{-3}) of A in the fluid before the start of its injection through the grids and the concentration remains constant for regions far from the grid.

Since only component B flows before the injection of component A, ρ_{A0} is zero. This notation is maintained to allow a more general solution as in the case when component A and a neutral substance pass through the grid at which B is delivered. In this case, ρ_{A0} would be the concentration of A before the injection of B. In our case, to maintain the formalism as in Sections 13.3 and 14.3, ρ_{A0} can be understood as a reference concentration, which may be even set as 1 kg m^{-3}. The conditions given by Equations 15.40 and 15.41 follow assumption 4.

Let us consider the following dimensionless variables:

$$\psi = \frac{\rho_A - \rho_{A0}}{\rho_{A0}} \tag{15.42}$$

$$\chi = \frac{ax}{v_x} \tag{15.43}$$

$$\xi = \frac{ay}{v_y} \tag{15.44}$$

$$\tau = at \tag{15.45}$$

Here

$$a = \frac{k}{\rho_{A0}} \tag{15.46}$$

The unit for the above constant parameter a would be s^{-1}.

Using Equations 15.42 through 15.46, Equation 15.38 becomes

$$\frac{\partial \psi}{\partial \tau} + \frac{\partial \psi}{\partial \chi} + \frac{\partial \psi}{\partial \xi} = -1 \tag{15.47}$$

The boundary conditions are written as

$$\psi(0,\chi,\xi) = 0, \quad \chi > 0, \quad \xi > 0 \tag{15.48}$$

$$\left.\frac{\partial \psi}{\partial \chi}\right|_{\chi=0} = -b_\chi - c_\chi\,\psi(\chi=0), \quad \tau > 0, \quad \xi \geq 0 \tag{15.49}$$

$$\left.\frac{\partial \psi}{\partial \xi}\right|_{\xi=0} = -b_\xi - c_\xi\,\psi(\xi=0), \quad \tau > 0, \quad \chi \geq 0 \tag{15.50}$$

where

$$b_\chi = \frac{N_{A,\chi=0}v_\chi}{D_{AB}a\rho_{A0}}, \quad b_\xi = \frac{N_{A,y=0}v_y}{D_{AB}a\rho_{A0}} \tag{15.51}$$

$$c_\chi = \frac{\beta v_\chi}{aD_{AB}}, \quad c_\xi = \frac{\beta v_y}{aD_{AB}} \tag{15.52}$$

The conditions given by Equations 15.49 and 15.50 are third-kind boundary conditions.

15.3.1 SOLUTION BY LAPLACE TRANSFORM

Considering the condition given by Equation 15.45, the transform in variable τ seems attractive, and

$$\Psi(s;\chi,\xi) = L\{\psi(\tau,\chi,\xi)\} \tag{15.53}$$

After applying this in Equation 15.48, and which when combined with this condition, one gets

$$s\Psi + \frac{\partial \Psi}{\partial \chi} + \frac{\partial \Psi}{\partial \xi} - \frac{1}{s} \tag{15.54}$$

The transforms of the remaining boundary conditions are

$$\left.\frac{\partial \Psi}{\partial \chi}\right|_{\chi=0} = -\frac{b_\chi}{s} - c_\chi\Psi(s;0,\xi), \quad \xi \geq 0 \tag{15.55}$$

$$\left.\frac{\partial \Psi}{\partial \xi}\right|_{\xi=0} = -\frac{b_\xi}{s} - c_\xi\Psi(s;\chi,0), \quad \chi \geq 0 \tag{15.56}$$

As described before, Equation 15.54 might be solved by several methods, one among them being the Laplace transform. Let the second transform be written as

$$\Xi(s,r;\xi) = L\{\Psi(s;\chi,\xi)\} \tag{15.57}$$

Applying Equation 15.57 to Equation 15.54, one gets

$$(s+r)\Xi - f(s;\xi) + \frac{d\Xi}{d\xi} = -\frac{1}{sr} \tag{15.58}$$

Again, f is a function of ξ and represents $\Psi(\chi=0,\xi)$. The semicolon is set to remind that unlike the variable ξ, s is just a parameter involved in function f. Of course, for now this function is unknown.

The transform of the condition given by Equation 15.56 is

$$\frac{d\Xi}{d\xi}\bigg|_{\xi=0} = -\frac{b_\xi}{rs} - c_\xi \Xi(s,r;0) \tag{15.59}$$

Equation 15.58 is linear and the solution can be sought by using the method of parameter variation, as shown in Appendix B. Therefore, the following is set:

$$\Xi(\xi) = \sigma(\xi)\omega(\xi) \tag{15.60}$$

Applying Equation 15.60 to Equation 15.58 allows writing

$$\omega\left[(s+r)\sigma + \frac{d\sigma}{d\xi}\right] - f(s;\xi) + \sigma\frac{d\omega}{d\xi} = -\frac{1}{sr} \tag{15.61}$$

Without any loss of generality, the term inside the bracket can be set equal to zero, which leads to

$$\sigma = \exp[-(s+r)\xi] \tag{15.62}$$

Consequently, Equation 15.61 yields

$$\omega = -\frac{e^{(s+r)\xi}}{sr(s+r)} + \int_0^\xi f(s;\xi)e^{(s+r)\xi}d\xi + C_1 \tag{15.63}$$

Substituting Equations 15.62 and 15.63 into Equation 15.60 gives

$$\Xi = e^{-(s+r)\xi}\int_0^\xi f(s;\xi)e^{(s+r)\xi}d\xi - \frac{1}{rs(s+r)} + C_1 e^{-(s+r)\xi} \tag{15.64}$$

The application of the condition given by Equation 15.59 requires the derivative of the former expression, which is given by

$$\frac{d\Xi}{d\xi} = -(s+r)e^{-(s+r)\xi}\int_0^\xi f(s;\xi)e^{(s+r)\xi}\,d\xi + f(s;\xi) - (s+r)C_1 e^{-(s+r)\xi} \tag{15.65}$$

The following is obtained after application of the condition given by Equation 15.59:

$$C_1 = \frac{1}{r+s-c_\xi}\left[f(s;0) + \frac{b_\xi}{sr} - \frac{c_\xi}{sr(s+r)}\right] \tag{15.66}$$

Using Equation 15.66 in Equation 15.64, one arrives at

$$\Xi = e^{-(s+r)\xi}\int_0^\xi f(s;\xi)e^{(s+r)\xi}\,d\xi - \frac{1}{rs(s+r)} + \frac{e^{-(s+r)\xi}}{r+s-c_\xi}\left[f(s;0) + \frac{b_\xi}{sr} - \frac{c_\xi}{sr(s+r)}\right] \tag{15.67}$$

As described before, the first inversion should be performed involving parameter r and variable χ. Similarly, it is assumed that the condition $\Psi(\chi=0,\xi)$ is not a function of variable ξ. This will be tested below, and if true, the new unknown function $f(s;\xi)$ is just a relation involving parameter s, which will be called $g(s)$. After this, the above equation becomes

$$\Xi = \frac{g(s)}{r+s} - \frac{g(s)e^{-(s+r)\xi}}{r+s} - \frac{1}{sr(r+s)} + \frac{e^{-(s+r)\xi}}{r+s-c_\xi}\left[g(s) + \frac{b_\xi}{sr} - \frac{c_\xi}{sr(r+s)}\right] \tag{15.68}$$

The inversion regarding parameter r results into

$$\Psi = g(s)e^{-s\chi} - g(s)e^{-s\chi}u(\chi-\xi) - \frac{1-e^{-s\chi}}{s^2}$$
$$+ g(s)e^{-(s-c_\xi)\chi-c_\xi\xi}u(\chi-\xi) + b_\xi\frac{e^{-s\xi}-e^{-(s-c_\xi)\chi-c_\xi\xi}}{s(s-c_\xi)}u(\chi-\xi)$$
$$+ \left[\frac{e^{-(s-c_\xi)\chi-c_\xi\xi}}{s(s-c_\xi)} - \frac{c_\xi e^{-s\xi}}{s^2(s-c_\xi)} - \frac{e^{-s\chi}}{s^2}\right]u(\chi-\xi) \tag{15.69}$$

Before inverting the above equation, some information regarding function $g(s)$ might be obtained by applying the condition given by Equation 15.55. This provides the following:

$$g(s) = -\frac{1-b_\chi}{s(s-c_\chi)} \tag{15.70}$$

Therefore, the assumption of $f(s;\xi)$ equal to $g(s)$, or independence of f on ξ, leads to a possible relation. Using the above, Equation 15.69 is rewritten as

$$\Psi = -\frac{(1-b_\chi)e^{-s\chi}}{s(s-c_\chi)} + \frac{(1-b_\chi)e^{-s\chi}}{s(s-c_\chi)}u(\chi-\xi) - \frac{1-e^{-s\chi}}{s^2}$$
$$- \frac{1-b_\chi}{s(s-c_\chi)}e^{-(s-c_\xi)\chi-c_\xi\xi}u(\chi-\xi) + b_\xi\frac{e^{-s\xi}-e^{-(s-c_\xi)\chi-c_\xi\xi}}{s(s-c_\xi)}u(\chi-\xi)$$
$$+ \left[\frac{e^{-(s-c_\xi)\chi-c_\xi\xi}}{s(s-c_\xi)} - \frac{c_\xi e^{-s\xi}}{s^2(s-c_\xi)} - \frac{e^{-s\chi}}{s^2}\right]u(\chi-\xi) \tag{15.71}$$

The inverse, regarding parameter s, is

$$\psi = (1-b_\chi)\frac{1-e^{c_\chi(\tau-\chi)}}{c_\chi}u(\tau-\chi) - (1-b_\chi)\frac{1-e^{c_\chi(\tau-\chi)}}{c_\chi}u(\tau-\chi)u(\chi-\xi)$$

$$- \tau + (\tau-\chi)u(\tau-\chi) + (1-b_\chi)\frac{1-e^{c_\chi(\tau-\chi)}}{c_\chi}e^{c_\xi(\chi-\xi)}u(\tau-\chi)u(\chi-\xi)$$

$$- b_\xi\frac{1-e^{c_\xi(\tau-\xi)}}{c_\xi}u(\tau-\xi)u(\chi-\xi) + b_\xi\frac{e^{c_\xi(\chi-\xi)}-e^{c_\xi(\tau-\xi)}}{c_\xi}u(\tau-\chi)u(\chi-\xi)$$

$$+ \left[-\frac{1-e^{c_\xi(\tau-\chi)}}{c_\xi}e^{c_\xi(\chi-\xi)}u(\tau-\chi) \right.$$

$$\left. + \frac{1+c_\xi(\tau-\xi)-e^{c_\xi(\tau-\xi)}}{c_\xi}u(\tau-\xi) - (\tau-\chi)u(\tau-\chi) \right]u(\chi-\xi)$$

This can be put in the following form:

$$\psi = -\tau + (\tau-\chi)u(\tau-\chi) + (1-b_\chi)\frac{1-e^{c_\chi(\tau-\chi)}}{c_\chi}u(\tau-\chi)$$

$$+ \left[\begin{array}{l} (1-b_\chi)\frac{1-e^{c_\chi(\tau-\chi)}}{c_\chi}\left[e^{c_\xi(\chi-\xi)}-1\right] \\ +b_\xi\frac{e^{c_\xi(\chi-\xi)}-e^{c_\xi(\tau-\xi)}}{c_\xi} - \frac{1-e^{c_\xi(\tau-\chi)}}{c_\xi}e^{c_\xi(\chi-\xi)}-(\tau-\chi) \end{array} \right]u(\tau-\chi)u(\chi-\xi)$$

$$+ \left[\frac{1+c_\xi(\tau-\xi)-e^{c_\xi(\tau-\xi)}}{c_\xi} - b_\xi\frac{1-e^{c_\xi(\tau-\xi)}}{c_\xi} \right]u(\tau-\xi)u(\chi-\xi) \qquad (15.72)$$

Figure 15.9 illustrates the case where $b_\chi=b_\xi=2$ and $c_\chi=c_\xi=1$ at dimensionless instant $\tau=1$ and Figure 15.10 illustrates the case where $b_\chi=b_\xi=4$ and $c_\chi=c_\xi=1$ at the same instant. Figure 15.11 illustrates the case where $b_\chi=2$, $b_\xi=4$, and $c_\chi=c_\xi=1$ at dimensionless instant $\tau=1$, and Figure 15.12 illustrates the case where $b_\chi=b_\xi=2$ and $c_\chi=1$, $c_\xi=2$ at the same instant.

As before, a wave front of component A travels from the grids toward the rest of the reacting region. Owing to the simplification of plug-flow, sharp modifications or changes in the derivative are observed at the wave front. Of course, the real situation is a more smooth variation, and this can be obtained by including the diffusive terms into the governing equations. As these are second-order derivatives, the effect is to anticipate the change and introduce a gradual variation of concentration at the wave front.

Despite an approximation, the present treatment provides some interesting features of the process. For instance, the rate of mass transfer by diffusion from the grid located at the $\chi=0$ plane is represented by increases in parameter b_χ, while for the one at $\xi=0$ by b_ξ. The rates of transfers by convection are represented by parameters c_χ and c_ξ. Therefore, increases on these parameters would provoke increases in the derivative of the concentration at these positions and ahead of the

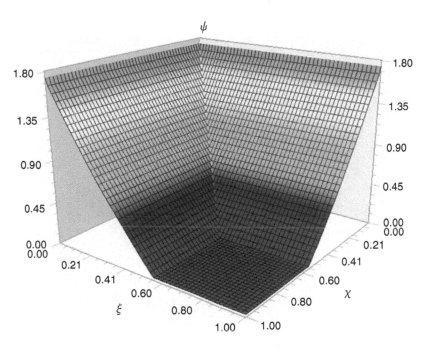

FIGURE 15.9 Dimensionless concentration (ψ) against dimensionless coordinates (χ,ξ) for the case where $b_\chi=b_\xi=2$ and $c_\chi=c_\xi=1$ at dimensionless instant $\tau=1$.

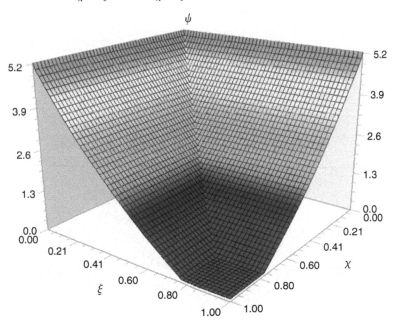

FIGURE 15.10 Dimensionless concentration (ψ) against dimensionless coordinates (χ,ξ) for the case where $b_\chi=b_\xi=4$ and $c_\chi=c_\xi=1$ at dimensionless instant $\tau=1$.

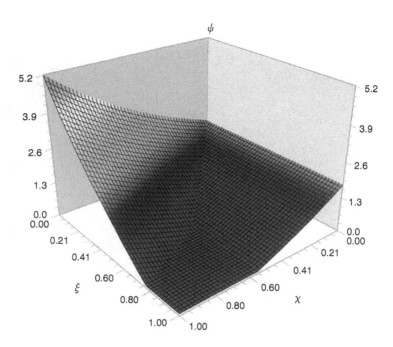

FIGURE 15.11 Dimensionless concentration (ψ) against dimensionless coordinates (χ, ξ) for the case where $b_\chi = 2$, $b_\xi = 4$, and $c_\chi = c_\xi = 1$ at dimensionless instant $\tau = 1$.

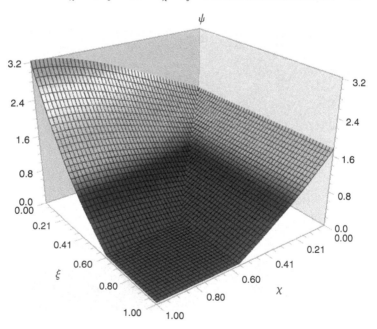

FIGURE 15.12 Dimensionless concentration (ψ) against dimensionless coordinates (χ, ξ) for the case where $b_\chi = b_\xi = 2$ and $c_\chi = 1$, $c_\xi = 2$ at dimensionless instant $\tau = 1$.

grid. In addition, to maintain such increases of derivatives, the concentration at the grids should be higher as well. The effect of increases in diffusion transfer can be seen by comparing Figures 15.9 and 15.10. The effect of diffusion transfer increases at just one grid is shown in Figure 15.11, while the effect of convective transfer can be observed by comparing Figures 15.9 and 15.12.

EXERCISES

1. Work on the details to arrive at Equation 15.33.
2. Work on the details of inversion to arrive at Equation 15.36.
3. Solve the problem in Section 15.2 using any branch of the method of weighted residuals. As a hint, try to depart from Equation 15.10 and the condition given by Equation 15.11 to write a possible choice for the approximations given by

$$\bar{\psi}_n(\tau,\chi,\xi) = \sum_{j=1}^{n} \tau^j F_j(\chi)G_j(\xi) \tag{15.73}$$

Here, F_j and G_j are trial functions to be defined by each approximation.
   The first approximation would be

$$\bar{\psi}_1(\tau,\chi,\xi) = \tau F_1(\chi)G_1(\xi) \tag{15.74}$$

Using Equation 15.10, the residue becomes

$$\Lambda_1 = F_1 G_1 + \tau F_1' G_1 + \tau F_1 G_1' \tag{15.75}$$

For the sake of simplicity, apply the collocation method at instant $\tau = a$. Therefore, after equating the residual to zero at this point, the following can be written:

$$1 + \frac{aF_1'}{F} + \frac{aG_1'}{G} = 0 \tag{15.76}$$

As F_1 and G_1 are functions of independent variables, it is possible to obtain

$$\frac{aF_1'}{F} = -1 - \frac{aG_1'}{G} = C_1 \tag{15.77}$$

Here C_1 is a constant.
   Solve the above two differential equations to obtain a form for the first approximation. The constants might be determined by the boundary conditions given by Equations 15.12 and 15.13.
4. Solve the problem in Section 15.3 for an irreversible first-order reaction, at which the source term is given by

$$-k\rho_A$$

16 Problems 321; Three Variables, 2nd Order, 1st Kind Boundary Condition

16.1 INTRODUCTION

This chapter presents methods to solve problems with three independent variables involving second-order differential equations and first-kind boundary conditions. Therefore, partial differential equations are involved. Mathematically, this class of cases can be summarized as $f\left(\phi, \omega_1, \omega_2, \omega_3, \frac{\partial \phi}{\partial \omega_i}, \frac{\partial^2 \phi}{\partial \omega_i^2}, \frac{\partial^2 \phi}{\partial \omega_i \partial \omega_j}\right)$, first-kind boundary condition.

This is also the last chapter. Of course, analytical and approximate solutions can be obtained for cases that are even more complex. However, when the complexity increases, it might be worthwhile or simpler to apply numerical solutions. As indicated in the preface, the book advances until a point where a compromise between complexity and applicability of analytical and approximated methods seemed reasonable.

16.2 TEMPERATURES IN A RECTANGULAR PLATE

Consider an insulated plate with prescribed temperatures at the borders. An initial given temperature profile $f(x,y)$ is imposed, as shown in Figure 16.1.

It is desired to obtain the temperature profile in the plate against time.

Let us consider the following assumptions:

1. Temperatures at the borders ($x = 0$ and $y = 0$) remain constant.
2. Plate is thin enough as well as insulated at the two main wide faces in such a way that no appreciable heat transfer occurs in the z direction.
3. At a given instant ($t = 0$), the temperature of the plate (except at the borders) is set according to a given function $f(x,y)$.

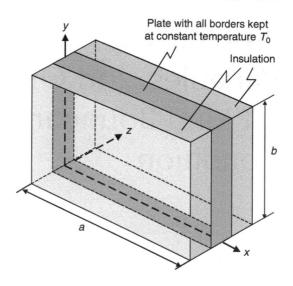

FIGURE 16.1 Rectangular plate with prescribed temperatures at the borders.

4. Plate is solid, and within the range of temperature of the present problem no phase change is possible. Therefore, there is no velocity field in the plate.
5. Range of temperature variation is small enough for the properties of plate material to be assumed as constants.
6. No chemical reaction, or any other form of energy generation, takes place in the plate.

From the above and applying Equation A.34, it is possible to write

$$\frac{\partial T}{\partial t} = D_{\mathrm{T}} \left(\frac{\partial^2 T}{\partial x^2} + \frac{\partial^2 T}{\partial y^2} \right) \tag{16.1}$$

Here D_{T} is the thermal diffusivity and is given by

$$D_{\mathrm{T}} = \frac{\lambda}{\rho C_{\mathrm{p}}} \tag{16.2}$$

Of course, there are various possibilities for a set of dimensionless variables. Here, the following is proposed:

$$\theta = \frac{T - T_0}{T_0} \tag{16.3}$$

$$\tau = \frac{D_{\mathrm{T}}}{a^2} t \tag{16.4}$$

$$\omega = \frac{x}{a} \tag{16.5}$$

$$\xi = \frac{y}{a} \tag{16.6}$$

Applying the above, Equation 16.1 can be written as

$$\frac{\partial\theta}{\partial\tau} = \frac{\partial^2\theta}{\partial\omega^2} + \frac{\partial^2\theta}{\partial\xi^2} \tag{16.7}$$

Let us call the ratio between the plate dimensions as

$$\gamma = \frac{b}{a} \tag{16.8}$$

The boundary conditions are

$$\theta(\tau,0,\xi) = 0, \quad 0 \le \xi \le \gamma, \quad \tau > 0 \tag{16.9}$$

$$\theta(\tau,\omega,\gamma) = 0, \quad 0 \le \omega \le 1, \quad \tau > 0 \tag{16.10}$$

$$\theta(\tau,1,\xi) = 0, \quad 0 \le \xi \le \gamma, \quad \tau > 0 \tag{16.11}$$

$$\theta(\tau,\omega,0) = 0, \quad 0 \le \omega \le 1, \quad \tau > 0 \tag{16.12}$$

The initial temperature profile is given by

$$\theta(0,\omega,\xi) = \varphi(\omega,\xi), \quad 0 < \omega < 1, \quad 0 < \xi < \gamma \tag{16.13}$$

Here

$$\varphi(\omega,\xi) = \frac{f(x,y) - T_0}{T_0} \tag{16.14}$$

16.2.1 SOLUTION BY SEPARATION OF VARIABLES

As shown in Appendix G, the assumption of the method makes it possible to write

$$\theta(\tau,\omega,\xi) = G(\tau)F(\omega,\xi) \tag{16.15}$$

Equation 16.15 is substituted into Equation 16.7 to give

$$\frac{1}{G}\frac{dG}{d\tau} = \frac{1}{F}\left(\frac{\partial^2 F}{\partial\omega^2} + \frac{\partial^2 F}{\partial\xi^2}\right) = -C^2, \tag{16.16}$$

where C is a real constant. This has been reached after verifying that complex or zero values for the constant C would lead to impossibilities or to a trivial solution.

Thus, it is possible to write

$$G(\tau) = K \exp(-C^2 \tau) \tag{16.17}$$

Let us now assume that

$$F(\omega, \xi) = H(\omega) I(\xi) \tag{16.18}$$

Using Equation 16.18 in Equation 16.16 would lead to

$$\frac{1}{H} \frac{d^2 H}{d\omega^2} = -\frac{1}{I} \frac{d^2 I}{d\xi^2} - C^2 = -\kappa^2 \tag{16.19}$$

According to Appendix B, the solutions of these two ordinary differential equations are given by

$$H = A_1 \sin(\kappa \omega) + A_2 \cos(\kappa \omega) \tag{16.20}$$

and

$$I = B_1 \sin(\zeta \xi) + B_2 \cos(\zeta \xi) \tag{16.21}$$

where

$$\zeta^2 = C^2 - \kappa^2 \tag{16.22}$$

The boundary conditions given by Equations 16.9 and 16.11 lead to

$$H(0) = H(1) = 0 \tag{16.23}$$

Similarly, conditions given by Equations 16.10 and 16.12 lead to

$$I(0) = I(\gamma) = 0 \tag{16.24}$$

From these boundary conditions, it is possible to obtain

$$H_n = A_n \sin(\kappa_n \omega) \tag{16.25}$$

and

$$I_m = B_m \sin(\zeta_m \xi) \tag{16.26}$$

Here the eigenvalues are

$$\kappa_n = n\pi, \quad \zeta_m = \frac{m\pi}{\gamma} \tag{16.27}$$

Let

$$C_{m,n} = \sqrt{\kappa_n^2 + \zeta_m^2} \qquad (16.28)$$

Consequently, the solutions for $G(t)$ are

$$G_{m,n}(t) = K_{m,n} \exp(-C_{m,n}^2 \tau) \qquad (16.29)$$

With this, the general solution becomes

$$\theta(\tau,\omega,\xi) = \sum_{m=1}^{\infty} \sum_{n=1}^{\infty} E_{m,n} \sin(\kappa_n \omega) \sin(\zeta_m \xi) \exp(-C_{m,n}^2 \tau) \qquad (16.30)$$

Finally, the boundary condition given by Equation 16.13 can be applied to give

$$\varphi(\omega,\xi) = \sum_{n=1}^{\infty} R_n(\xi) \sin(\kappa_n \omega) \qquad (16.31)$$

where

$$R_n(\xi) = \sum_{m=1}^{\infty} E_{m,n} \sin(\zeta_m \xi) \qquad (16.32)$$

Therefore, a half-range odd expansion (see Appendix G) is set and

$$R_n(\xi) = 2 \int_0^1 \varphi(\omega,\xi) \sin(\kappa_n \omega) \, d\omega \qquad (16.33)$$

On the other hand, from Equation 16.32

$$E_{m,n} = \frac{2}{\gamma} \int_0^{\gamma} R_n(\xi) \sin(\zeta_m \xi) \, d\xi = \frac{4}{\gamma} \int_0^{\gamma} \int_0^1 \varphi(\omega,\xi) \sin(\kappa_n \omega) \sin(\zeta_m \xi) \, d\omega \, d\xi \qquad (16.34)$$

Hence, from the given function $\varphi(\omega,\xi)$, it is possible to determine $E_{m,n}$, and from Equation 16.30, it is possible to determine the temperature profile (θ) as a function of time and space coordinates.

16.2.2   EXAMPLE OF APPLICATION

To exemplify, let us consider a simple function describing a constant initial temperature of the plate, or

$$f(x,y) = T_1 \qquad (16.35)$$

Therefore, from Equation 16.14

$$\varphi(\omega,\xi) = \frac{T_1 - T_0}{T_0} \tag{16.36}$$

With this, Equation 16.34 becomes

$$
\begin{aligned}
E_{m,n} &= \frac{4(T_1 - T_0)}{T_0\gamma} \int_0^\gamma \int_0^1 \sin(\kappa_n\omega) \sin(\zeta_m\xi)\, d\omega\, d\xi \\
&= \frac{4(T_1 - T_0)}{T_0\gamma} \frac{(\cos\kappa_n - 1)[\cos(\zeta_m\gamma) - 1]}{\kappa_n\zeta_m}
\end{aligned}
\tag{16.37}
$$

Using the above, Equations 16.27 and 16.28 substituted into Equation 16.30 would provide the dimensionless temperature as

$$
\begin{aligned}
\theta(\tau,\omega,\xi) &= \frac{4(T_1 - T_0)}{\pi^2 T_0} \sum_{m=1}^\infty \sum_{n=1}^\infty \frac{1 - (-1)^n}{n} \frac{1 - (-1)^m}{m} \sin(n\pi\omega) \sin\left(\frac{m\pi\xi}{\gamma}\right) \\
&\quad \exp\left[-\pi^2\left(n^2 + \frac{m^2}{\gamma^2}\right)\tau\right]
\end{aligned}
\tag{16.38}
$$

Of course, the borders would remain at dimensionless temperature $\theta = 0$. If $T_1 = 600$ K and $T_0 = 300$ K, the initial value for the dimensionless temperatures inside the plate would be 1.0. From this, the heat transfer by conduction would lead to decreased temperatures inside the plate. Consider a case with carbon steel, with the following properties:

- Density $(\rho) = 7854$ kg m^{-3}.
- Specific heat $(C_p) = 434$ J kg^{-1} K^{-1}.
- Thermal conductivity $(\lambda) = 56.7$ W m^{-1} K^{-1}.
- According to Equation 16.2, the above values would provide a thermal diffusivity $(D_T) = 1.66 \times 10^{-5}$ m^2 s^{-1}.

For a square plate $(\gamma = 1)$ with 1 m at each side, the dimensionless time variable Equation 16.4 would be related to the real time (measured in seconds) $\tau = 1.66 \times 10^{-5}$ t. Therefore, after approximately 600 s, the dimensionless time would be 1.0×10^{-3}, and the temperature profile in the plate is illustrated in Figure 16.2. One should remember that the plate is perfectly insulated at its faces and the temperatures of the borders remain constant.

After approximately 6000 s (or $\tau = 0.01$), the temperature profile is shown in Figure 16.3, and after 60,000 s (or $\tau = 0.1$) in Figure 16.4.

Of course, materials with thermal diffusivity lower than that of carbon steel would require more time to match the same dimensionless times of the above example.

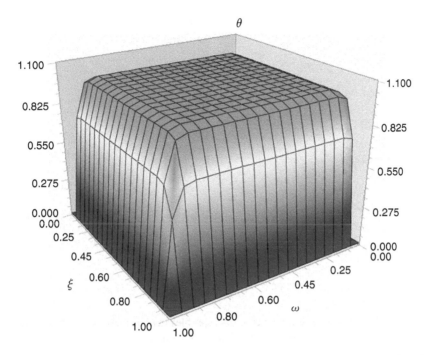

FIGURE 16.2 Dimensionless temperature (θ) profiles in a square plate ($\gamma = 1$) against the dimensionless coordinates (ω and ξ) at dimensionless time $\tau = 0.001$.

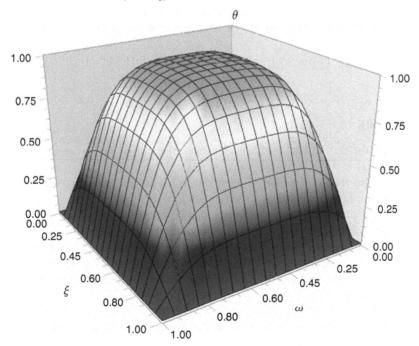

FIGURE 16.3 Dimensionless temperature (θ) profiles in a square plate ($\gamma = 1$) against the dimensionless coordinates (ω and ξ) at dimensionless time $\tau = 0.01$.

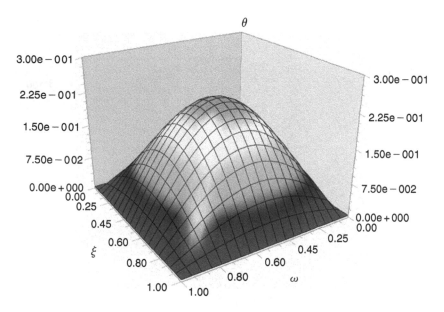

FIGURE 16.4 Dimensionless temperature (θ) profiles in a square plate ($\gamma = 1$) against the dimensionless coordinates (ω and ξ) at dimensionless time $\tau = 0.1$.

From the above solution, the reader is invited to obtain the flux of heat transfer between the plate and the environment at the borders ($x = 0$), or

$$q_x(x = 0) = -\lambda \left. \frac{\partial T}{\partial x} \right|_{x=0} \tag{16.39}$$

EXERCISES

1. Obtain the flux of heat transfer between the plate and the environment at the borders ($x = 0$).
2. Apply the relationship for the heat flux achieved in the last problem to the particular case presented in Section 16.2.2.
3. A porous plate with prescribed concentration of component A at the borders is shown in Figure 16.5. An initial given concentration profile $f(x,y)$ is imposed. The profile describes the concentration of component A in a mixture with component B. Components A and B react according to a zero-order reaction.

    It is desired to obtain the profile of component A concentration in the plate against time.

    Assume the following:

    (a) Concentration of species A at the borders ($x = 0$ and $y = 0$) remains constant and equal to ρ_{A0}.

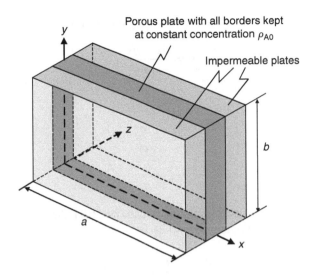

FIGURE 16.5 Porous plate with borders at constant concentration of a substance that diffuses into the plate interior.

(b) Porous plate is thin enough as well as sandwiched between two impermeable plates at the main wide faces in such a way that no mass transfer occurs in the z direction.

(c) At a given instant ($t = 0$), the composition of a mixture of components A and B is known in the plate (except at the borders) and described by function $f(x,y)$.

(d) A zero-order and irreversible reaction occurs between components A and B.

(e) Despite the mass transfer, no appreciable velocity field exists in the plate.

(f) Reaction between components A and B is neither endothermic nor exothermic and no mixture heat is developed. Therefore, the temperature in the plate remains constant.

(g) All physical properties of fluid mixture in the plate are similar and remain approximately constant.

From the above and applying Equation A.40, show that it is possible to write

$$\frac{\partial \rho_A}{\partial t} = D_e \left(\frac{\partial^2 \rho_A}{\partial x^2} + \frac{\partial^2 \rho_A}{\partial y^2} \right) - k \qquad (16.40)$$

Here D_e is the effective diffusivity of component A in the porous plate with any concentration of component B and reaction products. The reaction parameter is given in terms of concentration per unit time. In the SI system, it is represented in kmol m^{-3} s^{-1}.

4. In the above Exercise 3, use the following changes of variables:

$$\theta = \frac{\rho_A - \rho_{A_0}}{\rho_{A0}} \quad\quad (16.41)$$

$$\tau = \frac{D_e}{a^2} t \quad\quad (16.42)$$

$$\omega = \frac{x}{a} \quad\quad (16.43)$$

$$\xi = \frac{y}{a} \quad\quad (16.44)$$

To demonstrate that after applying Equations 16.41 through 16.44, it is possible to write Equation 16.40 as

$$\frac{\partial \theta}{\partial \tau} = \frac{\partial^2 \theta}{\partial \omega^2} + \frac{\partial^2 \theta}{\partial \xi^2} - \delta \quad\quad (16.45)$$

Here

$$\delta = \frac{ka^2}{\rho_{A0} D_e} \quad\quad (16.46)$$

5. In the previous problem, use the ratio between the plate dimensions as

$$\gamma = \frac{b}{a} \quad\quad (16.47)$$

The boundary conditions can be written as

$$\theta(\tau, 0, \xi) = 0, \quad 0 \le \xi \le \gamma, \quad \tau > 0 \quad\quad (16.48)$$

$$\theta(\tau, \omega, \gamma) = 0, \quad 0 \le \omega \le 1, \quad \tau > 0 \quad\quad (16.49)$$

$$\theta(\tau, 1, \xi) = 0, \quad 0 \le \xi \le \gamma, \quad \tau > 0 \quad\quad (16.50)$$

$$\theta(\tau, \omega, 0) = 0, \quad 0 \le \omega \le 1, \quad \tau > 0 \quad\quad (16.51)$$

The initial temperature profile is given by

$$\theta(0, \omega, \xi) = \varphi(\omega, \xi), \quad 0 < \omega < 1, \quad 0 < \xi < \gamma \quad\quad (16.52)$$

where

$$\varphi(\omega, \xi) = \frac{f(x,y) - \rho_{A0}}{\rho_{A0}} \quad\quad (16.53)$$

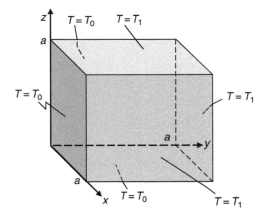

FIGURE 16.6 Solid cube with prescribed temperatures at the faces.

6. Apply the method of separation of variables, to solve the above boundary value problem

$$\theta(\tau,\omega,\xi) - G(\tau)F(\omega,\xi) \tag{16.54}$$

7. Apply Laplace transform to solve Problem 4.
8. A solid cube is illustrated in Figure 16.6. The faces are kept at prescribed temperatures, as indicated in the figure. Set the governing differential equations and boundary conditions that would allow obtaining the temperature profile inside this solid.

Appendix A
Fundamental Equations
of Transport Phenomena

A.1 INTRODUCTION

The basic equations of transport phenomena are summarized below. The reader can find the complete deductions in several texts [1–4].

The rectangular, cylindrical, and spherical coordinate systems are illustrated in Figure A.1.

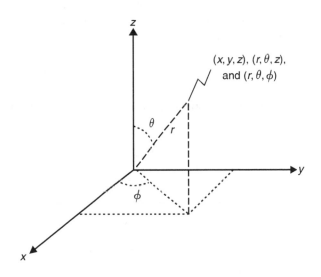

FIGURE A.1 Systems of rectangular (x, y, z), cylindrical (r, θ, z), and spherical (r, θ, ϕ) coordinates.

TABLE A.1

Equations of Mass Conservation in Several Coordinate Systems

Coordinate System	Equation	Number
Rectangular	$\dfrac{\partial \rho}{\partial t} + \dfrac{\partial \rho v_x}{\partial x} + \dfrac{\partial \rho v_y}{\partial y} + \dfrac{\partial \rho v_z}{\partial z} = 0$	A.1
Cylindrical	$\dfrac{\partial \rho}{\partial t} + \dfrac{1}{r}\dfrac{\partial \rho r v_r}{\partial r} + \dfrac{1}{r}\dfrac{\partial \rho v_\theta}{\partial \theta} + \dfrac{\partial \rho v_z}{\partial z} = 0$	A.2
Spherical	$\dfrac{\partial \rho}{\partial t} + \dfrac{1}{r^2}\dfrac{\partial \rho r^2 v_r}{\partial r} + \dfrac{1}{r \sin \theta}\dfrac{\partial \rho v_\theta \sin \theta}{\partial \theta} + \dfrac{1}{r \sin \theta}\dfrac{\partial \rho v_\phi}{\partial \phi} = 0$	A.3

A.2 GLOBAL CONTINUITY

Table A.1 shows the various global continuity equations written for each kind of coordinate system. The global continuity should be differentiated from those that can be written for each chemical species, as shown ahead in Section A.5.

A.3 MOMENTUM TRANSFER

The momentum conservation or momentum transfer equations are shown below in Table A.2. These equations are also referred to as equations of motion.

The same equations are presented in Table A.3 in the case of Newtonian fluid with constant density and viscosity.

TABLE A.2

Equations of Momentum Conservation in Rectangular Coordinates

Component	Equation	Number
x	$\rho\left(\dfrac{\partial v_x}{\partial t} + v_x\dfrac{\partial v_x}{\partial x} + v_y\dfrac{\partial v_x}{\partial y} + v_z\dfrac{\partial v_x}{\partial z}\right)$ $= -\dfrac{\partial p}{\partial x} - \left(\dfrac{\partial \tau_{xx}}{\partial x} + \dfrac{\partial \tau_{yx}}{\partial y} + \dfrac{\partial \tau_{zx}}{\partial z}\right) + \rho g_x + R_{Fx}$	A.4
y	$\rho\left(\dfrac{\partial v_y}{\partial t} + v_x\dfrac{\partial v_y}{\partial x} + v_y\dfrac{\partial v_y}{\partial y} + v_z\dfrac{\partial v_y}{\partial z}\right)$ $= -\dfrac{\partial p}{\partial y} - \left(\dfrac{\partial \tau_{xy}}{\partial x} + \dfrac{\partial \tau_{yy}}{\partial y} + \dfrac{\partial \tau_{zy}}{\partial z}\right) + \rho g_y + R_{Fy}$	A.5
z	$\rho\left(\dfrac{\partial v_z}{\partial t} + v_x\dfrac{\partial v_z}{\partial x} + v_y\dfrac{\partial v_z}{\partial y} + v_z\dfrac{\partial v_z}{\partial z}\right)$ $= -\dfrac{\partial p}{\partial z} - \left(\dfrac{\partial \tau_{xz}}{\partial x} + \dfrac{\partial \tau_{yz}}{\partial y} + \dfrac{\partial \tau_{zz}}{\partial z}\right) + \rho g_z + R_{Fz}$	A.6

Note: For Newtonian fluids, a list of stress tensors is found in Table A.8. R_F is the contribution $(+)$ or withdrawal $(-)$ of momentum due to force fields evenly imposed on the control volume at the indicated direction; units in SI system are kg m^{-2} s^{-2}.

TABLE A.3
Equations of Momentum Conservation in Rectangular Coordinates for a Newtonian Fluid with Constant Density and Viscosity

Component	Equation	Number
x	$$\rho\left(\frac{\partial v_x}{\partial t} + v_x\frac{\partial v_x}{\partial x} + v_y\frac{\partial v_x}{\partial y} + v_z\frac{\partial v_x}{\partial z}\right)$$ $$= -\frac{\partial p}{\partial x} + \mu\left(\frac{\partial^2 v_x}{\partial x^2} + \frac{\partial^2 v_x}{\partial y^2} + \frac{\partial^2 v_x}{\partial z^2}\right) + \rho g_x + R_{Fx}$$	A.7
y	$$\rho\left(\frac{\partial v_y}{\partial t} + v_x\frac{\partial v_y}{\partial x} + v_y\frac{\partial v_y}{\partial y} + v_z\frac{\partial v_y}{\partial z}\right)$$ $$= -\frac{\partial p}{\partial y} + \mu\left(\frac{\partial^2 v_y}{\partial x^2} + \frac{\partial^2 v_y}{\partial y^2} + \frac{\partial^2 v_y}{\partial z^2}\right) + \rho g_y + R_{Fy}$$	A.8
z	$$\rho\left(\frac{\partial v_z}{\partial t} + v_x\frac{\partial v_z}{\partial x} + v_y\frac{\partial v_z}{\partial y} + v_z\frac{\partial v_z}{\partial z}\right)$$ $$= -\frac{\partial p}{\partial z} + \mu\left(\frac{\partial^2 v_z}{\partial x^2} + \frac{\partial^2 v_z}{\partial y^2} + \frac{\partial^2 v_z}{\partial z^2}\right) + \rho g_z + R_{Fz}$$	A.9

Note: R_F is the contribution (+) or withdrawal (−) of momentum due to force fields evenly imposed on the control volume at the indicated direction; units in SI system are $kg\ m^{-2}\ s^{-2}$.

Tables A.4 and A.5 are equivalent to Tables A.2 and A.3, however, for the case of cylindrical coordinates.

Table A.6 presents the momentum transfer equations for spherical coordinates and Table A.7 gives the same relationships for cases of Newtonian fluids with constant density and viscosity.

TABLE A.4
Equations of Momentum Conservation in Cylindrical Coordinates

Component	Equation	Number
r	$$\rho\left(\frac{\partial v_r}{\partial t} + v_r\frac{\partial v_r}{\partial r} + \frac{v_\theta}{r}\frac{\partial v_r}{\partial \theta} - \frac{v_\theta^2}{r} + v_z\frac{\partial v_r}{\partial z}\right)$$ $$= -\frac{\partial p}{\partial r} - \left(\frac{1}{r}\frac{\partial(r\tau_{rr})}{\partial r} + \frac{1}{r}\frac{\partial \tau_{r\theta}}{\partial \theta} - \frac{\tau_{\theta\theta}}{r} + \frac{\partial \tau_{rz}}{\partial z}\right) + \rho g_r + R_{Fr}$$	A.10
θ	$$\rho\left(\frac{\partial v_\theta}{\partial t} + v_r\frac{\partial v_\theta}{\partial r} + \frac{v_\theta}{r}\frac{\partial v_\theta}{\partial \theta} + \frac{v_r v_\theta}{r} + v_z\frac{\partial v_\theta}{\partial z}\right)$$ $$= -\frac{1}{r}\frac{\partial p}{\partial \theta} - \left(\frac{1}{r^2}\frac{\partial r^2\tau_{r\theta}}{\partial r} + \frac{1}{r}\frac{\partial \tau_{\theta\theta}}{\partial \theta} + \frac{\partial \tau_{\theta z}}{\partial z}\right) + \rho g_\theta + R_{F\theta}$$	A.11
z	$$\rho\left(\frac{\partial v_z}{\partial t} + v_r\frac{\partial v_z}{\partial r} + \frac{v_\theta}{r}\frac{\partial v_z}{\partial \theta} + v_z\frac{\partial v_z}{\partial z}\right)$$ $$= -\frac{\partial p}{\partial z} - \left(\frac{1}{r}\frac{\partial r\tau_{rz}}{\partial r} + \frac{1}{r}\frac{\partial \tau_{\theta z}}{\partial \theta} + \frac{\partial \tau_{zz}}{\partial z}\right) + \rho g_z + R_{Fz}$$	A.12

Note: For Newtonian fluids, a list of stress tensors is found in Table A.8. R_F is the contribution (+) or withdrawal (−) of momentum due to force fields evenly imposed on the control volume at the indicated direction; units in SI system are $kg\ m^{-2}\ s^{-2}$.

TABLE A.5
Equations of Momentum Conservation in Cylindrical Coordinates for a Newtonian Fluid with Constant Density and Viscosity

Component	Equation	Number
r	$\rho\left(\dfrac{\partial v_r}{\partial t}+v_r\dfrac{\partial v_r}{\partial r}+\dfrac{v_\theta}{r}\dfrac{\partial v_r}{\partial \theta}-\dfrac{v_\theta^2}{r}+v_z\dfrac{\partial v_r}{\partial z}\right)$ $$=-\dfrac{\partial p}{\partial r}+\mu\left[\dfrac{\partial}{\partial r}\left(\dfrac{1}{r}\dfrac{\partial r v_r}{\partial r}\right)+\dfrac{1}{r^2}\dfrac{\partial^2 v_r}{\partial \theta^2}-\dfrac{2}{r^2}\dfrac{\partial v_\theta}{\partial \theta}+\dfrac{\partial^2 v_r}{\partial z^2}\right]+\rho g_r+R_{Fr}$$	A.13
θ	$\rho\left(\dfrac{\partial v_\theta}{\partial t}+v_r\dfrac{\partial v_\theta}{\partial r}+\dfrac{v_\theta}{r}\dfrac{\partial v_\theta}{\partial \theta}+\dfrac{v_r v_\theta}{r}+v_z\dfrac{\partial v_\theta}{\partial z}\right)$ $$=-\dfrac{1}{r}\dfrac{\partial p}{\partial \theta}+\mu\left[\dfrac{\partial}{\partial r}\left(\dfrac{1}{r}\dfrac{\partial r v_\theta}{\partial r}\right)+\dfrac{1}{r^2}\dfrac{\partial^2 v_\theta}{\partial \theta^2}+\dfrac{2}{r^2}\dfrac{\partial v_r}{\partial \theta}+\dfrac{\partial^2 v_\theta}{\partial z^2}\right]+\rho g_\theta+R_{F\theta}$$	A.14
z	$\rho\left(\dfrac{\partial v_z}{\partial t}+v_r\dfrac{\partial v_z}{\partial r}+\dfrac{v_\theta}{r}\dfrac{\partial v_z}{\partial \theta}+v_z\dfrac{\partial v_z}{\partial z}\right)$ $$=-\dfrac{\partial p}{\partial z}+\mu\left[\dfrac{1}{r}\dfrac{\partial}{\partial r}\left(r\dfrac{\partial v_z}{\partial r}\right)+\dfrac{1}{r^2}\dfrac{\partial^2 v_z}{\partial \theta^2}+\dfrac{\partial^2 v_z}{\partial z^2}\right]+\rho g_z+R_{Fz}$$	A.15

Note: R_F is the contribution (+) or withdrawal (−) of momentum due to force fields evenly imposed on the control volume at the indicated direction; units in SI system are kg m^{-2}s^{-2}.

TABLE A.6
Equations of Momentum Conservation in Spherical Coordinates

Component	Equation	Number
r	$\rho\left(\dfrac{\partial v_r}{\partial t}+v_r\dfrac{\partial v_r}{\partial r}+\dfrac{v_\theta}{r}\dfrac{\partial v_r}{\partial \theta}-\dfrac{v_\theta^2+v_\phi^2}{r}+\dfrac{v_\phi}{r\sin\theta}\dfrac{\partial v_r}{\partial \phi}\right)$ $$=-\dfrac{\partial p}{\partial r}-\left(\dfrac{1}{r^2}\dfrac{\partial r^2\tau_{rr}}{\partial r}+\dfrac{1}{r\sin\theta}\dfrac{\partial \tau_{r\theta}\sin\theta}{\partial \theta}-\dfrac{\tau_{\theta\theta}+\tau_{\phi\phi}}{r}+\dfrac{1}{r\sin\theta}\dfrac{\partial \tau_{r\phi}}{\partial \phi}\right)+\rho g_r+R_{Fr}$$	A.16
θ	$\rho\left(\dfrac{\partial v_\theta}{\partial t}+v_r\dfrac{\partial v_\theta}{\partial r}+\dfrac{v_\theta}{r}\dfrac{\partial v_\theta}{\partial \theta}+\dfrac{v_r v_\theta}{r}+\dfrac{v_\phi}{r\sin\theta}\dfrac{\partial v_\theta}{\partial \phi}-\dfrac{v_\phi^2\cot\theta}{r}\right)$ $$=-\dfrac{1}{r}\dfrac{\partial p}{\partial \theta}-\left(\dfrac{1}{r^2}\dfrac{\partial r^2\tau_{r\theta}}{\partial r}+\dfrac{1}{r\sin\theta}\dfrac{\partial \tau_{\theta\theta}\sin\theta}{\partial \theta}+\dfrac{1}{r\sin\theta}\dfrac{\partial \tau_{\theta\phi}}{\partial \phi}+\dfrac{\tau_{r\theta}}{r}-\dfrac{\cot\theta}{r}\tau_{\phi\phi}\right)+\rho g_\theta+R_{F\theta}$$	A.17
ϕ	$\rho\left(\dfrac{\partial v_\phi}{\partial t}+v_r\dfrac{\partial v_\phi}{\partial r}+\dfrac{v_\theta}{r}\dfrac{\partial v_\phi}{\partial \theta}+\dfrac{v_\phi}{r\sin\theta}\dfrac{\partial v_\phi}{\partial \phi}+\dfrac{v_\phi v_r}{r}+\dfrac{v_\phi v_\theta}{r}\cot\theta\right)$ $$=-\dfrac{1}{r\sin\theta}\dfrac{\partial p}{\partial \phi}-\left(\dfrac{1}{r^2}\dfrac{\partial r^2\tau_{r\phi}}{\partial r}+\dfrac{1}{r}\dfrac{\partial \tau_{\theta\phi}}{\partial \theta}+\dfrac{1}{r\sin\theta}\dfrac{\partial \tau_{\phi\phi}}{\partial \phi}+\dfrac{\tau_{r\phi}}{r}+\dfrac{2\cot\theta}{r}\tau_{\theta\phi}\right)$$ $$+\rho g_\phi+R_{F\phi}$$	A.18

Note: For Newtonian fluids, a list of stress tensors is found in Table A.8. R_F is the contribution (+) or withdrawal (−) of momentum due to force fields evenly imposed on the control volume at the indicated direction; units in SI system are kg m^{-2}s^{-2}.

TABLE A.7

Equations of Momentum Conservation in Spherical Coordinates for a Newtonian Fluid with Constant Density and Viscosity

Component	Equation	Number
r	$\rho\left(\dfrac{\partial v_r}{\partial t}+v_r\dfrac{\partial v_r}{\partial r}+\dfrac{v_\theta}{r}\dfrac{\partial v_r}{\partial \theta}-\dfrac{v_\theta^2+v_\phi^2}{r}+\dfrac{v_\phi}{r\sin\theta}\dfrac{\partial v_r}{\partial \phi}\right)$ $=-\dfrac{\partial p}{\partial r}+\mu\left(\nabla^2 v_r-\dfrac{2}{r^2}v_r-\dfrac{2}{r^2}\dfrac{\partial v_\theta}{\partial \theta}-\dfrac{2}{r^2}\dfrac{\partial v_\theta}{\partial \theta}-\dfrac{2}{r^2}v_\theta\cot\theta-\dfrac{2}{r^2\sin\theta}\dfrac{\partial v_\phi}{\partial \phi}\right)$ $+\rho g_r+R_{Fr}$	A.19
θ	$\rho\left(\dfrac{\partial v_\theta}{\partial t}+v_r\dfrac{\partial v_\theta}{\partial r}+\dfrac{v_\theta}{r}\dfrac{\partial v_\theta}{\partial \theta}+\dfrac{v_r v_\theta}{r}+\dfrac{v_\phi}{r\sin\theta}\dfrac{\partial v_\theta}{\partial \phi}-\dfrac{v_\phi^2\cot\theta}{r}\right)$ $=-\dfrac{1}{r}\dfrac{\partial p}{\partial \theta}+\mu\left(\nabla^2 v_\theta+\dfrac{2}{r^2}\dfrac{\partial v_r}{\partial \theta}-\dfrac{v_\theta}{r^2\sin\theta}-\dfrac{2\cos\theta}{r^2\sin^2\theta}\dfrac{\partial v_\phi}{\partial \phi}\right)+\rho g_\theta+R_{F\theta}$	A.20
ϕ	$\rho\left(\dfrac{\partial v_\phi}{\partial t}+v_r\dfrac{\partial v_\phi}{\partial r}+\dfrac{v_\theta}{r}\dfrac{\partial v_\phi}{\partial \theta}+\dfrac{v_\phi}{r\sin\theta}\dfrac{\partial v_\phi}{\partial \phi}+\dfrac{v_\phi v_r}{r}+\dfrac{v_\phi v_\theta}{r}\cot\theta\right)$ $=-\dfrac{1}{r\sin\theta}\dfrac{\partial p}{\partial \phi}+\mu\left(\nabla^2 v_\phi-\dfrac{v_\phi}{r^2\sin^2\theta}+\dfrac{2}{r^2\sin\theta}\dfrac{\partial v_r}{\partial \phi}+\dfrac{2\cos\theta}{r^2\sin^2\theta}\dfrac{\partial v_\theta}{\partial \phi}\right)$ $+\rho g_\phi+R_{F\phi}$	A.21

Note: In these equations $\nabla^2=\frac{1}{r^2}\frac{\partial}{\partial r}\left(r^2\frac{\partial}{\partial r}\right)+\frac{1}{r^2\sin\theta}\frac{\partial}{\partial \theta}\left(\sin\theta\frac{\partial}{\partial \theta}\right)+\frac{1}{r^2\sin^2\theta}\frac{\partial^2}{\partial \phi^2}$, R_F is the contribution (+) or withdrawal (−) of momentum due to force fields evenly imposed on the control volume at the indicated direction; units in SI system are kg m^{-2} s^{-2}.

The terms of shear stress shown in the above tables are provided in detail in Table A.8.

A.4 ENERGY TRANSFER

The equations describing the conservation of energy—also called equations of energy transfer—are presented below.

Table A.9 presents the components of energy fluxes for several coordinate systems.

Table A.10 presents the equations of energy conservation in terms of fluxes and involving shear stress components.

The same equations are presented in Table A.11 in the case of Newtonian fluid with constant density, viscosity, and thermal conductivity.

A.5 MASS TRANSFER

Table A.12 shows the continuity of individual chemical species in terms of mass fluxes at various coordinate systems.

Table A.13 shows the same equations, however, in terms of velocities involving fluids with constant density and diffusivity.

TABLE A.8
Components of the Shear Stress for Newtonian Fluid at Various Coordinate Systems

Rectangular

$$\tau_{xx} = -\mu\left[2\frac{\partial v_x}{\partial x} - \frac{2}{3}\left(\frac{\partial v_x}{\partial x} + \frac{\partial v_y}{\partial y} + \frac{\partial v_z}{\partial z}\right)\right] \quad \text{A.4a}$$

$$\tau_{yy} = -\mu\left[2\frac{\partial v_y}{\partial y} - \frac{2}{3}\left(\frac{\partial v_x}{\partial x} + \frac{\partial v_y}{\partial y} + \frac{\partial v_z}{\partial z}\right)\right] \quad \text{A.5a}$$

$$\tau_{zz} = -\mu\left[2\frac{\partial v_z}{\partial z} - \frac{2}{3}\left(\frac{\partial v_x}{\partial x} + \frac{\partial v_y}{\partial y} + \frac{\partial v_z}{\partial z}\right)\right] \quad \text{A.6a}$$

$$\tau_{xy} = \tau_{yx} = -\mu\left[\frac{\partial v_x}{\partial y} + \frac{\partial v_y}{\partial x}\right] \quad \text{A.7a}$$

$$\tau_{yz} = \tau_{zy} = -\mu\left[\frac{\partial v_y}{\partial z} + \frac{\partial v_z}{\partial y}\right] \quad \text{A.8a}$$

$$\tau_{xz} = \tau_{zx} = -\mu\left[\frac{\partial v_x}{\partial z} + \frac{\partial v_z}{\partial x}\right] \quad \text{A.9a}$$

Cylindrical

$$\tau_{rr} = -\mu\left\{2\frac{\partial v_r}{\partial r} - \frac{2}{3}\left[\frac{1}{r}\frac{\partial(rv_r)}{\partial r} + \frac{1}{r}\frac{\partial v_\theta}{\partial \theta} + \frac{\partial v_z}{\partial z}\right]\right\} \quad \text{A.10a}$$

$$\tau_{\theta\theta} = -\mu\left\{2\frac{\partial v_\theta}{\partial \theta} + 2\frac{v_r}{r} - \frac{2}{3}\left[\frac{1}{r}\frac{\partial(rv_r)}{\partial r} + \frac{1}{r}\frac{\partial v_\theta}{\partial \theta} + \frac{\partial v_z}{\partial z}\right]\right\} \quad \text{A.11a}$$

$$\tau_{zz} = -\mu\left\{2\frac{\partial v_z}{\partial z} - \frac{2}{3}\left[\frac{1}{r}\frac{\partial(rv_r)}{\partial r} + \frac{1}{r}\frac{\partial v_\theta}{\partial \theta} + \frac{\partial v_z}{\partial z}\right]\right\} \quad \text{A.12a}$$

$$\tau_{r\theta} = \tau_{\theta r} = -\mu\left[r\frac{\partial}{\partial r}\left(\frac{v_\theta}{r}\right) + \frac{1}{r}\frac{\partial v_r}{\partial \theta}\right] \quad \text{A.13a}$$

$$\tau_{\theta z} = \tau_{z\theta} = -\mu\left[\frac{\partial v_\theta}{\partial z} + \frac{1}{r}\frac{\partial v_z}{\partial \theta}\right] \quad \text{A.14a}$$

$$\tau_{zr} = \tau_{rz} = -\mu\left[\frac{\partial v_z}{\partial r} + \frac{\partial v_r}{\partial z}\right] \quad \text{A.15a}$$

Spherical

$$\tau_{rr} = -\mu\left\{2\frac{\partial v_r}{\partial r} - \frac{2}{3}\left[\frac{1}{r^2}\frac{\partial(r^2 v_r)}{\partial r} + \frac{1}{r\sin\theta}\frac{\partial(v_\theta\sin\theta)}{\partial\theta} + \frac{1}{r\sin\theta}\frac{\partial v_\phi}{\partial\phi}\right]\right\} \quad \text{A.16a}$$

$$\tau_{\theta\theta} = -\mu\left\{2\left[\frac{1}{r}\frac{\partial v_\theta}{\partial\theta} + \frac{v_r}{r}\right] - \frac{2}{3}\left[\frac{1}{r^2}\frac{\partial(r^2 v_r)}{\partial r} + \frac{1}{r\sin\theta}\frac{\partial(v_\theta\sin\theta)}{\partial\theta} + \frac{1}{r\sin\theta}\frac{\partial v_\phi}{\partial\phi}\right]\right\} \quad \text{A.17a}$$

$$\tau_{\phi\phi} = -\mu\left\{2\left(\frac{1}{r\sin\theta}\frac{\partial v_\phi}{\partial\phi} + \frac{v_r}{r} + \frac{v_\theta\cot\theta}{r}\right) - \frac{2}{3}\left[\frac{1}{r^2}\frac{\partial(r^2 v_r)}{\partial r} + \frac{1}{r\sin\theta}\frac{\partial(v_\theta\sin\theta)}{\partial\theta} + \frac{1}{r\sin\theta}\frac{\partial v_\phi}{\partial\phi}\right]\right\} \quad \text{A.18a}$$

$$\tau_{r\theta} = \tau_{\theta r} = -\mu\left[r\frac{\partial}{\partial r}\left(\frac{v_\theta}{r}\right) + \frac{1}{r}\frac{\partial v_r}{\partial\theta}\right] \quad \text{A.19a}$$

$$\tau_{\theta\phi} = \tau_{\phi\theta} = -\mu\left[\frac{\sin\theta}{r}\frac{\partial}{\partial\theta}\left(\frac{v_\phi}{\sin\theta}\right) + \frac{1}{r\sin\theta}\frac{\partial v_\theta}{\partial\phi}\right] \quad \text{A.20a}$$

$$\tau_{\phi r} = \tau_{r\phi} = -\mu\left[\frac{1}{r\sin\theta}\frac{\partial v_r}{\partial\phi} + r\frac{\partial}{\partial r}\left(\frac{v_\phi}{r}\right)\right] \quad \text{A.21a}$$

TABLE A.9
Components of Energy Fluxes

Coordinate System	Equations[a]	Numbers
Rectangular	$q_x = -\lambda \dfrac{\partial T}{\partial x}, \quad q_y = -\lambda \dfrac{\partial T}{\partial y}, \quad q_z = -\lambda \dfrac{\partial T}{\partial z}$	A.22, 23, 24
Cylindrical	$q_r = -\lambda \dfrac{\partial T}{\partial r}, \quad q_\theta = -\lambda \dfrac{1}{r}\dfrac{\partial T}{\partial \theta}, \quad q_z = -\lambda \dfrac{\partial T}{\partial z}$	A.25, 26, 27
Spherical	$q_r = -\lambda \dfrac{\partial T}{\partial r}, \quad q_\theta = -\lambda \dfrac{1}{r}\dfrac{\partial T}{\partial \theta}, \quad q_\phi = -\lambda \dfrac{1}{r \sin \theta}\dfrac{\partial T}{\partial \phi}$	A.28, 29, 30

[a] The above equations represent the heat fluxes only in terms of conduction.

A.5.1 IMPORTANT CORRELATIONS ON MASS TRANSFER

Like other phenomena, mass transfer problems involve fluxes and main variables, or concentrations. To help on discussion throughout the book, some of these relations are listed below:

A.5.1.1 Molar and Mass Concentrations

$$\tilde{\rho}_j = \frac{\rho_j}{M_j} \tag{A.43}$$

$$\tilde{\rho} = \sum_{j=1}^{n} \tilde{\rho}_j \tag{A.44}$$

$$\rho = \sum_{j=1}^{n} \rho_j \tag{A.45}$$

A.5.1.2 Average Molecular Mass

$$M = \frac{\rho}{\tilde{\rho}} \tag{A.46}$$

$$M = \sum_{j=1}^{n} x_j M_j \tag{A.47}$$

A.5.1.3 Sums of Molar and Mass Fractions

$$1 = \sum_{j=1}^{n} x_j = \sum_{j=1}^{n} w_j \tag{A.48}$$

TABLE A.10
Equations of Energy Conservation in Terms of Energy Fluxes and Involving Shear Stress Components

Coordinate System	Equation[a]	Number

Rectangular

$$\rho C_v \left(\frac{\partial T}{\partial t} + v_x \frac{\partial T}{\partial x} + v_y \frac{\partial T}{\partial y} + v_z \frac{\partial T}{\partial z} \right) = -\left(\frac{\partial q_x}{\partial x} + \frac{\partial q_y}{\partial y} + \frac{\partial q_z}{\partial z} \right)$$

$$- T \left(\frac{\partial p}{\partial T} \right)_\rho \left(\frac{\partial v_x}{\partial x} + \frac{\partial v_y}{\partial y} + \frac{\partial v_z}{\partial z} \right) - \left(\tau_{xx} \frac{\partial v_x}{\partial x} + \tau_{yy} \frac{\partial v_y}{\partial y} + \tau_{zz} \frac{\partial v_z}{\partial z} \right)$$

$$- \left[\tau_{xy} \left(\frac{\partial v_x}{\partial y} + \frac{\partial v_y}{\partial x} \right) + \tau_{xz} \left(\frac{\partial v_x}{\partial z} + \frac{\partial v_z}{\partial x} \right) + \tau_{yz} \left(\frac{\partial v_y}{\partial z} + \frac{\partial v_z}{\partial y} \right) \right] + R_Q$$

A.31

Cylindrical

$$\rho C_v \left(\frac{\partial T}{\partial t} + v_r \frac{\partial T}{\partial r} + \frac{v_\theta}{r} \frac{\partial T}{\partial \theta} + v_z \frac{\partial T}{\partial z} \right) = -\left[\frac{1}{r} \frac{\partial}{\partial r}(r q_r) + \frac{1}{r} \frac{\partial q_\theta}{\partial \theta} + \frac{\partial q_z}{\partial z} \right]$$

$$- T \left(\frac{\partial p}{\partial T} \right)_\rho \left[\frac{1}{r} \frac{\partial}{\partial r}(r v_r) + \frac{1}{r} \frac{\partial v_\theta}{\partial \theta} + \frac{\partial v_z}{\partial z} \right] - \left[\tau_{rr} \frac{\partial v_r}{\partial r} + \tau_{\theta\theta} \frac{1}{r} \left(\frac{\partial v_\theta}{\partial \theta} + v_r \right) + \tau_{zz} \frac{\partial v_z}{\partial z} \right]$$

$$- \left\{ \tau_{r\theta} \left[r \frac{\partial}{\partial r}\left(\frac{v_\theta}{r}\right) + \frac{1}{r} \frac{\partial v_r}{\partial \theta} \right] + \tau_{rz} \left(\frac{\partial v_r}{\partial z} + \frac{\partial v_z}{\partial r} \right) + \tau_{\theta z} \left(\frac{1}{r} \frac{\partial v_z}{\partial \theta} + \frac{\partial v_\theta}{\partial z} \right) \right\} + R_Q$$

A.32

Spherical

$$\rho C_v \left(\frac{\partial T}{\partial t} + v_r \frac{\partial T}{\partial r} + \frac{v_\theta}{r} \frac{\partial T}{\partial \theta} + \frac{v_\phi}{r\sin\theta} \frac{\partial T}{\partial \phi} \right)$$

$$= -\left[\frac{1}{r^2} \frac{\partial}{\partial r}(r^2 q_r) + \frac{1}{r\sin\theta} \frac{\partial(q_\theta \sin\theta)}{\partial \theta} + \frac{1}{r\sin\theta} \frac{\partial q_\phi}{\partial \phi} \right]$$

$$- T \left(\frac{\partial p}{\partial T} \right)_\rho \left[\frac{1}{r^2} \frac{\partial}{\partial r}(r^2 v_r) + \frac{1}{r\sin\theta} \frac{\partial(v_\theta \sin\theta)}{\partial \theta} + \frac{1}{r\sin\theta} \frac{\partial v_\phi}{\partial \phi} \right]$$

$$- \left[\tau_{rr} \frac{\partial v_r}{\partial r} + \tau_{\theta\theta} \frac{1}{r} \left(\frac{\partial v_\theta}{\partial \theta} + \frac{v_r}{r} \right) + \tau_{\phi\phi} \left(\frac{1}{r\sin\theta} \frac{\partial v_\phi}{\partial \phi} + \frac{v_r}{r} + \frac{v_\theta \cot\theta}{r} \right) \right]$$

$$- \left[\tau_{r\theta} \left(\frac{1}{r} \frac{\partial v_r}{\partial \theta} + \frac{\partial v_\theta}{\partial r} - \frac{v_\theta}{r} \right) + \tau_{r\phi} \left(\frac{1}{r\sin\theta} \frac{\partial v_r}{\partial \phi} + \frac{\partial v_\phi}{\partial r} - \frac{v_\phi}{r} \right) \right.$$

$$\left. + \tau_{\theta\phi} \left(\frac{1}{r} \frac{\partial v_\phi}{\partial \theta} + \frac{1}{r\sin\theta} \frac{\partial v_\theta}{\partial \phi} - \frac{\cot\theta}{r} v_\phi \right) \right] + R_Q$$

A.33

[a] R_Q is the rate of generation (+) or consumption (−) of energy per unit volume. Therefore, in SI system its unit is W m^{-3}.

TABLE A.11
Equations of Energy Conservation for a Newtonian Fluid with Constant Density, Thermal Conductivity, and Viscosity

Coordinate System	Equation[a]	Number
Rectangular	$\rho C_p\left(\dfrac{\partial T}{\partial t}+v_x\dfrac{\partial T}{\partial x}+v_y\dfrac{\partial T}{\partial y}+v_z\dfrac{\partial T}{\partial z}\right)$ $=\lambda\left(\dfrac{\partial^2 T}{\partial x^2}+\dfrac{\partial^2 T}{\partial y^2}+\dfrac{\partial^2 T}{\partial z^2}\right)+2\mu\left[\left(\dfrac{\partial v_x}{\partial x}\right)^2+\left(\dfrac{\partial v_y}{\partial y}\right)^2+\left(\dfrac{\partial v_z}{\partial z}\right)^2\right]$ $+\mu\left[\left(\dfrac{\partial v_x}{\partial y}+\dfrac{\partial v_y}{\partial x}\right)^2+\left(\dfrac{\partial v_x}{\partial z}+\dfrac{\partial v_z}{\partial x}\right)^2+\left(\dfrac{\partial v_y}{\partial z}+\dfrac{\partial v_z}{\partial y}\right)^2\right]+R_Q$	A.34
Cylindrical	$\rho C_p\left(\dfrac{\partial T}{\partial t}+v_r\dfrac{\partial T}{\partial r}+\dfrac{v_\theta}{r}\dfrac{\partial T}{\partial \theta}+v_z\dfrac{\partial T}{\partial z}\right)=\lambda\left[\dfrac{1}{r}\dfrac{\partial}{\partial r}\left(r\dfrac{\partial T}{\partial r}\right)+\dfrac{1}{r^2}\dfrac{\partial^2 T}{\partial \theta^2}+\dfrac{\partial^2 T}{\partial z^2}\right]$ $+2\mu\left\{\left(\dfrac{\partial v_r}{\partial r}\right)^2+\left[\dfrac{1}{r}\left(\dfrac{\partial v_\theta}{\partial \theta}+v_r\right)\right]^2+\left(\dfrac{\partial v_z}{\partial z}\right)^2\right\}$ $+\mu\left\{\left(\dfrac{\partial v_\theta}{\partial z}+\dfrac{1}{r}\dfrac{\partial v_z}{\partial \theta}\right)^2+\left(\dfrac{\partial v_z}{\partial r}+\dfrac{\partial v_r}{\partial z}\right)^2+\left[\dfrac{1}{r}\dfrac{\partial v_r}{\partial \theta}+r\dfrac{\partial}{\partial r}\left(\dfrac{v_\theta}{r}\right)\right]^2\right\}+R_Q$	A.35
Spherical	$\rho C_p\left(\dfrac{\partial T}{\partial t}+v_r\dfrac{\partial T}{\partial r}+\dfrac{v_\theta}{r}\dfrac{\partial T}{\partial \theta}+\dfrac{v_\phi}{r\sin\theta}\dfrac{\partial T}{\partial \phi}\right)$ $=\lambda\left[\dfrac{1}{r^2}\dfrac{\partial}{\partial r}\left(r^2\dfrac{\partial T}{\partial r}\right)+\dfrac{1}{r^2\sin\theta}\dfrac{\partial}{\partial \theta}\left(\sin\theta\dfrac{\partial T}{\partial \theta}\right)+\dfrac{1}{r^2\sin^2\theta}\dfrac{\partial^2 T}{\partial \phi^2}\right]$ $+2\mu\left[\left(\dfrac{\partial v_r}{\partial r}\right)^2+\left(\dfrac{1}{r}\dfrac{\partial v_\theta}{\partial \theta}+\dfrac{v_r}{r}\right)^2+\left(\dfrac{1}{r\sin\theta}\dfrac{\partial v_\phi}{\partial \phi}+\dfrac{v_r}{r}+\dfrac{v_\theta\cot\theta}{r}\right)^2\right]$ $+\mu\left\{\left[r\dfrac{\partial}{\partial r}\left(\dfrac{v_\theta}{r}\right)+\dfrac{1}{r}\dfrac{\partial v_r}{\partial \theta}\right]^2+\left[\dfrac{1}{r\sin\theta}\dfrac{\partial v_r}{\partial \phi}+r\dfrac{\partial}{\partial r}\left(\dfrac{v_\phi}{r}\right)\right]^2 \right.$ $\left. +\left[\dfrac{\sin\theta}{r}\dfrac{\partial}{\partial \theta}\left(\dfrac{v_\phi}{\sin\theta}\right)+\dfrac{1}{r\sin\theta}\dfrac{\partial v_\theta}{\partial \phi}\right]^2\right\}+R_Q$	A.36

[a] R_Q is the rate of generation (+) or consumption (−) of energy per unit of volume.

TABLE A.12
Equations for Mass Conservation of Chemical Species in Terms of Fluxes

Coordinate System	Equations[a]	Numbers
Rectangular	$\dfrac{\partial \rho_A}{\partial t} + \dfrac{\partial N_{Ax}}{\partial x} + \dfrac{\partial N_{Ay}}{\partial y} + \dfrac{\partial N_{Az}}{\partial z} = R_{M,A}$	A.37
Cylindrical	$\dfrac{\partial \rho_A}{\partial t} + \dfrac{1}{r}\dfrac{\partial}{\partial r}(rN_{Ar}) + \dfrac{1}{r}\dfrac{\partial N_{A\theta}}{\partial \theta} + \dfrac{\partial N_{Az}}{\partial z} = R_{M,A}$	A.38
Spherical	$\dfrac{\partial \rho_A}{\partial t} + \dfrac{1}{r^2}\dfrac{\partial}{\partial r}(r^2 N_{Ar})$ $+ \dfrac{1}{r\sin\theta}\dfrac{\partial}{\partial \theta}(N_{A\theta}\sin\theta) + \dfrac{1}{r\sin\theta}\dfrac{\partial N_{A\phi}}{\partial \phi} = R_{M,A}$	A.39

[a] $R_{M,A}$ is the uniform rate of generation (+) or consumption (−) per unit volume throughout the entire control volume.

TABLE A.13
Equations for Mass Conservation of Chemical Species in Terms of Velocities and for Fluid with Constant Density and Diffusivity

Coordinate System	Equation[a]	Number
Rectangular	$\dfrac{\partial \rho_A}{\partial t} + v_x\dfrac{\partial \rho_A}{\partial x} + v_y\dfrac{\partial \rho_A}{\partial y} + v_z\dfrac{\partial \rho_A}{\partial z}$ $= D_{AB}\left(\dfrac{\partial^2 \rho_A}{\partial x^2} + \dfrac{\partial^2 \rho_A}{\partial y^2} + \dfrac{\partial^2 \rho_A}{\partial z^2}\right) + R_{M,A}$	A.40
Cylindrical	$\dfrac{\partial \rho_A}{\partial t} + v_r\dfrac{\partial \rho_A}{\partial r} + \dfrac{v_\theta}{r}\dfrac{\partial \rho_A}{\partial \theta} + v_z\dfrac{\partial \rho_A}{\partial z}$ $= D_{AB}\left[\dfrac{1}{r}\dfrac{\partial}{\partial r}\left(r\dfrac{\partial \rho_A}{\partial r}\right) + \dfrac{1}{r^2}\dfrac{\partial^2 \rho_A}{\partial \theta^2} + \dfrac{\partial^2 \rho_A}{\partial z^2}\right] + R_{M,A}$	A.41
Spherical	$\dfrac{\partial \rho_A}{\partial t} + v_r\dfrac{\partial \rho_A}{\partial r} + \dfrac{v_\theta}{r}\dfrac{\partial \rho_A}{\partial \theta} + \dfrac{v_\phi}{r\sin\theta}\dfrac{\partial \rho_A}{\partial \phi}$ $= D_{AB}\left[\dfrac{1}{r^2}\dfrac{\partial}{\partial r}\left(r^2\dfrac{\partial \rho_A}{\partial r}\right) + \dfrac{1}{r^2\sin\theta}\dfrac{\partial}{\partial \theta}\left(\sin\theta\dfrac{\partial \rho_A}{\partial \theta}\right) + \dfrac{1}{r^2\sin^2\theta}\dfrac{\partial^2 \rho_A}{\partial \phi^2}\right]$ $+ R_{M,A}$	A.42

[a] $R_{M,A}$ is the uniform rate of generation (+) or consumption (−) per unit volume throughout the entire control volume.

A.5.1.4   Molar and Mass Fractions

$$x_j = \frac{\dfrac{w_j}{M_j}}{\displaystyle\sum_{j=1}^{n}\dfrac{w_j}{M_j}} \tag{A.49}$$

$$w_j = \frac{x_j M_j}{\sum\limits_{j=1}^{n} x_j M_j} \tag{A.50}$$

A.5.1.5 Mass and Molar Average Velocities

$$v = \frac{\sum\limits_{j=1}^{n} \rho_j v_j}{\sum\limits_{j=1}^{n} \rho_j} = \sum\limits_{j=1}^{n} w_j v_j \tag{A.51}$$

$$\tilde{v} = \frac{\sum\limits_{j=1}^{n} \tilde{\rho}_j v_j}{\sum\limits_{j=1}^{n} \tilde{\rho}_j} = \sum\limits_{j=1}^{n} x_j v_j \tag{A.52}$$

A.5.1.6 Binary (A and B Species) Mass and Molar Fluxes in Relation to Inertial Coordinates

$$\mathbf{N}_A = \rho_A v_A = -D_{AB}\rho \nabla w_A + w_A(\mathbf{N}_A + \mathbf{N}_B) \tag{A.53}$$

$$\tilde{\mathbf{N}}_A = \tilde{\rho}_A v_A = -D_{AB}\tilde{\rho} \nabla x_A + x_A(\tilde{\mathbf{N}}_A + \tilde{\mathbf{N}}_B) \tag{A.54}$$

A.5.1.7 Sums of Mass and Molar Fluxes

$$\sum\limits_{j=1}^{n} N_j = \rho v \tag{A.55}$$

$$\sum\limits_{j=1}^{n} \tilde{N}_j = \tilde{\rho}\tilde{v} \tag{A.56}$$

A.6 PARABOLIC, ELLIPTIC, AND HYPERBOLIC PARTIAL DIFFERENTIAL EQUATIONS

A classification of partial differential equations often found in the literature is based on the concepts of quadratic parabolic, elliptic, and hyperbolic functions.

The definitions are sometimes useful because this classification may indicate the difficulty of analytical solution for partial differential equations.

All are based on a more general form given by

$$a_1 \frac{\partial^2 \varphi}{\partial x^2} + 2a_2 \frac{\partial^2 \varphi}{\partial x \partial y} + a_3 \frac{\partial^2 \varphi}{\partial y^2} + a_4 \frac{\partial \varphi}{\partial x} + a_5 \frac{\partial \varphi}{\partial y} + a_6 = 0 \tag{A.57}$$

Here a_i $(i = 1, 2, \ldots, 6)$ are constants.

Let a matrix M be given by

$$M = \begin{bmatrix} a_1 & a_2 \\ a_2 & a_3 \end{bmatrix} \tag{A.58}$$

The differential equation (Equation A.57) would be

- Parabolic, if $M = 0$
- Elliptic, if $M > 0$
- Hyperbolic, if $M < 0$

The following are examples of parabolic equations, which are related to conductive transfers:

$$\frac{\partial^2 T}{\partial x^2} + \frac{\partial^2 T}{\partial y^2} = 0 \tag{A.59}$$

Usually, solving parabolic differential equations are made simpler by using analytical methods, and these differential equations also allow the application of several methods.

Despite some usefulness of such classification, it is not applied in this work because the difficulty of finding an analytical solution for partial differential equations is not just a matter of the equation in itself, but depends also on the kind of boundary condition or conditions involved in the problem.

REFERENCES

1. Bird, R.B., Stewart, W.E., and Lightfoot, E.N., *Transport Phenomena*, John Wiley & Sons, New York, 1960.
2. Slattery, J.C., *Momentum, Energy, and Mass Transfer in Continua*, Robert E. Krieger, New York, 1978.
3. Brodkey, R.S., *The Phenomena of Fluid Motions*, Dover, New York, 1967.
4. Luikov, A.V., *Heat and Mass Transfer*, Mir, Moscow, 1980.

Appendix B
Fundamental Aspects
of Ordinary Differential
Equations

B.1 INTRODUCTION

A significant fraction of the differential equations derived from situations in transport phenomena lead to ordinary differential equations. Fortunately, most of these equations fall into the range of linear equations, and the main methods to solve these are presented here.

This is an overview on the subject, and more and deeper material as well as demonstrations of theorems stated here can be found throughout the literature [1–6].

A linear differential equation is an equation that can be written in the following form:

$$a_n(x)\frac{d^n y}{dx^n} + a_{n-1}(x)\frac{d^{n-1} y}{dx^{n-1}} + \cdots + a_1(x)\frac{dy}{dx} + a_0(x)y = b(x) \qquad (B.1)$$

where $a_i(x)$ and $b(x)$ are functions of the independent variable x. On the other hand, an example of nonlinear equation is

$$\frac{dy}{dx} + xy^2 = 2x$$

Equation B.1 is called homogeneous if the function $b(x)$ is identically zero; otherwise, it is heterogeneous.

B.2 FIRST-ORDER LINEAR EQUATIONS

This is the most simple form of differential equations and is given by

$$\frac{dy}{dx} + a(x)\,y = b(x) \qquad (B.2)$$

B.2.1 SEPARABLE EQUATIONS

If in Equation B.2, the function $a(x)$ is identical to zero, the equation is called separable and a simple integration is enough to arrive at the solution, or

$$y = \int b(x)\,dx + C \tag{B.3}$$

Here C is the integration constant, which can be determined by the boundary condition of a physical problem.

Another class of separable differential equation arises from homogeneous equations, or

$$\frac{dy}{dx} + a(x)\,y = 0 \tag{B.4}$$

Such equations can be integrated to give

$$\ln y = -\int a(x)\,dx + \ln C \tag{B.5}$$

or

$$y(x) = C \exp\left(\int a(x)\,dx \right) \tag{B.6}$$

B.2.2 VARIATION OF PARAMETERS

Variation of parameters is a method applicable to a wide range of differential equations. It starts by assuming the solution as the product of two functions in the form

$$y(x) = y_1(x)\,y_2(x) \tag{B.7}$$

This does not imply on loss of generality. However, an additional degree of freedom is introduced.

Combining Equations B.7 and B.2, one gets

$$y_1' y_2 + y_1 y_2' + a y_1 y_2 = b \tag{B.8}$$

In turn, this can be written as

$$y_1(y_2' + a y_2) + y_1' y_2 = b \tag{B.9}$$

The additionally introduced degree of freedom is now used to impose that

$$y_2' + ay_2 = 0 \tag{B.10}$$

This is a separable differential equation, with solution given by

$$y_2(x) = \exp\left(-\int a(x)\, dx\right) \tag{B.11}$$

Notice that the integration constant was arbitrarily set as zero. Owing to the additional degree of freedom introduced before, this is possible without any loss of generality. Besides, the boundary condition will be satisfied through the introduction of a constant in the following steps.

Equations B.9 through B.11 lead to

$$y_1(x) = \int \frac{b(x)}{y_2(x)}\, dx + C = \int b(x)\, e^{\int a(x)dx}\, dx + C \tag{B.12}$$

The final solution is obtained using the above last two equations in Equation B.7. Constant C would be obtained through the application of boundary condition.

B.3 FIRST-ORDER NONLINEAR EQUATIONS

There are various methods for solving nonlinear differential equations, some being approximated methods such as the method of weighted residuals, which is described in Appendix E. For the time being, a simple but effective method is shown. It is called Picard's method. It is an iterative procedure as explained below.

B.3.1 PICARD'S METHOD

Consider a first-order equation in the general form

$$\frac{dy}{dx} = f(x,y) \tag{B.13}$$

This form includes a wide range of nonlinear equations. If a boundary condition

$$y(x_0) = y_0 \tag{B.14}$$

is imposed, the integration of Equation B.13 leads to

$$y(x) = y_0 + \int_{x_0}^{x} f[z, y(z)]\, dz \tag{B.15}$$

Now, if $y(z)$ is replaced by its value at boundary condition or the one given by Equation B.14, an approximation is obtained and is named $y_1(x)$, which is given by

$$y_1(x) = y_0 + \int_{x_0}^{x} f(z, y_0)\, dz \qquad (B.16)$$

The next step is to assume the new level of approximation as

$$y_2(x) = y_0 + \int_{x_0}^{x} f[z, y_1(z)]\, dz \qquad (B.17)$$

The nth step would be

$$y_n(x) = y_0 + \int_{x_0}^{x} f[z, y_{n-1}(z)]\, dz \qquad (B.18)$$

Of course, the deviation between the computed values using the approximation $y_n(x)$ and the respective exact solution $y(x)$ would depend on the number of steps or n.

B.3.2 EXISTENCE AND UNIQUENESS OF SOLUTIONS

A first-order differential boundary problem given by Equations B.13 and B.14 has at least one solution in the domain (x,y) in which $f(x,y)$ is continuous and finite.

In addition to the previous properties that some boundary condition problems have a single solution exists in the above region if the derivative $\dfrac{\partial f}{\partial y}$ is also finite.

The reader can find rigorous demonstrations in proper texts [1].

B.4   SECOND-ORDER LINEAR EQUATIONS

A second-order linear equation can be put in the form

$$\frac{d^2 y}{dx^2} + a_1(x)\frac{dy}{dx} + a_0(x)\, y = b(x) \qquad (B.19)$$

The easiest or preferred method for the solution of this class of equation depends on the forms of functions $a_1(x)$, $a_0(x)$, and $b(x)$. As long as these functions are continuous and bounded, Equation B.19 allows two independent or particular solutions: $y_1(x)$ and $y_2(x)$. A general solution is written as a linear combination of these two solutions, or

$$y(x) = C_1 y_1(x) + C_2 y_2(x) \qquad (B.20)$$

Therefore, an infinite number of solutions are possible and any physical problem should involve two and only two boundary conditions to allow a particular solution.

B.4.1 SECOND-ORDER LINEAR EQUATION WITH CONSTANT COEFFICIENTS

In this case since a_0 and a_1 are constants, the method of solving Equation B.19 would depend on whether the function $b(x)$ is identical to zero or not.

B.4.1.1 Homogeneous Linear Second Order

If in Equation B.19, the function $b(x)$ is identical to zero, the differential equation becomes homogeneous, or

$$\frac{d^2y}{dx^2} + a_1\frac{dy}{dx} + a_0y = 0 \tag{B.21}$$

The solutions can be sought in the form

$$y(x) = e^{Bx} \tag{B.22}$$

Here B is a constant that can be determined by Equations B.21 and B.22, or

$$B^2 + a_1(x)B + a_0(x) = 0 \tag{B.23}$$

Equation B.23 is a second-order polynomial with the following solutions:

$$B_1 = \frac{-a_1 + \sqrt{a_1^2 - 4a_0}}{2} \quad \text{and} \quad B_2 = \frac{-a_1 - \sqrt{a_1^2 - 4a_0}}{2} \tag{B.24a,b}$$

Therefore, the general solution for Equation B.19 is a linear combination of the two solutions, or

$$y(x) = C_1e^{B_1x} + C_2e^{B_2x} \tag{B.25}$$

Of course, this assumes two independent solutions. However, this is not always the case. To cover all possibilities, three cases are to be addressed:

1. Two real and distinct roots, which occur if $a_1^2 - 4a_0 > 0$
2. A real double root, which occurs if $a_1^2 - 4a_0 = 0$
3. Complex conjugate roots, which occur if $a_1^2 - 4a_0 < 0$

B.4.1.1.1 Case 1: Two Distinct Real Roots
The general solution is given by Equation B.25, and the constants C_1 and C_2 can be determined if two boundary conditions are available.

B.4.1.1.2 Case 2: A Real Double Root

In this case the following occurs:

$$B_1 = B_2 = -\frac{a_1}{2} = B \tag{B.26}$$

This would lead to one particular solution

$$y_1(x) = e^{Bx} \tag{B.27}$$

The other solution can be found through the application of variation of parameters, also called reduction of order. With no loss of generality, a second solution is set in the form

$$y_2(x) = u(x)\,y_1(x) \tag{B.28}$$

Using Equation B.28 in Equation B.21, one reaches at

$$u''y_1 + 2u'y_1' + uy_1'' + a_1(uy_1' + u'y_1) + ua_0y_1 = 0 \tag{B.29}$$

Collecting the terms, the above equation becomes

$$u(y_1'' + a_1y_1' + a_0y_1) + u''y_1 + u'(2y_1' + a_1y_1) = 0 \tag{B.30}$$

It should be noticed that the content inside the first parenthesis is equal to zero because y_1 is a solution of Equation B.21. In addition, the one inside the second parenthesis is also zero. This can be verified by using Equations B.26 and B.27. As y_1 is not identical to zero, Equation B.30 leads to

$$\frac{d^2u}{dx^2} = 0 \tag{B.31}$$

Therefore, with a particular solution given by

$$u(x) = x \tag{B.32}$$

Finally, the general solution becomes

$$y(x) = e^{Bx}(C_1 + C_2x) \tag{B.33}$$

The integration constants would be determined by boundary conditions.

B.4.1.1.3 Case 3: Conjugate Complex Roots

In this case, the roots are

$$B_1 = -\frac{a_1}{2} + iC \text{ and } B_2 = -\frac{a_1}{2} - iC \tag{B.34a,b}$$

Here

$$C = \frac{\sqrt{4a_0 - a_1^2}}{2} \tag{B.35}$$

Therefore, the particular solutions of Equation B.25 are

$$y_1(x) = e^{-\frac{a_1}{2}x} e^{iCx} \quad \text{and} \quad y_2(x) = e^{-\frac{a_1}{2}x} e^{-iCx} \tag{B.36}$$

If the Euler formula

$$e^{-ai} = \cos a + i \sin a \tag{B.37}$$

is applied, the above solutions can be combined to give a general solution as

$$y(x) = e^{-\frac{a_1}{2}x} [C_1 \sin(Cx) + C_2 \cos(Cx)] \tag{B.38}$$

The details are left as an exercise to the reader. Again, C_1 and C_2 can be determined with two boundary conditions.

B.4.1.2   Nonhomogeneous Linear Second Order

A nonhomogeneous linear second-order differential equation with constant coefficients is written as

$$\frac{d^2y}{dx^2} + a_1 \frac{dy}{dx} + a_0 y = b(x) \tag{B.39}$$

The solution of this class of differential equations requires the following principles:

1. Difference between the two solutions of Equation B.39 is the solution of Equation B.21.
2. Sum of the solution for Equation B.39 and the solution for Equation B.21 is also the solution for Equation B.39.

   Therefore, a general solution for Equation B.39 can be written as

$$y(x) = y_h(x) + y_p(x) \tag{B.40}$$

Here y_h is the general solution of homogeneous equation (Equation B.21), and y_p is a particular solution for the heterogeneous equation (Equation B.39).

   We have already seen how to obtain the solutions y_1 and y_2 for the homogeneous equation. Let us now concentrate on how a particular solution y_p for the heterogeneous equation can be found.

TABLE B.1
Some Functions for the Method of Undetermined Coefficients

$b(x)$	$y_p(x)$
Ae^{ax}	Ce^{ax}
Ax^n $(n=0,1,\ldots)$	$A_{n+2}x^{n+2} + A_{n+1}x^{n+1} + A_n x^n + A_{n-1}x^{n-1} + \cdots + A_1 x + A_0$
$A\cos Bx$	$A_1 \cos Bx + A_2 \sin Bx$
$A\sin Bx$	$A_1 \cos Bx + A_2 \sin Bx$
$Ae^{ax}\cos Cx$	$e^{ax}(A_1 \cos Cx + A_2 \sin Cx)$
$Ae^{ax}\sin Cx$	$e^{ax}(A_1 \cos Cx + A_2 \sin Cx)$
$A\cosh Bx$	$A_1 \cosh Bx + A_2 \sinh Bx$
$A\sinh Bx$	$A_1 \cosh Bx + A_2 \sinh Bx$
$Ae^{ax}\cosh Cx$	$e^{ax}(A_1 \cosh Cx + A_2 \sinh Cx)$
$Ae^{ax}\sinh Cx$	$e^{ax}(A_1 \cosh Cx + A_2 \sinh Cx)$

For this, the method of undetermined coefficients may be applied. It is simply based on the following rules:

1. Basic rule: If the function $b(x)$ at Equation B.39 is one of the functions in the first column of Table B.1, the particular function $y_p(x)$ is given in the form shown in the respective second column. The coefficients for y_p are found by applying this solution into Equation B.39.
2. Modification rule: If the above choice of function y_p coincides with a solution of the homogenous equation (Equation B.21), the true solution can be obtained by multiplying the function y_p by the independent variable x. In case where y_p coincides with the double root, multiply it by x^2. Again, the coefficients for y_p are found by applying it in Equation B.39.
3. Sum rule: If $b(x)$ is a sum of the functions listed in the first column of Table B.1, the function y_p will be given by the sum of respective functions in the second column. Again, the coefficients for y_p are found by applying it in Equation B.39.

B.4.1.2.1 Wronskian

As seen before, a general solution of Equation B.19 is obtained by linear combinations of two independent solutions $y_1(x)$ and $y_2(x)$. These two functions form a basis for the solutions of Equation B.39. To ensure the independence and linearity of these two functions, the following identity:

$$A_1 y_1(x) + A_2 y_2(x) = 0$$

should be possible if and only if

$$A_1 = A_2 = 0$$

Alternatively, the independence of y_1 and y_2 can be tested using Wronskian, which is a determinant defined by

$$W(y_1, y_2) = \begin{vmatrix} y_1 & y_2 \\ y_1' & y_2' \end{vmatrix} \tag{B.41}$$

If the above Wronskian is identical to zero at any value x within the domain for which $a_1(x)$ and $a_0(x)$ are continuous, then the functions $y_1(x)$ and $y_2(x)$ are linearly dependent.

B.4.1.3 How to Obtain the Particular Solution for Any Case

In Section B.4.1.2, a practical set of rules was shown to allow obtaining a particular solution for a nonhomogeneous Equation B.39. On the other hand, the functions in Table B.1 are limited to a range of possibilities. The method shown below overcomes such limitations. In addition, the method is applicable to all linear equations, such as Equation B.19. Therefore, it is not bounded to equations with constant coefficients.

Consider $y_1(x)$ and $y_2(x)$ as a basis for solutions for the homogeneous differential equation

$$\frac{d^2 y}{dx^2} + a_1(x) \frac{dy}{dx} + a_0(x) y = 0 \tag{B.42}$$

The particular solution $y_p(x)$ for Equation B.19 can be obtained by

$$y_p(x) = -y_1(x) \int \frac{y_2(x) b(x)}{W} \, dx + y_2(x) \int \frac{y_1(x) b(x)}{W} \, dx \tag{B.43}$$

and the Wronskian computed by Equation B.41.

This method assumes that functions $a_1(x)$, $a_2(x)$, and $b(x)$ are analytical (or continuous, at least by parts) within the domain of solution.

B.4.2 EULER–CAUCHY EQUATION

Few particular differential equations with variable coefficients allow simple solutions. Among them, the Euler–Cauchy equation is given by

$$x^2 \frac{d^2 y}{dx^2} + a_1 x \frac{dy}{dx} + a_0 y = 0 \tag{B.44}$$

The solutions can be found in the form

$$y(x) = x^B \tag{B.45}$$

where B can be determined after combining Equations B.45 and B.44 to arrive at the following second-order polynomial:

$$B^2 + (a_1 - 1)B + a_0 = 0 \tag{B.46}$$

The roots are

$$B_1 = -\frac{a_1 - 1}{2} + \frac{\sqrt{(a_1 - 1)^2 - 4a_0}}{2} \quad \text{and} \quad B_2 = -\frac{a_1 - 1}{2} - \frac{\sqrt{(a_1 - 1)^2 - 4a_0}}{2} \tag{B.47}$$

Again, three possible cases are discussed ahead.

B.4.2.1 Different Real Roots

In this case, the general solution for Equation B.44 is

$$y(x) = C_1 x^{B_1} + C_2 x^{B_2} \tag{B.48}$$

where B_1 and B_2 are the distinct real roots and constants C_1 and C_2 may be now determined from boundary conditions.

B.4.2.2 Double Root

Similar to the case of homogeneous constant coefficient equations, the variation of parameters can be applied to lead to the following general solution:

$$y(x) = (C_1 + C_2 \ln x) x^{(1-a_1)/2} \tag{B.49}$$

The details are left as an exercise to the reader.

B.4.2.3 Complex Conjugate Roots

In this case the roots are

$$B_1 = -\frac{a_1 - 1}{2} + i \frac{\sqrt{-(a_1 - 1)^2 + 4a_0}}{2} = -\frac{a_1 - 1}{2} + iC \quad \text{and} \quad B_2 = -\frac{a_1 - 1}{2} - iC \tag{B.50}$$

The general solution is written as

$$y(x) = [C_1 \cos (C \ln x) + C_2 \sin (C \ln x)] x^{(1-a_1)/2} \tag{B.51}$$

Again, the details are left as an exercise to the reader.

B.4.3 Existence and Uniqueness

Like the first-order differential equations, the existence and uniqueness of second-order equations should be assured.

The following is a homogeneous linear differential equation:

$$\frac{d^2 y}{dx^2} + a_1(x)\frac{dy}{dx} + a_0(x)\,y = 0 \tag{B.52}$$

with initial boundary conditions given by

$$y(x_0) = C_0 \quad \text{and} \quad y'(x_1) = C_1 \tag{B.53a,b}$$

Equation B.52 has a unique solution in a given domain if functions $a_1(x)$ and $a_0(x)$ are continuous in that domain [1].

REFERENCES

1. Ince, E.L., *Ordinary Differential Equations*, Dover, New York, 1956.
2. Kreyszig, E., *Advanced Engineering Mathematics*, 7th ed., John Wiley & Sons, New York, 1993.
3. Hildebrand, F.B., *Advanced Calculus for Applications*, 2nd ed., Prentice-Hall, Englewood Cliffs, New Jersey, 1976.
4. Bajpai, A.C., Mustoe, L.R., and Walker, D., *Advanced Engineering Mathematics*, John Wiley & Sons, Chichester, 1977.
5. Pipes, L.A. and Harvill, L.R., *Applied Mathematics for Engineers and Physicists*, McGraw-Hill Kogakusha, Tokyo, 1970.
6. Sokolnikoff, I.S. and Redheffer, R.M., *Mathematics of Physics and Modern Engineering*, 2nd ed., McGraw-Hill, New York, 1966.

Appendix C
Method of Power Series
and Special Functions

C.1 INTRODUCTION

The method of power series is useful to enlarge the range of boundary value problems solvable by analytical procedures. The present text introduces these methods as well as a few classical functions, which are derived from solutions of differential equations. Among these are the Bessel functions.

The basic principle behind the method comes from the fact that all continuous or continuous by parts functions, with finite derivatives, can be represented by series. Since functions representing parameters of heat, mass, and momentum transfers fall in this category, it is fair to expect that a method seeking solutions in the form of series could be useful to solve transport phenomena problems. More details on the material presented here can be found in the literature [1–6].

C.2 POWER SERIES

Most of the common functions can be represented by series. Examples are

$$e^x = 1 + x + \frac{x^2}{2!} + \frac{x^3}{3!} + \cdots = \sum_{m=0}^{\infty} \frac{x^m}{m!} \tag{C.1}$$

$$\frac{1}{1-x} = 1 + x + x^2 + \cdots = \sum_{m=0}^{\infty} x^m \tag{C.2}$$

$$\sin x = x - \frac{x^3}{3!} + \frac{x^5}{5!} - \cdots = \sum_{m=0}^{\infty} (-1)^m \frac{x^{2m+1}}{(2m+1)!} \tag{C.3}$$

$$\cos x = 1 - \frac{x^2}{2!} + \frac{x^4}{4!} - \cdots = \sum_{m=0}^{\infty} (-1)^m \frac{x^{2m}}{(2m)!} \tag{C.4}$$

$$\sinh x = x + \frac{x^3}{3!} + \frac{x^5}{5!} + \cdots = \sum_{m=0}^{\infty} \frac{x^{2m+1}}{(2m+1)!} \tag{C.5}$$

$$\cosh x = 1 + \frac{x^2}{2!} + \frac{x^4}{4!} + \cdots = \sum_{m=0}^{\infty} \frac{x^{2m}}{(2m)!} \tag{C.6}$$

Consider D, the dominium at which a function $f(x)$ is continuous or piecewise continuous and in which $f(x)$ has finite derivatives. Taylor series generalizes the possibility of writing such a function by

$$f(x) = f(a) + \frac{df}{dx}\bigg|_{x=a} (x-a) + \frac{d^2 f}{dx^2}\bigg|_{x=a} \frac{(x-a)^2}{2!}$$
$$+ \cdots + \frac{d^{(m-1)} f}{dx^{(m-1)}}\bigg|_{x=a} \frac{(x-a)^{(m-1)}}{(m-1)!} + R_m \tag{C.7}$$

Here, a is a value within the dominium D and R_m is the residue or remainder after the term m of series given by Equation C.7, which is given by

$$R_m = \frac{d^m f}{dx^m}\bigg|_{x=\xi} \frac{(x-a)^m}{m!} \tag{C.8}$$

Parameter ξ can be taken as any value between x and a. If a series is convergent, the remainder should satisfy the condition

$$\lim_{x \to \infty} R_m = 0 \tag{C.9}$$

In this case, the function $f(x)$ is called an analytical function with center at a.

If, in the above equations $a = 0$, the series is also called the MacLaurin series.

C.2.1　An Example

Consider the following differential simple equation:

$$\frac{d^2 y}{dx^2} - y = 0 \tag{C.10}$$

If it were assumed that the solution $y(x)$ is an analytical function, it would be represented by a convergent power series such as

$$y(x) = \sum_{m=0}^{\infty} a_m x^m \tag{C.11}$$

Therefore

$$\frac{dy}{dx} = \sum_{m=0}^{\infty} m a_m x^{m-1} \tag{C.12}$$

and

$$\frac{d^2y}{dx^2} = \sum_{m=0}^{\infty} m(m-1) \, a_m x^{m-2} \tag{C.13}$$

Equation C.10 becomes

$$\sum_{m=0}^{\infty} m(m-1) \, a_m x^{m-2} - \sum_{m=0}^{\infty} a_m x^m = 0 \tag{C.14}$$

Now, to satisfy the above equation, it is necessary to obtain the relationships between the coefficients of terms with same power of x, or

- From those with x^0 (terms of x^{-2} or x^{-1} cannot be obtained because the coefficient of the first series is zero)

$$2a_2 - a_0 = 0 \quad \text{or} \quad a_2 = \frac{a_0}{2} \tag{C.15}$$

- From those with x^1

$$6a_3 - a_1 = 0 \quad \text{or} \quad a_3 = \frac{a_1}{6} \tag{C.16}$$

- From those with x^2

$$12a_4 - a_2 = 0 \quad \text{or} \quad a_4 = \frac{a_2}{12} = \frac{a_0}{4!} \tag{C.17}$$

- From those with x^3

$$20a_5 - a_3 = 0 \quad \text{or} \quad a_5 = \frac{a_3}{20} = \frac{a_1}{5!} \tag{C.18}$$

As seen, all even terms can be represented by fractions of a_0 and the odd by fractions of a_1, or

$$a_{2n} = \frac{a_0}{(2n)!} \quad \text{and} \quad a_{2n+1} = \frac{a_1}{(2n+1)!} \quad n = 0, 1, 2, \ldots \tag{C.19}$$

Therefore, the series given by Equation C.11, or solution, can be written as

$$y(x) = a_0 \sum_{n=0}^{\infty} \frac{x^{2n}}{(2n)!} + a_1 \sum_{n=0}^{\infty} \frac{x^{2n+1}}{(2n+1)!} \tag{C.20}$$

Using Equations C.5 and C.6, the above can also be expressed as

$$y(x) = a_0 \cosh x + a_1 \sinh x \tag{C.21}$$

This is the general form of solution for Equation C.10.

Despite the apparent simplicity of the method, the search for correlations between the coefficients may become strenuous. A simpler procedure to search for series coefficients is the Frobenius method, as described below.

C.3 FROBENIUS METHOD

This is a specific method to solve second-order linear differential equations with nonconstant coefficients, which can be written as

$$\frac{d^2 y}{dx^2} + \frac{b(x)}{x}\frac{dy}{dx} + \frac{c(x)}{x^2}y = 0 \tag{C.22}$$

where $b(x)$ and $c(x)$ are analytical functions of x with center at zero. The above equation has at least one solution in the form

$$y(x) = x^r \sum_{m=0}^{\infty} a_m x^m \tag{C.23}$$

Here, r is any number (real or complex) such that a_0 is not equal to zero. Equation C.23 also has another solution, which may be similar to it (with different r or a_m coefficients) or may contain a logarithmic term. This is given in detail ahead.

C.3.1 INDICIAL EQUATION

Equation C.22 can be written as

$$x^2\frac{d^2 y}{dx^2} + xb(x)\frac{dy}{dx} + c(x)y = 0 \tag{C.24}$$

Since $b(x)$ and $c(x)$ are analytical, these can be written as the following power series:

$$b(x) = \sum_{m=0}^{\infty} b_m x^m = b_0 + b_1 x + b_2 x^2 + \cdots \tag{C.25}$$

$$c(x) = \sum_{m=0}^{\infty} c_m x^m = c_0 + c_1 x + c_2 x^2 + \cdots \tag{C.26}$$

Now, the derivatives for function $y(x)$ are

$$\frac{dy}{dx} = \sum_{m=0}^{\infty} (m+r)a_m x^{m+r-1} \tag{C.27}$$

and

$$\frac{d^2y}{dx^2} = \sum_{m=0}^{\infty} (m+r-1)a_m x^{m+r-2} \tag{C.28}$$

Applying Equation C.23 along with Equations C.25 through C.28 into Equation C.24, collecting the terms with x^r, and assuming a_0 different from zero, one gets

$$r^2 + r(b_0 - 1) + c_0 = 0 \tag{C.29}$$

The above second-order polynomial given by Equation C.29 is called an indicial equation. Roots r_1 and r_2 determine the forms of the solutions, as proposed by Equation C.23. The following are the possibilities for the two roots:

1. Distinct and not differing by an integer. In this case, the two solutions are given by Equation C.23, each with the proper value for the root. Nonetheless, if the roots differ by an integer, one solution would be just a linear combination of the other.
2. Double root. In this case, in addition to solution $y_1(x)$ given by Equation C.23, the other would be

$$y_2(x) = Ay_1(x)\ln x + x^r \sum_{m=0}^{\infty} B_m x^m \tag{C.30}$$

3. Distinct roots differing by an integer. In this case, if the first solution y_1 is written as in Equation C.23 with the root r_1, the second would be

$$y_2(x) = Ay_1(x)\ln x + x^{r_2} \sum_{m=0}^{\infty} B_m x^m \tag{C.31}$$

In Equations C.30 and C.31, the coefficients A and B_m ($m = 0$, 1, 2, ...) are determined by replacing these solutions into the original differential equation and equating the coefficient of terms with the same power of x.

Several examples are presented throughout the text.

C.3.2 EXAMPLE

Consider the Euler–Cauchy equation:

$$x^2y'' - xy' + x = 0$$

Therefore

$$b(x) = -1, \quad c(x) = 1$$

From this one obtains $b_0 = -1$ and $c_0 = 1$. The indicial equation becomes

$$r^2 - 2r + 1 = 0$$

Therefore, a relation that provides double root equal to 1 is obtained. One solution would be given as

$$y_1(x) = \sum_{m=0}^{\infty} a_m x^{m+1}$$

Using this in the original equation, it is possible to write

$$\sum_{m=0}^{\infty} m(m+1)a_m x^{m+1} - \sum_{m=0}^{\infty} (m+1)a_m x^{m+1} + \sum_{m=0}^{\infty} a_m x^{m+1} = 0$$

This leads to

$$\sum_{m=0}^{\infty} m^2 a_m x^{m+1} = 0$$

Apart from the trivial solution, the only possibility is $m = 0$. Therefore, the first solution is

$$y_1(x) = x$$

Following the method, the other solution would be

$$y_2(x) = x \ln x + x \sum_{m=1}^{\infty} A_m x^m$$

Each part of the above solution is a solution of the original differential equation as well. In order to determine coefficients A_m, one should follow the same procedure as before. Of course, this would again give $m = 0$. Hence, the final solution is

$$y(x) = C_1 x + C_2 x \ln x$$

C.4 BESSEL EQUATIONS

There are some types of differential equations that frequently appear in problems related to transport phenomena. One such type is known as Bessel equations, which includes a whole class of equations. Among these, one branch is given by

$$x^2 \frac{d^2 y}{dx^2} + x \frac{dy}{dx} + (x^2 - a^2)y = 0 \tag{C.32}$$

Here, parameter a is a real number.

Following the Frobenius method, the proposed solution is given by Equation C.23 and the indicial equation would be

$$(r - a)(r + a) = 0 \qquad \text{(C.33)}$$

Therefore, the roots would be a and $-a$.

C.4.1 BESSEL FUNCTIONS OF THE FIRST KIND

In the cases, when parameter a is a real number and not an integer, the solution would be

$$y(x) = b_1 J_a(x) + b_2 J_{-a}(x) \qquad \text{(C.34)}$$

Here, b_1 and b_2 are constants, which can be determined from the boundary conditions. The functions $J(x)$ are known as Bessel functions of the first kind. The method described above allows one to obtain the coefficients of the series representing the solution as

$$J_a(x) = x^a \sum_{m=0}^{\infty} \frac{(-1)^m x^{2m}}{2^{2m+a} m! \; \Gamma(a + m + 1)} \qquad \text{(C.35)}$$

The gamma function is given by

$$\Gamma(b) = \int_0^{\infty} e^{-z} z^{a-1} \, dz \qquad \text{(C.36)}$$

It is possible to show that if n is an integer the following can be written:

$$\Gamma(n + 1) = n! \qquad \text{(C.37)}$$

Therefore, the gamma function can be seen as a generalization of the factorial operation to any number.

C.4.2 BESSEL FUNCTIONS OF THE SECOND KIND

The next two possibilities for roots of the indicial equation (Equation C.33) are

1. Zero or double root
2. Integer numbers

In any of these cases, the general solution is

$$y(x) = b_1 J_a(x) + b_2 Y_a(x) \qquad \text{(C.38)}$$

$Y(x)$ is called the Bessel function of the second kind and is written as

$$Y_n(x) = \frac{2}{n} J_n(x) \left(\ln \frac{x}{2} + \gamma \right) + \frac{x^n}{\pi} \sum_{m=0}^{\infty} \frac{(-1)^{m-1}(h_m + h_{m+n})}{2^{2m-n} m!(m+n)!} x^{2m}$$

$$- \frac{x^{-n}}{\pi} \sum_{m=0}^{n-1} \frac{(n-m-1)!}{2^{2m-n} m!} x^{2m} \tag{C.39}$$

Here, the parameter γ is the Euler constant, which is approximately equal to 0.57721566490. Function h_m is given by

$$h_m = 1 + \frac{1}{2} + \cdots + \frac{1}{m} \tag{C.40}$$

In addition, the second-order function can be written as

$$Y_a(x) = \frac{1}{\sin a\pi} [J_a(x) \cos a\pi - J_{-a}(x)] \tag{C.41}$$

C.4.3   MODIFIED BESSEL FUNCTIONS

If the differential equation (Equation C.32) is slightly changed to

$$x^2 \frac{d^2 y}{dx^2} + x \frac{dy}{dx} - (x^2 + a^2)y = 0 \tag{C.42}$$

this is called a modified Bessel equation and would generate the following general solution:

$$y(x) = b_1 I_a(x) + b_2 K_a(x) \tag{C.43}$$

$I(x)$ and $K(x)$ are the modified Bessel functions of the first and second kind, respectively. These are given by

$$I_a(x) = x^a \sum_{m=0}^{\infty} \frac{x^{2m+a}}{2^{2m+a} m! \, \Gamma(a+m+1)} \tag{C.44}$$

and

$$K_a(x) = \frac{\pi}{2 \sin a\pi} [I_{-a}(x) - I_a(x)] \tag{C.45}$$

C.4.4 Selected Relations

There are several relations between Bessel functions as well as commonly found functions [1–7], which are listed below:

$$J_{1/2}(x) = (\sin x)\sqrt{\frac{2}{\pi x}} \tag{C.46}$$

$$J_{-1/2}(x) = (\cos x)\sqrt{\frac{2}{\pi x}} \tag{C.47}$$

$$I_{1/2}(x) = (\sinh x)\sqrt{\frac{2}{\pi x}} \tag{C.48}$$

$$I_{-1/2}(x) = (\cosh x)\sqrt{\frac{2}{\pi x}} \tag{C.49}$$

$$J_{n+1}(x) = \frac{2n}{x}J_n(x) - J_{n-1}(x) \tag{C.50}$$

$$\frac{dJ_n(x)}{dx} = \frac{1}{2}[J_{n-1}(x) - J_{n+1}(x)] \tag{C.51}$$

$$\frac{d}{dx}[x^n J_n(x)] = x^n J_{n-1}(x) \tag{C.52}$$

$$\frac{d}{dx}[x^{-n} J_n(x)] = -x^{-n} J_{n+1}(x) \tag{C.53}$$

$$I_{-n}(x) = I_n(x) \tag{C.54}$$

$$K_{-n}(x) = K_n(x) \tag{C.55}$$

$$J_0'(x) = -J_1(x) \tag{C.56}$$

$$I_0'(x) = I_1(x) \tag{C.57}$$

$$K_0'(x) = -K_1(x) \tag{C.58}$$

$$J_1(0) = 0 \tag{C.59}$$

$$\lim_{x \to 0} Y_1 = -\infty \tag{C.60}$$

$$I_1(0) = 0 \tag{C.61}$$

$$\lim_{x \to 0} K_1(x) = \infty \tag{C.62}$$

Equations C.50 through C.53 are also applicable to functions $Y_n(x)$.

C.5   LEGENDRE FUNCTIONS

Likewise, Legendre functions arise from typical differential equations with the same name and are given by

$$(1 - x^2)\frac{d^2y}{dx^2} - 2x\frac{dy}{dx} + \left[a(a+1) - \frac{b^2}{1-x^2}\right]y = 0 \qquad \text{(C.63)}$$

The general solutions for this equation can also be found by applying the Frobenius method and are represented by $P_a^b(z)$ and $Q_a^b(z)$, which are the associated Legendre functions of the first and second kinds, respectively, or

$$y(x) = C_1 P_a^b(x) + C_2 P_a^b(x) \qquad \text{(C.64)}$$

Representations of these functions in series can be found in any manual on mathematics [7].

REFERENCES

1. Kreyszig, E., *Advanced Engineering Mathematics*, 7th ed., John Wiley & Sons, New York, 1993.
2. Hildebrand, F.B., *Advanced Calculus for Applications*, 2nd ed., Prentice-Hall, Englewood Cliffs, New Jersey, 1976.
3. Bajpai, A.C., Mustoe, L.R., and Walker, D., *Advanced Engineering Mathematics*, John Wiley & Sons, Chichester, 1977.
4. Ince, E.L., *Ordinary Differential Equations*, Dover, New York, 1956.
5. Pipes, L.A. and Harvill, L.R., *Applied Mathematics for Engineers and Physicists*, McGraw-Hill Kogakusha, Tokyo, 1970.
6. Sokolnikoff, I.S. and Redheffer, R.M., *Mathematics of Physics and Modern Engineering*, McGraw-Hill, 2nd ed., New York, 1966.
7. Abramovitz, M. and Stegun, I.A., *Handbook of Mathematical Functions*, Dover, New York, 1972.

Appendix D
Laplace Transform

D.1 INTRODUCTION

Differentiation and integration, as well as other mathematical procedures, are just operations that transform the values of involved variables into others. The Laplace transform is one of them. The branch of applied mathematics involving Laplace, Fourier, and other transforms is called operational mathematics and is extremely useful for solving a wide range of boundary-value problems involving ordinary and partial differential equations.

The objective of the material presented here is to help solve differential equations related to transport phenomena. Those interested in deeper studies of operational mathematics can find excellent texts in the literature [1–9].

D.2 BASIC PRINCIPLE

Consider an analytical function $f(t)$, i.e., continuous or at least piecewise continuous for $t \geq 0$. The following operation can be performed:

$$L\{f(t)\} = \int_0^\infty e^{-st}f(t) \, dt = F(s) \tag{D.1}$$

As seen, such operation transforms the function $f(t)$ into another function $F(s)$. In other words, function $F(s)$ is the "image" of $f(t)$ in a new space s.

For now, let s be a real and positive variable. Additionally, the existence of Laplace transform requires the following condition:

$$\lim_{s \to \infty} f(s) = 0 \tag{D.1a}$$

As an example, the transform of function $f(t) = 1$ is given by

$$L\{1\} = \int_0^\infty e^{-st} \, dt = -\frac{1}{s}e^{-st}\Big|_0^\infty = \frac{1}{s} \tag{D.2}$$

Equation D.2 was only possible after the condition of real and positive s. The transform also follows condition given by Equation D.1a.

Other examples are

$$L\{t\} = \frac{1}{s^2} \tag{D.3}$$

$$L\{e^{at}\} = \frac{1}{s-a}, \quad s > a \tag{D.4}$$

and

$$L\{\sin(at)\} = \frac{a}{s^2 + a^2} \tag{D.5}$$

D.3 A FEW BASIC PROPERTIES

Here, a few useful properties of the Laplace transform are shown. The importance and examples of their utilization are shown ahead.

D.3.1 FUNDAMENTAL OPERATIONS

As an integral, the transform is a linear operator, i.e.,

$$L\{af(t) + bg(t)\} = a L\{f(t)\} + b L\{g(t)\} \tag{D.6}$$

The inverse operation, or inverse transform, is indicated by

$$L^{-1}\{f(s)\} = F(t) \tag{D.7}$$

It is also a linear operator, or

$$L^{-1}\{af(s) + b g(s)\} = a L^{-1}\{f(s)\} + b L^{-1}\{g(s)\} \tag{D.8}$$

D.3.2 TRANSFORMS OF DERIVATIVES

The transform of a derivative can be obtained by applying integration by parts, or

$$L\{f'(t)\} = \int_0^\infty e^{-st} f'(t) dt = s \int_0^\infty e^{-st} f(t) dt + e^{-st} f(t)\big|_0^\infty \tag{D.9}$$

Therefore

$$L\{f'(t)\} = sF(s) - f(0) \tag{D.10}$$

The process can be repeated to obtain transforms of higher-order derivatives. For instance

$$L\{f''(t)\} = sL\{f'(t)\} - f'(0) = s[sL\{f(t)\} - f(0)] - f'(0)$$
$$= s^2 F(s) - sf(0) - f'(0)$$

(D.11)

In general

$$L\{f^{(n)}(t)\} = s^n F(s) - s^{n-1} f(0) - s^{n-2} f'(0) - \cdots - f^{(n-1)}(0)$$ (D.12)

D.3.3 TRANSFORMS OF INTEGRALS

It is possible to show that

$$L\left\{\int_0^t f(\tau)\ d\tau\right\} = \frac{1}{s} L\{f(t)\}$$

(D.13)

The proof for Equation D.13 is left as an exercise to the reader.

D.3.4 s-SHIFTING

It is possible to write

$$L\{e^{at} f(t)\} = F(s - a)$$

(D.14)

Hence, if function $f(t)$ is multiplied by an exponential of t, the result is a shift of the transform in the s-axis.

The proof of Equation D.14 is also left as an exercise.

D.3.5 t-SHIFTING

In order to understand the t-shifting property, it is necessary to introduce the unit step function, which is defined by

$$u(t - a) = \begin{cases} 0 & \text{if } t < a \\ 1 & \text{if } t > a \end{cases}$$

(D.15)

This is illustrated in Figure D.1.

From this definition, it is possible to write a new function $g(t)$ defined as follows:

$$g(t) = f(t - a)\,u(t - a) = \begin{cases} 0 & \text{if } t < a \\ f(t - a) & \text{if } t > a \end{cases}$$

(D.16)

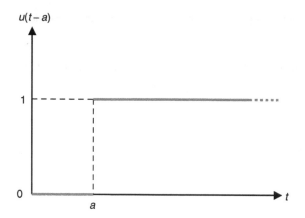

FIGURE D.1 Step function.

With this, and if $F(s)$ is the Laplace transform of $f(t)$, then the Laplace transform of $g(t)$ becomes

$$L\{g(t)\} = L\{f(t-a)\,u(t-a)\} = e^{-as}F(s) \qquad (D.17)$$

As seen, the last right-hand side represents the transform of $f(t)$ after a shift in the t-axis.

On this basis it is possible to write

$$L\{u(t-a)\} = \frac{e^{-as}}{s} \qquad (D.18)$$

Again, the proof for these relations is left as an exercise.

D.3.6 IMPULSE FUNCTION

Once the step function has been introduced, it is opportune to do the same for the impulse function (Figure D.2), which is defined as

$$I_a(h,t) = \begin{cases} \dfrac{1}{h} & \text{if } a < t < a+h \\ 0 & \text{if } t < a \text{ or } t > a+h \end{cases} \qquad (D.19)$$

The Laplace transform of such a function can be found by writing

$$I_a(h,t) = \frac{u(t-a) - u[t-(a+h)]}{h} \qquad (D.20)$$

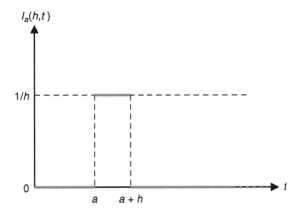

FIGURE D.2 The impulse function.

Applying the transform, this equation leads to

$$L\{I_a(h,t)\} = e^{-sa} \frac{1 - e^{-sh}}{hs}, \quad a \geq 0, \quad h > 0, \quad s > 0 \qquad \text{(D.21)}$$

Now, the following should be noticed:

$$\lim_{h \to 0} I_a(h,t) = 0, \quad \text{when} \quad t \neq a \qquad \text{(D.22)}$$

$$\lim_{h \to 0} I_a(h,t) = \infty, \quad \text{when} \quad t = a \qquad \text{(D.22a)}$$

This limit simulates the behavior of what is called the Dirac symbol $\delta_a(t)$, which is written as

$$\delta_a(t) = \lim_{h \to 0} I_a(h,t) = \begin{cases} 0, & t \neq a \\ \infty, & t = a \end{cases} \qquad \text{(D.22b)}$$

However, it is important to notice that the Dirac symbol is not a true function.
It is interesting to notice that

$$\lim_{h \to 0} \int_{-\infty}^{\infty} I_a(h,t) \, dt = 1 \qquad \text{(D.23)}$$

Therefore, from Equation D.21

$$\lim_{h \to 0} L\{I_a(h,t)\} = e^{-sa}, \quad a \geq 0, \quad h > 0, \quad s > 0 \qquad \text{(D.24)}$$

Thus, it is possible to write

$$\int_{-\infty}^{\infty} \delta_a(t) \, dt = 1 \tag{D.25}$$

and

$$L\{\delta_a(t)\} = e^{-sa} \tag{D.26}$$

One should notice that

$$\int_{-\infty}^{\infty} \delta_a(t) f(t) \, dt = f(a) \tag{D.27}$$

D.3.7 CONVOLUTION

The convolution is an operation involving two functions $f(t)$ and $g(t)$ and are defined by

$$f(t)*g(t) = \int_{0}^{t} f(\tau) g(t - \tau) \, d\tau \tag{D.28}$$

Now, if

$$F(s) = L\{f(t)\} \tag{D.29}$$

and

$$G(s) = L\{g(t)\} \tag{D.30}$$

the following can be shown:

$$L^{-1}\{F(s) \, G(s)\} = f(t)*g(t) \tag{D.31}$$

This property is very powerful, as it is shown in the main text.

D.3.8 DERIVATIVES OF TRANSFORMS

The derivative of the transform is given by

$$\frac{dF(s)}{ds} = F'(s) = \frac{d}{ds} \int_{0}^{\infty} e^{-st} f(t) \, dt = \int_{0}^{\infty} e^{-st}(-t) f(t) \, dt = L\{-tf(t)\} \tag{D.32}$$

This can be generalized for multiple differentiations to give

$$F^{(n)}(s) = L\{(-t)^n f(t)\} \tag{D.33}$$

From these properties, it is possible to write

$$L\{t^n f(t)\} = (-1)^n \frac{d^n}{ds^n} L\{f(t)\} = (-1)^n F^{(n)}(s) \tag{D.34}$$

With this, you see, for instance, that

$$L\{t^2 f'(t)\} = \frac{d^2}{ds^2} [sF(s) - f(0)] = sF''(s) + 2F'(s) \tag{D.35}$$

and

$$L\{t f''(t)\} = -\frac{d}{ds} [s^2 F(s) - sf(0) - f'(0)] = -s^2 F'(s) - 2sF(s) + f(0) \tag{D.36}$$

D.4   TRANSFORM OF A PARTIAL DERIVATIVE

Consider the transform of a partial derivative on an independent variable other than the transformed one (t), or

$$L\left\{\frac{\partial f(t,x)}{\partial x}\right\} = \int_0^\infty \frac{\partial f(t,x)}{\partial x} e^{-st} \, dt \tag{D.37}$$

Since x and t are independent variables, the above integral can be written as

$$\frac{\partial}{\partial x} \int_0^\infty f(t,x) e^{-st} \, dt$$

Of course, this is possible because x is an independent variable from t. Therefore,

$$L\left\{\frac{\partial f(t,x)}{\partial x}\right\} = \frac{d}{dx} L\{f(t,x)\} \tag{D.38}$$

This result is very useful for applications of transform to partial differential equations.

D.5   TABLES OF LAPLACE TRANSFORMS

Tables (Tables D.1 through D.5) that summarize the main properties of Laplace transforms as well as several transforms are presented below.

TABLE D.1
Summary of Table of Laplace Transforms

$f(t)$	$F(s) = L\{f(t)\}$
$f(t)$	$\int\limits_0^\infty e^{-st} f(t)\, dt$
$a\,f(t) + b\,g(t)$ (a and b are constants)	$aF(s) + bG(s)$
$\dfrac{df(t)}{dt}$	$sF(s) - f(0)$
$\dfrac{d^2 f(t)}{dt^2}$	$s^2 F(s) - sf(0) - \dfrac{df(t)}{dt}\bigg|_{t=0}$
$\dfrac{d^n f(t)}{dt^n}$	$s^n F(s) - s^{n-1} f(0) - s^{n-2}\dfrac{df(t)}{dt}\bigg|_{t=0} - \cdots - \dfrac{d^{n-1} f(t)}{dt^{n-1}}\bigg|_{t=0}$
$\int\limits_0^t f(x)\, dx$	$\dfrac{F(s)}{s}$
$t\,f(t)$	$-\dfrac{dF(s)}{ds}$
$t^n f(t)$	$(-1)^n \dfrac{d^n F(s)}{ds^n}$
$\dfrac{f(t)}{t}$	$\int\limits_s^\infty F(x)\, dx$
$e^{at} f(t)$	$F(s - a)$
$f(t-a)\,u(t-a) = \begin{cases} 0 & \text{if } t < a \\ f(t-a) & \text{if } t > a \end{cases}$	$e^{-as} F(s)$
$f(t)*g(t) = \int\limits_0^t f(\tau)g(t-\tau)\, d\tau$	$F(s)\,G(s)$
$\dfrac{\partial f(t,x)}{\partial x}$	$\dfrac{dF(s,x)}{dx}$

TABLE D.2
Laplace Transforms Involving Powers of t

$f(t)$	$F(s) = L\{f(t)\}$
1	$\dfrac{1}{s}$
t	$\dfrac{1}{s^2}$
$t^n, \quad n = 1, 2, 3, \ldots$	$\dfrac{n!}{s^{n+1}}$
$\dfrac{1}{\sqrt{t}}$	$\sqrt{\dfrac{1}{s}}$
$t^a, \quad a = \text{any real number}$	$\dfrac{\Gamma(a+1)}{s^{a+1}}$

TABLE D.3
Laplace Transforms Involving Exponentials of t

$f(t)$	$F(s) = L\{f(t)\}$
e^{at}	$\dfrac{1}{s-a}$
te^{at}	$\dfrac{1}{(s-a)^2}$
$t^{n-1}e^{at}, \quad n=1, 2, 3, \ldots$	$\dfrac{(n-1)!}{(s-a)^n}$
$e^{at} - e^{bt}$	$\dfrac{a-b}{(s-a)(s-b)}$
$ae^{at} - be^{bt}$	$\dfrac{s(a-b)}{(s-a)(s-b)}$
$\dfrac{e^{at}}{(a-b)(a-c)} + \dfrac{e^{bt}}{(b-a)(b-c)} + \dfrac{e^{ct}}{(c-a)(c-b)}$	$\dfrac{1}{(s-a)(s-b)(s-c)}$
$\dfrac{e^{at}}{(a-b)(a-c)(a-d)} + \dfrac{e^{bt}}{(b-a)(b-c)(b-d)}$ $+ \dfrac{e^{ct}}{(c-a)(c-b)(c-d)}$	$\dfrac{1}{(s-a)(s-b)(s-c)(s-d)}$
$e^{at} - at - 1$	$\dfrac{a^2}{s^2(s-a)}$
$e^{at} - 1 - at - \dfrac{a^2 t^2}{2}$	$\dfrac{a^3}{s^3(s-a)}$
$\dfrac{(1+2at)e^{at}}{\sqrt{t}}$	$\dfrac{s\sqrt{\pi}}{(s-a)\sqrt{s-a}}$
$\dfrac{e^{-bt} - e^{-at}}{t}$	$\ln\left(\dfrac{s+a}{s-a}\right)$
$\dfrac{e^{-bt} - e^{-at}}{t^{3/2}}$	$\dfrac{2(b-a)\sqrt{\pi}}{\sqrt{s+a}+\sqrt{s+b}}$
$\dfrac{\exp(-2\sqrt{at})}{\sqrt{\pi t}}$	$\dfrac{e^{a/s}}{\sqrt{s}}\operatorname{erfc}\left(\sqrt{\dfrac{a}{s}}\right)$
$\exp(-a^2 t^2)$	$\dfrac{\sqrt{\pi}\exp\left(\dfrac{s^2}{4a^2}\right)}{2a}\operatorname{erfc}\left(\dfrac{s}{2a}\right)$
$\dfrac{\exp\left(-\dfrac{a^2}{4t}\right)}{\sqrt{\pi t}}$	$\dfrac{\exp\left(-\dfrac{a}{s}\right)}{\sqrt{s}}$

TABLE D.4
Laplace Transforms Involving Trigonometric and Hyperbolic Trigonometric Terms on t

$f(t)$	$F(s) = L\{f(t)\}$
$\sin(at)$	$\dfrac{a}{s^2 + a^2}$
$\cos(at)$	$\dfrac{s}{s^2 + a^2}$
$\sinh(at)$	$\dfrac{a}{s^2 - a^2}$
$\cosh(at)$	$\dfrac{s}{s^2 - a^2}$
$t\sin(at)$	$\dfrac{2as}{(s^2 + a^2)^2}$
$t\cos(at)$	$\dfrac{s^2 - a^2}{(s^2 + a^2)^2}$
$\dfrac{\sin(at)}{t}$	$\tan^{-1}\left(\dfrac{a}{s}\right)$
$at - \sin(at)$	$\dfrac{a^3}{s^2(s^2 + a^2)}$
$1 - \cos(at)$	$\dfrac{a^2}{s(s^2 + a^2)}$
$\sin(at) - at\cos(at)$	$\dfrac{2a^3}{(s^2 + a^2)^2}$
$\sin(at) + at\cos(at)$	$\dfrac{2as^2}{(s^2 + a^2)^2}$
$\cos(at) - \cos(bt)$	$\dfrac{(b^2 - a^2)\,s}{(s^2 + a^2)(s^2 + b^2)}$
$\sin(at)\sinh(at)$	$\dfrac{2a^2s}{s^4 + 4a^4}$
$\cos(at)\cosh(at)$	$\dfrac{s^3}{s^4 + 4a^4}$
$\sinh(at) - \sin(at)$	$\dfrac{2a^3}{s^4 - a^4}$
$\sinh(at) + \sin(at)$	$\dfrac{2as^2}{s^4 - a^4}$
$\cosh(at) - \cos(at)$	$\dfrac{2a^2s}{s^4 - a^4}$
$\cosh(at) + \cos(at)$	$\dfrac{2s^3}{s^4 - a^4}$

TABLE D.4 (continued)
Laplace Transforms Involving Trigonometric
and Hyperbolic Trigonometric Terms on t

$f(t)$	$F(s) = L\{f(t)\}$
$\sin(at)\cosh(at) - \cos(at)\sinh(at)$	$\dfrac{4a^3}{s^4 + 4a^4}$
$\dfrac{1 - \cos(at)}{t}$	$\dfrac{1}{2}\ln\dfrac{s^2 + a^2}{s^2}$
$\dfrac{1 - \cosh(at)}{t}$	$\dfrac{1}{2}\ln\dfrac{s^2 - a^2}{s^2}$
$\dfrac{\cos(at) - \cos(bt)}{t}$	$\dfrac{1}{2}\ln\dfrac{s^2 + a^2}{s^2 + b^2}$
$t^2 \sin(at)$	$\dfrac{2a(3s^2 - a^2)}{(s^2 + a^2)^3}$
$t^2 \cos(at)$	$\dfrac{2(s^3 - 3a^2 s)}{(s^2 + a^2)^3}$
$t^3 \cos(at)$	$\dfrac{6(s^4 - 6a^2 s^2 + a^4)}{(s^2 + a^2)^4}$
$t^3 \sin(at)$	$\dfrac{24a(s^3 - a^2 s)}{(s^2 + a^2)^4}$
$t^2 \sinh(at)$	$\dfrac{2a(3s^2 + a^2)}{(s^2 - a^2)^3}$
$t^2 \cosh(at)$	$\dfrac{2(s^3 + 3a^2 s)}{(s^2 - a^2)^3}$
$t^3 \sinh(at)$	$\dfrac{24a(s^3 + a^2 s)}{(s^2 - a^2)^4}$
$t^3 \cosh(at)$	$\dfrac{6(s^4 + 6a^2 s^2 + a^4)}{(s^2 - a^2)^4}$
$t \sin(at) - at^2 \cos(at)$	$\dfrac{8a^3 s}{(s^2 + a^2)^3}$
$3t \sin(at) + at^2 \cos(at)$	$\dfrac{8as^3}{(s^2 + a^2)^3}$
$(1 + a^2 t^2) \sin(at) - at \cos(at)$	$\dfrac{8a^3 s^2}{(s^2 + a^2)^3}$
$(3 - a^2 t^2) \sin(at) - 3at \cos(at)$	$\dfrac{8a^5}{(s^2 + a^2)^3}$
$(3 - a^2 t^2) \sin(at) + 5at \cos(at)$	$\dfrac{8as^4}{(s^2 + a^2)^3}$

(continued)

TABLE D.4 (continued)
Laplace Transforms Involving Trigonometric and
Hyperbolic Trigonometric Terms on t

$f(t)$	$F(s) = L\{f(t)\}$
$(8 - a^2t^2)\cos(at) - 7at\sin(at)$	$\dfrac{8s^5}{(s^2 + a^2)^3}$
$3t\sinh(at) + at^2\cosh(at)$	$\dfrac{8as^3}{(s^2 - a^2)^3}$
$at\cosh(at) + (at^2 - 1)\sinh(at)$	$\dfrac{8a^3s^2}{(s^2 - a^2)^3}$
$(3 + a^2t^2)\sinh(at) - 3at\cosh(at)$	$\dfrac{8a^5}{(s^2 - a^2)^3}$
$(3 + a^2t^2)\sinh(at) + 5at\cosh(at)$	$\dfrac{8as^4}{(s^2 - a^2)^3}$
$(8 + a^2t^2)\cosh(at) + 7at\sinh(at)$	$\dfrac{8s^5}{(s^2 - a^2)^3}$
$\sin(2\sqrt{at})$	$\sqrt{\dfrac{\pi a}{s}}\,\dfrac{e^{-a/s}}{s}$
$\dfrac{\sin(2a\sqrt{t})}{\sqrt{t}}$	$\pi\,\mathrm{erf}\left(\dfrac{a}{\sqrt{s}}\right)$
$\dfrac{\cos(2\sqrt{at})}{\sqrt{t}}$	$\sqrt{\dfrac{\pi}{s}}\,e^{-a/s}$
$\sinh(2\sqrt{at})$	$\sqrt{\dfrac{\pi a}{s}}\,\dfrac{e^{a/s}}{s}$
$\dfrac{\cosh(2\sqrt{at})}{\sqrt{t}}$	$\sqrt{\dfrac{\pi}{s}}\,e^{a/s}$
$|\sin(at)|$	$\dfrac{a\coth\left(\dfrac{\pi s}{2a}\right)}{s^2 + a^2}$
$|\sin(t)| + \sin(t)$	$\dfrac{2}{(s^2 + 1)(1 - e^{-\pi s})}$
$e^{at}\sin(bt)$	$\dfrac{b}{(s - a)^2 + b^2}$
$e^{at}\cos(bt)$	$\dfrac{s - a}{(s - a)^2 + b^2}$
$\dfrac{e^{at/2}}{3a^2}\left[\sqrt{3}\sin\left(\dfrac{\sqrt{3}at}{2}\right) - \cos\left(\dfrac{\sqrt{3}at}{2}\right) + e^{-3at/2}\right]$	$\dfrac{1}{s^3 + a^3}$
$\dfrac{e^{at/2}}{3a^2}\left[\sqrt{3}\sin\left(\dfrac{\sqrt{3}at}{2}\right) + \cos\left(\dfrac{\sqrt{3}at}{2}\right) - e^{-3at/2}\right]$	$\dfrac{s}{s^3 + a^3}$

TABLE D.4 (continued)
Laplace Transforms Involving Trigonometric and
Hyperbolic Trigonometric Terms on t

$f(t)$	$F(s) = L\{f(t)\}$
$\dfrac{e^{-at/2}}{3a^2}\left[-\sqrt{3}\sin\left(\dfrac{\sqrt{3}at}{2}\right) - \cos\left(\dfrac{\sqrt{3}at}{2}\right) + e^{-3at/2}\right]$	$\dfrac{1}{s^3 - a^3}$
$\dfrac{e^{-at/2}}{3a^2}\left[\sqrt{3}\sin\left(\dfrac{\sqrt{3}at}{2}\right) - \cos\left(\dfrac{\sqrt{3}at}{2}\right) + e^{-3at/2}\right]$	$\dfrac{s}{s^3 - a^3}$
$\dfrac{1}{3}\left[2e^{at/2}\cos\left(\dfrac{\sqrt{3}at}{2}\right) + e^{-at}\right]$	$\dfrac{s^2}{s^3 + a^3}$
$\dfrac{1}{3}\left[2e^{-at/2}\cos\left(\dfrac{\sqrt{3}at}{2}\right) + e^{at}\right]$	$\dfrac{s^2}{s^3 - a^3}$
$\dfrac{x}{a} + \dfrac{2}{\pi}\sum_{j=1}^{\infty}\dfrac{(-1)^j}{j}\sin\left(\dfrac{j\pi x}{a}\right)\cos\left(\dfrac{j\pi t}{a}\right)$	$\dfrac{\sinh(xs)}{s\sinh(as)}$
$\dfrac{t}{a} + \dfrac{2}{\pi}\sum_{j=1}^{\infty}\dfrac{(-1)^j}{j}\cos\left(\dfrac{j\pi x}{a}\right)\sin\left(\dfrac{j\pi t}{a}\right)$	$\dfrac{\cosh(xs)}{s\sinh(as)}$
$\dfrac{xt}{a} + \dfrac{2a}{\pi^2}\sum_{j=1}^{\infty}\dfrac{(-1)^j}{j^2}\sin\left(\dfrac{j\pi x}{a}\right)\sin\left(\dfrac{j\pi t}{a}\right)$	$\dfrac{\sinh(xs)}{s^2\sinh(as)}$
$\dfrac{t^2}{2a} + \dfrac{2a}{\pi^2}\sum_{j=1}^{\infty}\dfrac{(-1)^j}{j^2}\cos\left(\dfrac{j\pi x}{a}\right)\left[1 - \cos\left(\dfrac{j\pi t}{a}\right)\right]$	$\dfrac{\cosh(xs)}{s^2\sinh(as)}$
$\dfrac{4}{\pi}\sum_{j=1}^{\infty}\dfrac{(-1)^j}{2j-1}\sin\left(\dfrac{(2j-1)\pi x}{2a}\right)\cos\left(\dfrac{(2j-1)\pi t}{2a}\right)$	$\dfrac{\sinh(xs)}{s\cosh(as)}$
$1 + \dfrac{4}{\pi}\sum_{j=1}^{\infty}\dfrac{(-1)^j}{2j-1}\cos\left(\dfrac{(2j-1)\pi x}{2a}\right)\sin\left(\dfrac{(2j-1)\pi t}{2a}\right)$	$\dfrac{\cosh(xs)}{s\cosh(as)}$
$x + \dfrac{8a}{\pi^2}\sum_{j=1}^{\infty}\dfrac{(-1)^j}{(2j-1)^2}\sin\left(\dfrac{(2j-1)\pi x}{2a}\right)\cos\left(\dfrac{(2j-1)\pi t}{2a}\right)$	$\dfrac{\sinh(xs)}{s^2\cosh(as)}$
$t + \dfrac{8a}{\pi^2}\sum_{j=1}^{\infty}\dfrac{(-1)^j}{(2j-1)^2}\cos\left(\dfrac{(2j-1)\pi x}{2a}\right)\sin\left(\dfrac{(2j-1)\pi t}{2a}\right)$	$\dfrac{\cosh(xs)}{s^2\cosh(as)}$
$\dfrac{t^2 + x^2 - a^2}{2} -$ $\dfrac{16a^2}{\pi^3}\sum_{j=1}^{\infty}\dfrac{(-1)^j}{(2j-1)^3}\cos\left(\dfrac{(2j-1)\pi x}{2a}\right)\sin\left(\dfrac{(2j-1)\pi t}{2a}\right)$	$\dfrac{\cosh(xs)}{s^3\cosh(as)}$
$\dfrac{2\pi}{a^2}\sum_{j=1}^{\infty}j(-1)^j\exp\left(-\dfrac{j^2\pi^2 t}{a^2}\right)\sin\left(\dfrac{j\pi x}{a}\right)$	$\dfrac{\sinh(x\sqrt{s})}{\sinh(a\sqrt{s})}$
$\dfrac{x}{a} + \dfrac{2}{\pi}\sum_{j=1}^{\infty}\dfrac{(-1)^j}{j}\exp\left(-\dfrac{j^2\pi^2 t}{a^2}\right)\sin\left(\dfrac{j\pi x}{2a}\right)$	$\dfrac{\sinh(x\sqrt{s})}{s\sinh(a\sqrt{s})}$

(continued)

TABLE D.4 (continued)
Laplace Transforms Involving Trigonometric and Hyperbolic Trigonometric Terms on t

$f(t)$	$F(s) = L\{f(t)\}$
$\dfrac{1}{a} + \dfrac{2}{a}\displaystyle\sum_{j=1}^{\infty}(-1)^j \exp\left(-\dfrac{j^2\pi^2 t}{a^2}\right)\cos\left(\dfrac{j\pi x}{2a}\right)$	$\dfrac{\cosh(x\sqrt{s})}{\sqrt{s}\sinh(a\sqrt{s})}$
$\dfrac{2}{a}\displaystyle\sum_{j=1}^{\infty}(-1)^{j-1}\exp\left(-\dfrac{(2j-1)^2\pi^2 t}{4a^2}\right)\sin\left(\dfrac{(2j-1)\pi x}{2a}\right)$	$\dfrac{\sinh(x\sqrt{s})}{\sqrt{s}\cosh(a\sqrt{s})}$
$\dfrac{\pi}{a^2}\displaystyle\sum_{j=1}^{\infty}(-1)^{j-1}(2j-1)\exp\left(-\dfrac{(2j-1)^2\pi^2 t}{4a^2}\right)$ $\cos\left(\dfrac{(2j-1)\pi x}{2a}\right)$	$\dfrac{\cosh(x\sqrt{s})}{\cosh(a\sqrt{s})}$
$\dfrac{xt}{a} + \dfrac{2a^2}{\pi^3}\displaystyle\sum_{j=1}^{\infty}\dfrac{(-1)^j}{j^3}\left[1-\exp\left(-\dfrac{j^2\pi^2 t}{a^2}\right)\right]\sin\left(\dfrac{j\pi x}{a}\right)$	$\dfrac{\sinh(x\sqrt{s})}{s^2\sinh(a\sqrt{s})}$
$1 + \dfrac{4}{\pi}\displaystyle\sum_{j=1}^{\infty}\dfrac{(-1)^j}{2j-1}\exp\left(-\dfrac{(2j-1)^2\pi^2 t}{4a^2}\right)\cos\left(\dfrac{(2j-1)\pi x}{2a}\right)$	$\dfrac{\cosh(x\sqrt{s})}{s\cosh(a\sqrt{s})}$
$\dfrac{x^2-a^2}{2} + t - \dfrac{16a^2}{\pi^3}\displaystyle\sum_{j=1}^{\infty}\dfrac{(-1)^j}{(2j-1)^3}\exp\left(-\dfrac{(2j-1)^2\pi^2 t}{4a^2}\right)$ $\cos\left(\dfrac{(2j-1)\pi x}{2a}\right)$	$\dfrac{\cosh(x\sqrt{s})}{s^2\cosh(a\sqrt{s})}$
$1 - 2\displaystyle\sum_{j=1}^{\infty}\dfrac{\sin b_j}{b_j+\sin b_j\cos b_j}\cos(b_j x)\exp\left(-b_j^2 t\right)$ where b_j are the roots of $b = b_j\tan b_j$	$\dfrac{1}{s}\dfrac{b\cosh(x\sqrt{s})}{\sqrt{s}\sinh(\sqrt{s})+b\cosh(\sqrt{s})}$
$x - 2\displaystyle\sum_{j=1}^{\infty}\dfrac{\sin b_j - b_j\cos b_j}{b_j - \sin b_j\cos b_j}\dfrac{\sin(b_j x)}{b_j}\exp\left(-b_j^2 t\right)$ where b_j are the roots of $a = 1 - b_j\cot b_j$	$\dfrac{a}{s}\dfrac{\sinh(x\sqrt{s})}{(a-1)\sinh(\sqrt{s})+\sqrt{s}\cosh(\sqrt{s})}$

TABLE D.5
Laplace Transforms Involving Special Functions of t

$f(t)$	$F(s) = L\{f(t)\}$
$\mathrm{erf}(at)$	$\dfrac{\exp\left(\dfrac{s^2}{4a^2}\right)\mathrm{erfc}\left(\dfrac{s}{2a}\right)}{s}$
$\mathrm{erf}\left(\dfrac{t}{2a}\right)$	$\dfrac{\exp(a^2 s^2)\mathrm{erfc}(as)}{s}$
$\mathrm{erf}(\sqrt{at})$	$\dfrac{\sqrt{a}}{s\sqrt{s+a}}$

TABLE D.5 (continued)
Laplace Transforms Involving Special Functions of t

$f(t)$	$F(s) = L\{f(t)\}$
$e^{at}\operatorname{erf}\left(\sqrt{at}\right)$	$\dfrac{\sqrt{a}}{(s-a)\sqrt{s}}$
$\operatorname{erf}\left(\dfrac{a}{2\sqrt{t}}\right)$	$\dfrac{1-\exp(-a\sqrt{s})}{s}$
$\operatorname{erfc}\left(\dfrac{a}{2\sqrt{t}}\right)$	$\dfrac{\exp(-a\sqrt{s})}{s}$
$\dfrac{\operatorname{erf}\left(\sqrt{at}\right)}{\sqrt{a}}$	$\dfrac{1}{s\sqrt{s+a}}$
$\dfrac{e^{at}\operatorname{erf}\left(\sqrt{at}\right)}{\sqrt{a}}$	$\dfrac{1}{(s-a)\sqrt{s}}$
$e^{at}\left[\dfrac{1}{\sqrt{\pi t}} - b\exp(b^2 t)\ \operatorname{erfc}(b\sqrt{t})\right]$	$\dfrac{1}{b+\sqrt{s-a}}$
$2\sqrt{\dfrac{t}{\pi}}\exp\left(-\dfrac{a^2}{4t}\right) - a\ \operatorname{erfc}\left(\dfrac{a}{2\sqrt{t}}\right)$	$\dfrac{\exp(-a\sqrt{s})}{s\sqrt{s}}$
$\exp(ab)\exp(a^2 t)\operatorname{erfc}\left(a\sqrt{t}+\dfrac{b}{a\sqrt{t}}\right)$	$\dfrac{\exp(-b\sqrt{s})}{(a+\sqrt{s})\sqrt{s}}$
$\exp(ab)\exp(a^2 t)\operatorname{erfc}\left(a\sqrt{t}+\dfrac{b}{a\sqrt{t}}\right) - \operatorname{erfc}\left(\dfrac{b}{2\sqrt{t}}\right)$	$-\dfrac{a\exp(-b\sqrt{s})}{s(a+\sqrt{s})}$
$J_0(at)$	$\dfrac{1}{\sqrt{s^2+a^2}}$
$I_0(at)$	$\dfrac{1}{\sqrt{s^2-a^2}}$
$a^b J_b(at),\quad b > -1$	$\dfrac{\left(\sqrt{s^2+a^2}-s\right)^b}{\sqrt{s^2+a^2}}$
$a^b I_b(at),\quad b > -1$	$\dfrac{\left(s-\sqrt{s^2-a^2}\right)^b}{\sqrt{s^2-a^2}}$
$t J_0(at)$	$\dfrac{s}{(s^2+a^2)^{3/2}}$
$t I_0(at)$	$\dfrac{s}{(s^2-a^2)^{3/2}}$
$t J_1(at)$	$\dfrac{a}{(s^2+a^2)^{3/2}}$
$t I_1(at)$	$\dfrac{a}{(s^2-a^2)^{3/2}}$

(continued)

TABLE D.5 (continued)
Laplace Transforms Involving Special Functions of *t*

$f(t)$	$F(s) = L\{f(t)\}$
$J_0(at) - at\, J_1(at)$	$\dfrac{s^2}{(s^2 + a^2)^{3/2}}$
$I_0(at) + at\, I_1(at)$	$\dfrac{s^2}{(s^2 - a^2)^{3/2}}$
$\left(\dfrac{t}{a}\right)^{b/2} J_b\left(2\sqrt{at}\right), \quad b > -1$	$\dfrac{\exp\left(-\dfrac{a}{s}\right)}{s^{b+1}}$

$$1 - 2\sum_{n=1}^{\infty} \frac{J_0\left(b_n \dfrac{x}{a}\right)}{b_n J_1(b_n)} \exp\left(-\frac{b_n^2 t}{a^2}\right) \qquad \frac{I_0(x\sqrt{s})}{s I_0(a\sqrt{s})}$$

b_n are the positive real roots of $J_0(b_n) = 0$

$$\frac{x^2 - a^2}{4} + t + 2a^2 \sum_{n=1}^{\infty} \frac{J_0\left(b_n \dfrac{x}{a}\right)}{b_n^3 J_1(b_n)} \exp\left(-\frac{b_n^2 t}{a^2}\right) \qquad \frac{I_0(x\sqrt{s})}{s^2 I_0(a\sqrt{s})}$$

b_n are the positive real roots of $J_0(b_n) = 0$

$$1 - 2\sum_{j=1}^{\infty} \frac{a}{b_j^2 + a^2}\, \frac{J_0(b_j x)}{J_0(b_j)} \exp\left(-b_j^2 t\right) \qquad \frac{a}{s}\, \frac{I_0(x\sqrt{s})}{I_1(\sqrt{s})\sqrt{s} + a I_0(\sqrt{s})}$$

b_j are the positive real roots of $a = \dfrac{b_j J_1(b_j)}{J_0(b_j)}$

REFERENCES

1. Churchill, R.V., *Operational Mathematics*, 3rd ed., McGraw-Hill Kogakusha, Tokyo, 1972.
2. Carslaw, H.S. and Jaeger, J.C., *Operational Methods in Applied Mathematics*, Dover, New York, 1948.
3. Hildebrand, F.B., *Advanced Calculus for Applications*, 2nd ed., Prentice-Hall, Englewood Cliffs, New Jersey, 1976.
4. Kreysizig, E., *Advanced Engineering Mathematics*, 7th ed., John Wiley, New York, 1993.
5. Bajpai, A.C., Mustoe, L.R., and Walter, D., *Advanced Engineering Mathematics*, John Wiley, Chichester, 1978.
6. Sokolnikoff, I.S. and Redheffer, R.M., *Mathematics of Physics and Modern Engineering*, 2nd ed., McGraw-Hill, New York, 1966.
7. Pipes, L.A. and Harvill, L.R., *Applied Mathematics for Engineers and Physicists*, McGraw-Hill Kogakusha, Tokyo, 1970.
8. Abramovitz, M. and Stegun, I.A., *Handbook of Mathematical Functions*, Dover, New York, 1972.
9. Oberhettinger, F. and Baddi, L., *Tables of Laplace Transforms*, Springer-Verlag, Berlin, 1973.

Appendix E
Method of Weighted
Residuals

E.1 INTRODUCTION

The method of weighted residuals (MWR) has been successfully applied to obtain approximate solutions for a wide range of boundary-value problems. The material presented here is just an introduction. More details and advanced material are available in the literature [1–4].

Probably, the idea of approximate solution for boundary value problems came from the fact that continuous, or at least piecewise continuous functions can be represented by series.

For instance, consider the simple differential equation

$$\frac{dy}{dx} = e^{2x} \tag{E.1}$$

with boundary condition

$$y(0) = 1 \tag{E.2}$$

The solution is given by

$$y(x) = \frac{1}{2}e^{2x} + \frac{1}{2} \tag{E.3}$$

This can be computed by

$$y(x) = 1 + \sum_{j=1}^{\infty} \frac{2^{j-1}}{j!}x^j \tag{E.4}$$

As one might notice, the first term on the right side obeys the boundary condition given by Equation E.2, whereas the terms in the summation would vanish at the boundary condition point.

An approximation $y_n(x)$ to the solution of a boundary value problem might be sought by employing the following relationship:

$$y(x) = y_n(x) + \Lambda_n(x) \tag{E.5}$$

where $\Lambda_n(x)$ would be the residual representing the difference between the approximation and the exact solution. Of course, if $y_n(x)$ were the exact solution, $\Lambda_n(x)$ would vanish.

In our example, if for instance

$$y(x) = 1 + \sum_{j=1}^{n} C_j x^j + \Lambda_n(x) \tag{E.6}$$

the task would be to determine constants C_j. Of course, one idea would be to optimize the search to achieve small or negligible residuals $\Lambda_n(x)$ with the minimal number n of approximations. The techniques or MWR presented here show possible routes for such approximations.

E.2   GENERAL THEORY

Consider the following differential equation:

$$D[y] = f(x) \tag{E.7}$$

where x is the dependent variable, y is the dependent one, and D is a differential operator and $f(x)$ is function of x in the domain Δ. Additionally, the boundary conditions of such an equation are given by

$$B_i[y] = g_i(x) \tag{E.8}$$

Here g_i are functions applicable in the same domain Δ of x.

One wishes to find, at least, approximated solutions $y_n(x)$ for the boundary value problem represented by Equations E.7 and E.8.

Successive approximations $(j = 1, 2, \ldots, n)$ of the exact solution may be found using linear combinations of trial functions ϕ given by

$$y_n(x) = \phi_0 + \sum_{j=1}^{n} C_j \phi_j \tag{E.9}$$

It would be convenient if ϕ_0 could satisfy the boundary conditions given by Equation E.8. In this way, it would be desired that functions ϕ_j become identical to zero at the boundary condition points. Summarizing, we would like that

$$B_i[\phi_0] = g_i(x), \quad i = 1, \ldots, p \tag{E.10}$$

$$B_i[\phi_j] = 0, \quad i = 1, \ldots, p, \quad j \neq 0 \tag{E.11}$$

After that, parameters C_j would be determined with the help of the original differential equation given by Equation E.7.

MWR is one of the techniques to accomplish this.

According to MWR, C_j are chosen with the aim of vanishing a weighted average deviation or residual. That residue represents the differences between the approximate solution in Equation E.9 and the exact solution of system given by Equations E.7 and E.8.

When the trial solution in Equation E.9 is substituted into Equation E.7, the residual Λ_n becomes

$$\Lambda_n(x) = f(x) - D[y_n] = f(x) - D\left[\phi_0 + \sum_{j=1}^{n} C_j \phi_j(x)\right] \tag{E.12}$$

As seen, Λ_n includes parameters $C_1, C_2, \ldots, C_n$. Function $y_n(x)$ is called nth approximation of $y(x)$. Of course, if y_n is the exact solution, Λ_n becomes identically zero.

Within a restricted amount of trials, a good approximation may be described as one in which Λ_n is minimized. Some minimization methods are described below.

E.3   VARIOUS METHODS

The way in which the residual Λ_n is small can be chosen. A proposal of the MWR is to impose that

$$\int_D W_k(x) \, \Lambda_n(x) \, dx = 0, \quad k = 1, 2, \ldots, n \tag{E.13}$$

This implies that Λ_n is "weighted" by functions W_k to provide a method for vanishing the product within the domain of the independent variable x. When the trial solution y_n (given by Equation E.9) is selected, Equation E.13 leads to n algebraic equations for $C_j (j = 1, \ldots, n)$.

The nature of $W_k(x)$ functions determines the particular method and the commonest cases are described below.

E.3.1   METHOD OF MOMENTS

According to the method of moments the weighting function is given by

$$W_k = x^k \tag{E.14}$$

Analytical and Approximate Methods in Transport Phenomena

Notice that the residue is forced to vanish under a crescent "moment" x^k. Actually, a more rigorous definition of the weight in this case is

$$W_k = P_k(x) \tag{E.15}$$

Here $P_k(x)$ are orthogonal* polynomials over a domain of variable x. Of course, functions given by Equation E.14 are not orthogonal polynomials on the interval $0 \le x \le 1$. In addition, smaller Λ_n for smaller number of terms in the series (Equation E.9) can be achieved if orthogonal functions were used.

E.3.2 Collocation

According to the method of collocation, n points, say p_i $(i = 1, 2, \ldots, n)$, are chosen in the domain Δ of x and the weighted function is defined by

$$W_k = \delta(p - p_k) \tag{E.16}$$

Here δ represents the unit impulse or Dirac delta function (see Appendix D). This function vanishes at all points of the domain, but at $p = p_k$. Therefore, applied as weighting functions, it leads to

$$\int_\Delta \delta(p - p_k)\Lambda_n \, dx = \Lambda_n(p_k) = 0 \tag{E.17}$$

It has been shown that the optimum points to impose zero residual are the roots of orthogonal polynomials. If such, the method is called orthogonal collocation.

In some cases, where a first simple choice would suffice, equidistant collocation might be applied. In this case, the dominium Δ of variable x is divided into equal spaces with collocation points as nodes separating those spaces.

E.3.3 Subdomain

The idea is to divide the domain Δ into smaller subdomains $\Delta_1, \Delta_2, \ldots, \Delta_n$. One might also select subdomains with intersections among them. The weighted functions W_k would be

$$W_k(\Delta_k) = 1, \quad W_k(\Delta_j) = 0, \quad j \ne k \tag{E.18}$$

Therefore, from Equation E.13

$$\int_{\Delta_k} \Lambda_n \, dx = 0 \tag{E.19}$$

---

* Appendix I presents more details about orthogonal functions.

E.3.4 Least Squares

In this case, the idea is to minimize the squares of the residuals regarding the parameters C_j, or

$$\frac{\partial}{\partial C_k} \int_\Delta \Lambda_n^2 \, dx = 2 \int_\Delta \frac{\partial \Lambda_n}{\partial C_k} \Lambda_n \, dx = 0, \quad k = 1, 2, \ldots, n \qquad \text{(E.20)}$$

Therefore, from Equation E.13

$$W_k = \frac{\partial \Lambda_n}{\partial C_k} \qquad \text{(E.21)}$$

E.3.5 Galerkin

In Galerkin's method, the weighting function is set as

$$W_k = \phi_k(x) \qquad \text{(E.22)}$$

ϕ_k are the functions used to build the trial solution given by Equation E.9. Thus, Galerkin's method asks for

$$\int_\Delta \phi_k \Lambda_n \, dx = 0, \quad k = 1, 2, \ldots, n \qquad \text{(E.23)}$$

Some authors define

$$W_k = \frac{\partial y_n}{\partial C_k} \qquad \text{(E.24)}$$

E.3.6 Comparing the Methods

At a given level n of approximation, the best method will be the one that provides the smallest residual (Equation E.12) for the same number of terms in the trial series (Equation E.9).

It has been possible to demonstrate [1–3] that Galerkin's method is the best for most of the situations. However, this does not ensure it as the easier method to apply, especially if analytical integrations of Equation E.13 are to be performed. Sometimes, a method that leads to simpler integrations might provide similar quality of solutions.

E.4 SELECTING THE APPROXIMATION ORDER

As seen, the order of approximation can increase indefinitely. Of course, a criterion should be established to determine at which level one might stop.

Usually, the professional faces a real problem where, for instance, a solution should be found within a given maximum deviation from the exact. For instance, this deviation might be the acceptable error in the temperature measured at a reactor. Industrial measurements of temperature normally fall into the precision of 1 K for combustion systems working in the order of 1000 K. Therefore, in this case a maximum deviation of 0.1% is allowed. Our approximate solution has to be within this deviation from the exact.

Of course, the exact solution is not known; otherwise, approximate methods would not be necessary. However, we do know the approximations until, say, order n. A comparison between that approximation and the next can be used through the following relation:

$$\frac{1}{x_2 - x_1} \int_{x_1}^{x_2} |\Lambda_n(x) - \Lambda_{n-1}(x)| dx \leq \text{maximum deviation} \qquad (E.25)$$

where x_1 and x_2 are the extremes of the dominium of variable x in which the solution has been sought. Thus, the nth approximation is acceptable if the average absolute deviation between its and the $(n-1)$th residue is below the maximum desired deviation between the approximate and the exact solution.

E.5 EXAMPLE OF APPLICATION

Consider the following differential equation:

$$\frac{dy}{dx} - y = x \qquad (E.26)$$

with the following boundary condition:

$$y(0) = 1 \qquad (E.27)$$

The dominium Δ of variable x is $0 \leq x \leq 1$.

This first-order linear differential equation is easily solved by several methods. For instance, the variation of parameters (see Appendix B) leads to

$$y = 2e^x - x - 1 \qquad (E.28)$$

Following the MWR, the approximate solutions given by Equation E.9 may be written upon choices for trial functions ϕ_0 and $\phi_j(j = 1, 2, \ldots, n)$. According to Equation E.10, the condition given by Equation E.27 would be satisfied by several functions. The simplest one is

$$\phi_0 = 1 \tag{E.29}$$

Several functions ϕ_j would satisfy the condition given by Equation E.11. Among the simplest choices there are polynomials on x, and

$$y_n = 1 + \sum_{j=1}^{n} C_j x^j \tag{E.30}$$

This is a good choice because of the following:

1. Simpler trial functions are easily integrated, as required by MWR.
2. There is no guarantee that more complex trial functions ϕ_j would lead to fewer steps or approximations n in order to achieve the acceptable deviation. Usually, orthogonal polynomials are the best choice. An example of orthogonal polynomials are the Legendre polynomials [4], and are given by

$$\frac{1}{2^j j!} \frac{d^j}{dx^j} (x^2 - 1)^j, \quad j = 0, 1, 2, \ldots \tag{E.31}$$

For instance, the fourth order Legendre polynomial is

$$1/8 \left(35^4 - 30x^2 + 3 \right)$$

with two real positive roots equal to 0.33998 and 0.86114. These values could be used as collocation points for a second approximation of a normalized function, i.e., one with dominium between 0 and 1.

E.5.1 FIRST APPROXIMATION

Therefore, the first approximation is

$$y_1 = 1 + C_1 x \tag{E.32}$$

From Equation E.12, the residue is

$$\Lambda_1 = C_1(1-x) - x - 1 \tag{E.33}$$

E.5.1.1 Method of Moments

According to Equation E.14

$$W_1 = x \tag{E.34}$$

Thus, the following should be imposed:

$$\int_0^1 x[C_1(1-x) - x - 1]\, dx = 0 \tag{E.35}$$

which leads to $C_1 = 5$. Therefore, by this method the first approximation is

$$y_1 = 1 + 5x \tag{E.36}$$

E.5.1.2   Collocation Method

Despite the optimum approach of orthogonal collocation, for the sake of simplicity we are going to use equidistant collocation. Consequently, $x = \frac{1}{2}$ would be the first collocation point, and

$$\Lambda_1\left(\frac{1}{2}\right) = 0 \tag{E.37}$$

Applying Equation E.33, one obtains $C_1 = 3$.

E.5.1.3   Subdomain Method

For the first approximation, there is just one subdomain, which is the same as the complete domain, and according to Equation E.19

$$\int_0^1 [C_1(1 - x) - x - 1] \mathrm{d}x = 0 \tag{E.38}$$

This leads to $C_1 = 3$.

E.5.1.4   Least Squares Method

Under this submethod the weights are given by Equation E.20, or

$$W_1 = \frac{\partial \Lambda_1}{\partial C_1} = 1 - x \tag{E.39}$$

Therefore

$$\int_0^1 (1 - x)[C_1(1 - x) - x - 1] \, \mathrm{d}x = 0 \tag{E.40}$$

which leads to $C_1 = 2$.

E.5.1.5   Galerkin's Method

From Equation E.22

$$W_1 = \phi_1(x) = x \tag{E.41}$$

Hence, the result will be the same as obtained from the method of moments. Note that this is just a coincidence due to the particular choice of trial functions.

Of course, the first approximation is far from the exact solution. In addition, we need at least two approximations in order to apply the criteria given by Equation E.25.

E.5.2 SECOND APPROXIMATION

From Equation E.30

$$y_2 = 1 + C_1 x + C_2 x^2 \qquad \text{(E.42)}$$

and

$$\Lambda_2 = C_1(1 - x) + C_2 x(2 - x) - x - 1 \qquad \text{(E.43)}$$

The above routine should be repeated to calculate C_1 and C_2. Obviously, two algebraic equations will be obtained after the application of each method. Let us just apply the collocation method.

E.5.2.1 Collocation

At collocation points $1/3$ and $2/3$, the residual should be set as zero and

$$6C_1 + 5C_2 = 12 \qquad \text{(E.44)}$$

and

$$3C_1 + 8C_2 = 15 \qquad \text{(E.45)}$$

The solution of the above system is $C_1 = 7/11$ and $C_2 = 18/11$. Therefore, by this method, the second approximation is

$$y_2 = 1 + 7/11x + 18/11x^2 \qquad \text{(E.46)}$$

E.5.2.2 Application of Approximation Criterion

It is now possible to apply the criterion indicated by Equation E.25. Employing Equation E.32 (with C_1 obtained by collocation method) and Equation E.46, one gets

$$\int_0^1 |\Lambda_1 - \Lambda_2| dx = \int_0^1 \left| 1 + 3x - 1 - \frac{7}{11}x - \frac{18}{11}x^2 \right| dx = \frac{7}{11} = 0.6363 \qquad \text{(E.47)}$$

Thus, a deviation around 64% between the exact and approximated solutions would be obtained. Probably, this precision would not be satisfactory for any engineering application.

E.5.3 THIRD APPROXIMATION

From Equation E.30, the third approximation is given by

$$y_3 = 1 + C_1 x + C_2 x^2 + C_3 x^3 \tag{E.48}$$

Equation E.12 is applied to obtain the residue as

$$\Lambda_3 = C_1(1-x) + C_2 x(2-x) + C_3 x^2(3-x) - x - 1 \tag{E.49}$$

E.5.3.1 Collocation

At collocation points $1/4$, $1/2$, and $3/4$, the residual again equated to zero results in the following system:

$$48C_1 + 28C_2 + 11C_3 - 80 = 0 \tag{E.50}$$

$$4C_1 + 6C_2 + 5C_3 - 12 = 0 \tag{E.51}$$

$$16C_1 + 60C_2 + 81C_3 - 112 = 0 \tag{E.52}$$

The solution is $C_1 = 1.32094$, $C_2 = 0.122924$, $C_3 = 1.19574$.

With that, the deviation between the second and third approximations can be computed by

$$\int_0^1 |\Lambda_2 - \Lambda_3|\, dx = \int_0^1 |(0.63636 - 1.32094)x + (1.63636 - 0.122924)x^2 - 1.19574x^3|\, dx$$

$$= 0.1367$$

TABLE E.1
Comparison between Computed Values Using the Exact Solution and Third Approximation by Equidistant Collocation Method

x	$y(x)$ (exact)	$y_3(x)$	Error (%)
0.1	1.11034	1.13452	2.18
0.3	1.39972	1.43963	2.85
0.5	1.79744	1.84067	2.41
0.7	2.32751	2.39503	2.90
0.9	3.01921	3.16011	4.67

Therefore, a deviation around 13.7% has been achieved. This is a significant improvement in relation to the last level of error (64%). Table E.1 compares the exact and the third approximated solution.

The reader is invited to continue the process in order to achieve the desired level of precision.

REFERENCES

1. Villadsen, J. and Michelsen, M.L. *Solution of Differential Equation Models by Polynomial Approximation*, Prentice-Hall, Englewood Cliffs, New Jersey, 1978.
2. Finlayson, B.A. *The Method of Weighted Residuals and Variational Principles*, Academic Press, New York, 1972.
3. Ames, W.F. *Nonlinear Ordinary Differential Equations in Transport Phenomena*, Academic Press, New York, 1968.
4. Abramovitz, M. and Stegun, I.A. *Handbook of Mathematical Functions*, Dover, New York, 1972.

Appendix F
Method of Similarity

F.1 INTRODUCTION

Consider the following partial differential equation:

$$f\left(x_i, y, \frac{\partial^j y}{\partial x_i^j}\right) = 0, \quad i = 1, 2, \ldots, n \quad j = 1, 2, \ldots, m \tag{F.1}$$

The method of similarity is based on the fact that, for several cases, the solution of the above equation is given by $n-1$ combinations of the variables x_i. That is why this method is also known as method of combination of variables.

As an example, consider the partial differential equation (Equation 10.52), or

$$\frac{\partial \psi}{\partial t} = \nu \frac{\partial^2 \psi}{\partial z^2} \tag{F.2}$$

The boundary conditions are

$$\psi(0,y) = 0, \quad -\infty \leq y \leq \infty \tag{F.3}$$

$$\psi(t,0) = 1, \quad t > 0 \tag{F.4}$$

$$\psi(t,\infty) = 0, \quad t > 0 \tag{F.5}$$

As seen, in Chapter 10 the solution is given by

$$\psi = \text{erfc}\left(\frac{z}{2\sqrt{\nu t}}\right) \tag{F.6}$$

From that, one would notice that the two independent variables (t and z) are combined into $\dfrac{z}{2\sqrt{\nu t}}$. Therefore, it is possible to imagine a new single variable given by

$$\omega = \frac{z}{2\sqrt{\nu t}} \tag{F.7}$$

This could replace t and z, and the dimensionless temperature ψ would be a function of only one variable, or

$$\psi = \psi(\omega) \tag{F.8}$$

In this way, the second-order partial differential equation given in Equation F.2 could be transformed into a single second-order ordinary differential equation.

Of course, a second-order ordinary differential equation would require two and only two boundary conditions. Therefore, Equation F.2 would be transformed into the ordinary differential equation only if the three boundary conditions (Equations F.3 through F.5) could be coalesced into just two. Equation F.7 allows us to see that both limits for $t = 0$ or $z \to \infty$ lead to $\omega \to \infty$. In addition, conditions given in Equations F.3 and F.5 impose the same value for ψ. This allows the coalescence and the boundary conditions for the ordinary differential equation. One boundary condition would be given by

$$\psi(\omega \to \infty) = 0 \tag{F.9}$$

The other boundary condition would be obtained from Equation F.4, or

$$\psi(\omega = 0) = 1 \tag{F.10}$$

Of course, the combination given by Equation F.7 could be seen only after arriving at the solution of the partial boundary problem. Instead, the method to be shown ahead would allow verifying if a combination of variables is possible in order to simplify the problem of solving a boundary value problem.

Before that—however not rigorous or always effective—a simple method of finding the possible combination of variables is to write a new independent variable as

$$\omega = t^{b_1} z^{b_2} \tag{F.11}$$

Here b_1 and b_2 are constants to be found. The constants should be such as to allow writing Equation F.2 as an ordinary differential equation. Moreover, the boundary conditions of the original equation should be such as to permit appropriate coalescence of boundary conditions.

Just for the sake of an example, let us apply Equation F.11. The changes of variables are given by

$$\frac{\partial \psi}{\partial t} = \frac{\partial \omega}{\partial t} \frac{d\psi}{d\omega} = b_1 t^{b_1 - 1} z^{b_2} \frac{d\psi}{d\omega} \tag{F.12}$$

$$\frac{\partial \psi}{\partial z} = \frac{\partial \omega}{\partial z} \frac{d\psi}{d\omega} = b_2 z^{b_2 - 1} t^{b_1} \frac{d\psi}{d\omega} \tag{F.13}$$

$$\frac{\partial^2 \psi}{\partial z^2} = \frac{\partial}{\partial z}\left(\frac{\partial \psi}{\partial z}\right) = \frac{\partial}{\partial z}\left(b_2 z^{b_2-1} t^{b_1} \frac{d\psi}{d\omega}\right)$$

$$= b_2(b_2-1)z^{b_2-2}t^{b_1}\frac{d\psi}{d\omega} + b_2 z^{b_2-1}t^{b_1}\frac{\partial \omega}{\partial z}\frac{d^2\psi}{d\omega^2}$$

$$= b_2(b_2-1)z^{b_2-2}t^{b_1}\frac{d\psi}{d\omega} + b_2^2 z^{2(b_2-1)}t^{2b_1}\frac{d^2\psi}{d\omega^2} \qquad \text{(F.14)}$$

Applying Equations F.12 and F.14 into Equation F.2, one gets

$$b_1 t^{-1} z^2 \frac{d\psi}{d\omega} = \nu b_2(b_2-1)\frac{d\psi}{d\omega} + \nu b_2^2 t^{b_1} z^{b_2}\frac{d^2\psi}{d\omega^2} \qquad \text{(F.15)}$$

Without any loss of generality, the following values could be chosen: $b_1 = -1/2$ and $b_2 = 1$. Therefore

$$\omega = t^{-\frac{1}{2}}z \qquad \text{(F.16)}$$

Hence, Equation F.15 can be written as

$$-\frac{1}{2}\omega\frac{d\psi}{d\omega} = \nu\frac{d^2\psi}{d\omega^2} \qquad \text{(F.17)}$$

Calling $y = \frac{d\psi}{d\omega}$, it is possible to integrate the above and to write

$$y = C_1 \exp\left(-\frac{\omega^2}{4\nu}\right) \qquad \text{(F.18)}$$

After another integration

$$\psi = C_1 \int \exp\left(-\frac{\omega^2}{4\nu}\right) d\omega + C_2 \qquad \text{(F.19)}$$

Due to Equation F.16, the boundary conditions given by Equations F.3 through F.5 can be replaced by Equations F.9 and F.10.

Finally, Equation F.19 can also be written as

$$\psi = C_1 \int_\omega^\infty \exp\left(-\frac{\omega^2}{4\nu}\right) d\omega + C_2 \qquad \text{(F.20)}$$

Notice that the introduction of integration limits does not particularize the solution. The advantage of such a maneuver rests on the fact that the condition given by Equation F.9 would be satisfied for $C_2 = 0$. In addition, if the following variable

$$\eta = \frac{\omega}{2\sqrt{\nu}} \qquad \text{(F.21)}$$

is used, Equation F.20 becomes

$$\psi = C_3 \int_{\eta}^{\infty} \exp(-\eta^2)\,\mathrm{d}\eta = C_3 \mathrm{erfc}(\eta) \tag{F.22}$$

Here C_3 is just another constant, or

$$C_3 = C_1 \frac{\mathrm{d}\omega}{\mathrm{d}\eta} = C_1 \frac{1}{2\sqrt{\nu}} \tag{F.23}$$

The application of the condition given by Equation F.10 would lead to $C_1 = 1$, and therefore

$$\psi = \mathrm{erfc}(\eta) = \mathrm{erfc}\left(\frac{\omega}{2\sqrt{\nu}}\right) = \mathrm{erfc}\left(\frac{z}{2\sqrt{\nu t}}\right) \tag{F.24}$$

This reproduces Equation F.6.

Despite the success, the approach presented here is not a reliable procedure. For instance, Equation F.11 assumes a given form of combination. In the case of the above example, the variables are to be combined by multiplication of respective powers. This might not work for several situations.

A more successful procedure to find the correct combination of variables was developed during the 1950s and 1960s by several contributions [1–5]. It is called the generalized method of similarity and is presented below for the case of a single partial differential equation. It can also be applied to systems of such equations.

F.2　GENERALIZED METHOD OF SIMILARITY

F.2.1　Definitions

Let there be a function $f(x)$ to which a transformation is defined by

$$T\{f(x)\} = \varphi(\bar{x};a)f(\bar{x}) = g(\bar{x}) \tag{F.25}$$

As seen, the independent variable x is transformed into a new variable $\bar{x}$. If the function $\varphi(\bar{x};a)$—involving parameter a—is such that function $g(\bar{x})$ has exactly the same form of dependence on $\bar{x}$ than $f(x)$ on x, $f(x)$ is said to be "conformally invariant" under that transformation.

Additionally, $f(x)$ would be called a "conformally invariant constant" if $\varphi(\bar{x};a)$ is independent of variable $\bar{x}$.

F.2.2 SIMILARITY

Let there be a differential equation in which x_i $(i = 1, 2, \ldots, n)$ and y are, respectively, independent and dependent variables. Additionally, the following transformations are set:

$$\bar{x}_i = a^{b_i} x_i \ (i = 1, 2, 3, \ldots, n) \tag{F.26}$$

$$\bar{y} = a^c y \tag{F.27}$$

Here the parameter a is real and different from zero. Parameters b_i and c should be determined in such a way that $y(x_i)$ would be conformally invariant under the above transformations. This would provide a system of equations with nontrivial solutions, which may generate similar variables ω_j and f. These variables would be called the invariants. In this way, it would be possible to eliminate variables and to reduce the degree of the partial differential equation and, eventually, transforming it into a system of ordinary differential equations. For instance, if it is desired to eliminate variable x_1, two possibilities arise:

1. $b_1 \neq 0$. In this case, the invariants would be

$$\omega_j = \frac{x_j}{x_1^{b_j/b_1}} \quad (j = 2, 3, \ldots, n) \tag{F.28}$$

and

$$f(\omega_2, \omega_3, \ldots, \omega_n) = \frac{y(x_1, x_2, \ldots, x_n)}{x_1^{c/b_1}} \tag{F.29}$$

2. $b_1 = 0$. In this case, the invariants would be

$$\omega_j = \frac{x_j}{\exp(b_j x_1)} \quad (j = 2, 3, \ldots, n) \tag{F.30}$$

and

$$f(\omega_2, \omega_3, \ldots, \omega_n) = \frac{y(x_1, x_2, \ldots, x_n)}{\exp(c x_1)} \tag{F.31}$$

Applications of these concepts are found in the main text.

REFERENCES

1. Morgan, A.J.A., *Quart. J. Math. (Oxford)*, 2, 250, 1952.
2. Morgan, A.J.A., *Trans. ASME*, 80, 1559, 1958.

3. Hansen, A.G., *Trans. ASME*, 80, 1553, 1958.
4. Birkhoff, G., *Hydrodynamics*, Princeton Univ. Press, Princeton, New Jersey, 1960.
5. Manobar, R., Some similarity solutions of partial differential equations of boundary layer equations, Mathematics Research Center (University of Wisconsin) Technical Summary Report, No. 375, 1963.

Appendix G
Fourier Series and Method
of Separation of Variables

G.1 FOURIER SERIES

As has been seen in Appendix C, an analytical function can be always represented by a series. The Fourier series is one among these representations.

The objective of the material presented here is to help solve differential equations related to transport phenomena. Those interested in deeper aspects of the Fourier series would find excellent texts in the literature [1–6].

G.1.1 PERIODIC FUNCTIONS

A periodic function follows the relation

$$f(x+p) = f(x+jp) = f(x) \tag{G.1}$$

Here x is the independent variable, p is a constant—also called the period of function $f(x)$—and j is any integer number. An example of a trigonometric series is

$$\sum_{j=0}^{\infty} a_j \cos(jx) + b_j \sin(jx) \tag{G.2}$$

In this case, the period is 2π, and this series is an example of a Fourier series.

A periodic function, with period equal to 2π, can be represented by the Fourier series given by Equation G.2. In the case of a periodic function with any period p, it is given by

$$f(x) = a_0 + \sum_{j=1}^{\infty} \left[a_j \cos\left(\frac{2\pi}{p}jx\right) + b_j \sin\left(\frac{2\pi}{p}jx\right) \right] \tag{G.3}$$

If the function to be represented is piecewise continuous, the value of the Fourier representation is the average between the left and right limits of the function at the discontinuities.

G.1.2 COEFFICIENTS OF FOURIER SERIES

Given the function $f(x)$ with period p, the problem is how to determine the coefficients a_j and b_j of the Fourier series. This can be answered by integration maneuvers, which lead to the so-called Euler formulas.

The first step is to integrate both sides of Equation G.3 between $-p/2$ and $p/2$ to obtain

$$a_0 = \frac{1}{p} \int_{-p/2}^{p/2} f(x)\,dx \tag{G.4}$$

The second step is to multiply Equation G.3 by $\cos\left(\frac{2\pi}{p} ix\right)$ (where i is any integer) and again integrate both sides between $-p/2$ and $p/2$ to obtain

$$a_j = \frac{2}{p} \int_{-p/2}^{p/2} f(x)\cos\left(\frac{2\pi}{p}jx\right)dx, \quad j = 1, 2, 3, \ldots \tag{G.5}$$

The last step is to repeat the previous step using $\sin\left(\frac{2\pi}{p} ix\right)$ instead of cosine in order to get

$$b_j = \frac{2}{p} \int_{-p/2}^{p/2} f(x)\sin\left(\frac{2\pi}{p}jx\right)dx, \quad j = 1, 2, 3, \ldots \tag{G.6}$$

The details are left as exercises to the reader.

G.1.3 EVEN AND ODD FUNCTIONS

An even function satisfies the relation

$$f(x) = f(-x) \tag{G.7}$$

On the other hand, the following is valid for an odd function:

$$f(x) = -f(-x) \tag{G.8}$$

Sine and cosine, respectively, are simple examples of odd and even functions.

The integrations indicated in Equations G.4 through G.6 involve cosine and sine functions. On the other hand, it is known that

- Product of two even functions results in another even function.
- Product of two odd functions results in an even function.
- Product of an even and an odd functions results in an odd function.

In addition, the value of the integral of an odd function over a symmetric interval (such as from $-p/2$ to $p/2$) would be zero because the area of half period compensates the area of the other half. Therefore

- If $f(x)$ is odd, the coefficients a_j ($j \geq 0$) in Equations G.4 and G.5 would be equal to zero.
- If $f(x)$ is even, the coefficients b_j ($j \geq 1$) in Equation G.6 would be equal to zero.

These rules simplify the determination of coefficients for the Fourier series.

G.1.4 HALF-RANGE EXPANSIONS

Until now, only periodic functions have been represented by Fourier series. However, the method can be extended to nonperiodic functions.

Let there be a function with a limited range, say 0 to c. If this last limit could be imagined as the half period ($p/2$), two possible extensions can also be devised: even and odd, as shown by Figure G.1.

These even and odd extensions are periodic functions and therefore possible to represent by even and odd Fourier series, respectively. As seen, such a method can be applied to any function defined at a limited interval.

In particular, the cosine (or even) half-expansion is given by

$$f(x) = a_0 + \sum_{j=1}^{\infty} \left[a_j \cos\left(\frac{2\pi}{p}jx\right) \right] \tag{G.9}$$

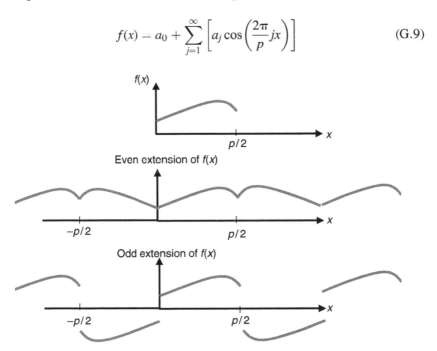

FIGURE G.1 Even and odd extensions of a function.

Here

$$a_0 = \frac{2}{p} \int_0^{p/2} f(x)\, dx \tag{G.10}$$

$$a_j = \frac{4}{p} \int_0^{p/2} f(x) \cos\left(\frac{2\pi}{p} jx\right) dx, \quad j = 1, 2, 3, \ldots \tag{G.11}$$

A sine (or odd) half-range expansion is given by

$$f(x) = \sum_{j=1}^{\infty} \left[b_j \sin\left(\frac{2\pi}{p} jx\right) \right] \tag{G.12}$$

with

$$b_j = \frac{4}{p} \int_0^{p/2} f(x) \sin\left(\frac{2\pi}{p} jx\right) dx, \quad j = 1, 2, 3, \ldots \tag{G.13}$$

G.1.5   FOURIER INTEGRALS

The representation of a function by Fourier series can be extended for cases where the function is not limited or periodic. The series now takes the form of integrals that are called Fourier integrals.

Let there be the function $f(x)$, which is piecewise continuous in a finite interval and has finite left as well as right derivatives at every point. Such a function can be represented by the Fourier integral given by

$$f(x) = \int_0^{\infty} [A(y) \cos yx + B(y) \sin yx]\, dy \tag{G.14}$$

where

$$A(y) = \frac{1}{\pi} \int_{-\infty}^{\infty} f(z) \cos yz\, dz \tag{G.15}$$

and

$$B(y) = \frac{1}{\pi} \int_{-\infty}^{\infty} f(z) \sin yz\, dz \tag{G.16}$$

Here the concept of even and odd functions $f(x)$ can also be applied. This would lead to

- Terms $A(y) = 0$ for odd functions $f(x)$
- Terms $B(y) = 0$ for even functions $f(x)$

G.2　METHOD OF SEPARATION OF VARIABLES

The method of separation of variables is among the most used for solving boundary value problems involving partial differential equations (PDE). It is based on the assumption that a function of two variables can be written as the product of two functions of single variable

$$f(x,y) = F(x)\,G(y) \qquad (G.17)$$

Of course, this cannot be guaranteed for all functions $f(x)$. However, there is a wide range of PDE problems, which allow solutions such as those given by Equation G.17.

G.2.1　PROCEDURE

For the simplest case of PDE involving two independent variables, the method of separation of variables is always applied through the following steps:

1. Assume the form given by Equation G.17 to obtain two ordinary differential equations.
2. Solve these ordinary differential equations.
3. Apply Fourier series in order to compose solutions that satisfy the boundary conditions of the original PDE problem.

The main text shows various examples of applications.

REFERENCES

1. Churchill, R.V., *Operational Mathematics*, 3rd ed., McGraw-Hill Kogakusha, Tokyo, 1972.
2. Hildebrand, F.B., *Advanced Calculus for Applications*, 2nd ed., Prentice-Hall, Englewood Cliffs, New Jersey, 1976.
3. Kreysizig, E., *Advanced Engineering Mathematics*, 7th ed., John Wiley, New York, 1993.
4. Sokolnikoff, I.S. and Redheffer, R.M., *Mathematics of Physics and Modern Engineering*, 2nd ed., McGraw-Hill, New York, 1966.
5. Pipes, L.A. and Harvill, L.R., *Applied Mathematics for Engineers and Physicists*, McGraw-Hill, Tokyo, 1970.
6. Bajpai, A.C., Mustoe, L.R., and Walter, D., *Advanced Engineering Mathematics*, John Wiley, Chichester, U.K., 1978.

Appendix H
Fourier Transforms

H.1 INTRODUCTION

As seen in Appendix G, the Fourier series is an important tool of applied mathematics and allows solving partial differential equations. Similar to Laplace transforms, the concepts of Fourier series can be used to define new transforms.

Fourier transforms are classified in the field of operational mathematics. Actually, along with Laplace transforms and others, it is within the range of integral operators. Those interested in deeper studies of operational mathematics would find very good texts in the literature [1–6]. The objective here is to just present material helpful for solving differential equations related to transport phenomena.

H.2 FOURIER COSINE AND SINE TRANSFORMS

As seen in Appendix G, the Fourier integrals allow representations of even and odd functions. An even function $f(x)$ would be represented by

$$f(x) = \int_0^\infty A(s) \cos (sx) \, ds \tag{H.1}$$

where

$$A(s) = \frac{2}{\pi} \int_0^\infty f(z) \cos (sz) \, dz \tag{H.2}$$

The Fourier cosine transform is related to these representations, and is defined as*

---

* Some authors include the factor $(\sqrt{2/\pi})$ before the integral. This is optional and does not influence applications of Fourier transforms.

$$F_c\{f(x)\} = F_c(s) = \int_0^\infty f(z)\cos(sz)\,dz \qquad (H.3)$$

Therefore, the original function $f(x)$ can be restored, or the inversion of $F_c(s)$ is

$$f(x) = \frac{2}{\pi}\int_0^\infty F_c(s)\cos(sx)\,ds \qquad (H.4)$$

An odd function $f(x)$ would be represented by

$$f(x) = \int_0^\infty A(s)\sin(sx)\,ds \qquad (H.5)$$

where

$$A(s) = \frac{2}{\pi}\int_0^\infty f(z)\sin(sz)\,dz \qquad (H.6)$$

The Fourier sine transform is related to these representations, and is defined as

$$F_s\{f(x)\} = F_s(s) = \int_0^\infty f(z)\sin(sz)\,dz \qquad (H.7)$$

Therefore, the original function $f(x)$ can be restored, or the inversion is

$$f(x) = \frac{2}{\pi}\int_0^\infty F_s(s)\sin(sx)\,ds \qquad (H.8)$$

H.2.1 EXAMPLES OF TRANSFORMS

Consider the following function:

$$f(x) = e^{-ax}, \quad x > 0 \qquad (H.9)$$

No matter if it is even or odd, the cosine or sine transforms can be found for such a function.

The Fourier cosine transform would be given by Equation H.3, or

$$F_c\{x\} = F_c(s) = \int_0^\infty e^{-az} \cos(sz)\, dz = \frac{a}{s^2 + a^2} \qquad (H.10)$$

The Fourier sine transform would be given by Equation H.7, or

$$F_s\{x\} = F_s(s) = \int_0^\infty e^{-az} \sin(sz)\, dz = \frac{s}{s^2 + a^2} \qquad (H.11)$$

H.3 COMPLEX OR EXPONENTIAL FOURIER TRANSFORM

In addition to the sine and cosine variants, the Fourier transform also has what is called the complex form. The complex Fourier transform—or just the Fourier transform—of a function $f(x)$ is defined by

$$F_e\{f(x)\} = F_e(s) = \int_{-\infty}^\infty f(z) e^{-isz}\, dz \qquad (H.12)$$

Its inverse is given by

$$f(x) = \frac{1}{2\pi} \int_{-\infty}^\infty F_e(s) e^{isx}\, ds \qquad (H.13)$$

H.4 EXISTENCE OF FOURIER TRANSFORMS

The following conditions should be met in order to ensure the existence of Fourier transform of a function $f(x)$:

1. $f(x)$ should be continuous, or at least piecewise continuous. In other words, $f(x)$ cannot have discontinuities at which it tends to positive or negative infinite values. Therefore, function $f(x)$ must have finite right-hand and left-hand derivatives at each point of its dominium, even if at certain points the values of those derivatives do not coincide.
2. Within its dominium, function $f(x)$ should be absolutely integrable. This means that the integrals of $f(x)$ should acquire finite values, no matter what the integral limits. In other words, the following limits should exist:

$$\lim_{a \to -\infty} \int_a^0 |f(x)|\, dx \quad \text{and} \quad \lim_{a \to \infty} \int_0^a |f(x)|\, dx$$

H.5   PROPERTIES OF FOURIER TRANSFORMS

Like Laplace transforms, Fourier transforms present several important properties. These are very convenient because they facilitate the transformation as well as the inversions. As several properties are common to the various forms of transforms, from now on if no index to the symbol F for the Fourier transform operator is indicated, the formula is valid for any kind of transform, whether cosine, sine, or exponential.

H.5.1   Linearity of Fourier Transforms

Similar to all integral transforms, the Fourier transforms maintain the property of linearity. Therefore, if a and b are constants, $f(x)$ and $g(x)$ are absolutely integrable and piecewise continuous, and the following can be written for any sort of Fourier transforms (sine, cosine, or complex):

$$F(af + bg) = aF(f) + bF(g) \tag{H.14}$$

H.5.2   Transform of Derivatives

Given a function $f(x)$ that follows the requirements of integrability and piecewise continuity, the following can be easily shown:*

- For cosine Fourier transform

$$F_c\left\{\frac{df}{dx}\right\} = sF_s\{f(x)\} - f(0) \tag{H.15}$$

- In the case of sine transform

$$F_s\left\{\frac{df}{dx}\right\} = -sF_c\{f(x)\} \tag{H.16}$$

- For exponential Fourier transform

$$F_e\left\{\frac{df}{dx}\right\} = -isF_e\{f(x)\} \tag{H.17}$$

The second derivatives are given by

$$F_c\{f''(x)\} = -s^2F_c\{f(x)\} - \frac{2}{\pi}f'(0) \tag{H.18}$$

---

* Such demonstrations are left as exercises to the reader. In any case, demonstrations can be found in the literature [1–6].

$$F_s\{f''(x)\} = -s^2 F_s\{f(x)\} + \sqrt{\frac{2}{\pi}} sf(0) \qquad (H.19)$$

$$F_e\{f''(x)\} = -s^2 F_e\{f(x)\} \qquad (H.20)$$

As seen, in all cases the transform of derivatives leads to a decrease in the level of derivation, and that is extremely useful for the solution of differential equations.

Despite this, the range of application of Fourier transform cannot be compared with those possible by Laplace transform. Examples of such applications can be found in various chapters.

H.5.3 CONVOLUTION

Convolution can be applied to complex Fourier transform and the following can be demonstrated [1–5]:

$$F\{f(x)*g(x)\} = F\{f(x)\}F\{g(x)\} \qquad (H.21)$$

The convolution operation is defined by

$$f(x)*g(x) = \int_{-\infty}^{\infty} f(z)g(x-z)\,dz = \int_{-\infty}^{\infty} f(x-z)g(z)\,dz \qquad (H.22)$$

H.6 GENERALIZED INTEGRAL TRANSFORMS

The various integral transforms, including Laplace and Fourier, can be written in a more generalized form

$$I\{f(x)\} = I(s) = \int_{a}^{b} f(z)g(z,s)\,dz \qquad (H.23)$$

Therefore, the following transforms define:

- In the case of Laplace: $a \to 0$, $b \to \infty$, $g(z,s) = e^{-sz}$
- In the case of Fourier exponential: $a \to -\infty$, $b \to \infty$, $g(z,s) = e^{-isz}$
- In the case of Fourier cosine: $a \to 0$, $b \to \infty$, $g(z,s) = \cos(sz)$
- In the case of Fourier sine: $a \to 0$, $b \to \infty$, $g(z,s) = \sin(sz)$
- In the case of finite Fourier cosine: $a \to -\pi$, $b \to \pi$, $g(z,s) = \cos(sz)$
- In the case of finite Fourier sine: $a \to -\pi$, $b \to \pi$, $g(z,s) = \sin(sz)$

Other forms can be found throughout the literature [1–6]. Obviously, depending on the problem at hand, some might be found more useful than others.

REFERENCES

1. Churchill, R.V., *Operational Mathematics*, 3rd ed., McGraw-Hill Kogakusha, Tokyo, 1972.
2. Hildebrand, F.B., *Advanced Calculus for Applications*, 2nd ed., Prentice-Hall, Englewood Cliffs, New Jersey, 1976.
3. Kreysizig, E., *Advanced Engineering Mathematics*, 7th ed., John Wiley, New York, 1993.
4. Bajpai, A.C., Mustoe, L.R., and Walter, D., *Advanced Engineering Mathematics*, John Wiley, Chichester, U.K., 1978.
5. Pipes, L.A. and Harvill, L.R., *Applied Mathematics for Engineers and Physicists*, McGraw-Hill Kogakusha, Tokyo, 1970.
6. Sokolnikoff, I.S. and Redheffer, R.M., *Mathematics of Physics and Modern Engineering*, 2nd ed., McGraw-Hill, New York, 1966.

Appendix I
Generalized Representation by Series

I.1 INTRODUCTION

The various concepts that allow generalizations for representations of functions by series based on various other functions are presented here. Therefore, it is a generalization of the concept of Fourier series.

Besides trigonometric functions, as used for Fourier series, several commonly used functions, such as Bessel and Legendre polynomials, can be used to build a series that represents a given function.

This is just an introduction to the subject and those interested in deeper discussions would find plenty of material in the literature [1–4].

I.2 ORTHOGONAL FUNCTIONS

Within a given interval $a \leq x \leq b$, two different functions $f_1(x)$ and $f_2(x)$ are considered orthogonal regarding a weight $w(x)$ when it is possible to write

$$\int_a^b w(x) f_1(x) f_2(x) \, dx = 0 \qquad (I.1)$$

Having in mind that orthogonality, the norm of function $f_1(x)$ is written as

$$\|f_1\| = \sqrt{\int_a^b w(x) f_1^2(x) \, dx} \qquad (I.2)$$

In addition to being orthogonal, the above functions f_1 and f_2 are called orthonormal if, within the same interval $a - b$, both have norms equal to 1.

If the weight function $w(x)$ is equal to 1, the functions f_1 and f_2 are just said to be orthogonal.

Examples of orthogonal functions are

- Trigonometric; such as $\sin(c_j\, x)$, with integer c_j. It is easy to verify that

$$\int_{-\pi}^{\pi} \sin(c_1 x)\, \sin(c_2 x)\, dx = 0 \tag{I.3}$$

- Their norm is

$$\|\sin(c_1 x)\| = \sqrt{\int_{-\pi}^{\pi} \sin^2(c_1 x)\, dx} = \pi \tag{I.4}$$

- Bessel functions of integer order n regarding weight $w(x) = x$ in the interval $0 \le x \le \kappa$, where κ is a real positive constant

$$\int_{0}^{\kappa} x J_n(c_{in} x)\, J_n(c_{jn} x)\, dx = 0, \quad i \ne j \tag{I.5}$$

where c_i ($i = 1,\ 2,\ 3 \ldots$) are given by $c_{in} = \alpha_{in}/\kappa$, and α_{in} are the roots of the equation

$$J_n(\alpha_{in}) = 0 \tag{I.6}$$

I.3   ORTHOGONAL SERIES

It is possible to show that a given function $f(x)$ can be represented by an orthogonal set of functions $g_j(x)$ ($j = 1,\ 2,\ 3, \ldots$), with respect to a weighted function $w(x)$, within an interval $a \le x \le b$, or

$$f(x) = \sum_{i}^{\infty} a_i g_i(x) \tag{I.7}$$

The coefficients a_i are given by

$$a_i = \frac{1}{\|g_i\|^2} \int_{a}^{b} w(x) f(x)\, g_i(x)\, dx \tag{I.8}$$

Let us exemplify with Bessel function of integer order (n). Following the above, if the weight function is $w(x) = x$, it is possible to write

$$f(x) = \sum_{i}^{\infty} a_i J_n(c_{in}x) \tag{I.9}$$

Since

$$\|J_n(c_{in}x)\|^2 = \int_0^{\kappa} x J_n^2(c_{in}x)\, dx = \frac{\kappa^2}{2} J_{n+1}^2(c_{in}\kappa) \tag{I.10}$$

Then

$$a_i = \frac{2}{\kappa^2 J_{n+1}^2(\alpha_{in})} \int_0^{\kappa} x f(x)\, J_n(c_{in}x)\, dx, \quad i = 1, 2, 3, \ldots \tag{I.11}$$

The coefficients $\alpha_{in} = c_{in}\,\kappa$ are given by Equation I.6.

EXERCISES

1. Demonstrate Equations I.3 and I.4.
2. For $\kappa = 1$, demonstrate Equation I.5.

REFERENCES

1. Hildebrand, F.B., *Advanced Calculus for Applications*, 2nd ed., Prentice-Hall, Englewood Cliffs, New Jersey, 1976.
2. Kreysizig, E., *Advanced Engineering Mathematics*, 7th ed., John Wiley, New York, 1993.
3. Bajpai, A.C., Mustoe, L.R., and Walter, D., *Advanced Engineering Mathematics*, John Wiley, Chichester, U.K., 1978.
4. Pipes, L.A. and Harvill, L.R., *Applied Mathematics for Engineers and Physicists*, McGraw-Hill Kogakusha, Tokyo, 1970.

Index